Waste Treatment and Environmental Sustainability: Current Trends, Challenges and Management Strategies

Waste Treatment and Environmental Sustainability: Current Trends, Challenges and Management Strategies

Editors

Sunil Kumar
Pooja Sharma
Deblina Dutta

MDPI • Basel • Beijing • Wuhan • Barcelona • Belgrade • Manchester • Tokyo • Cluj • Tianjin

Editors

Sunil Kumar
Waste Reprocessing Division
CSIR-NEERI
Nagpur
India

Pooja Sharma
NUS, Environmental Research Institute
National University of Singapore
Singapore

Deblina Dutta
Environmental Science and Engineering
SRM University-AP
Amaravati
India

Editorial Office
MDPI
St. Alban-Anlage 66
4052 Basel, Switzerland

This is a reprint of articles from the Special Issue published online in the open access journal *Sustainability* (ISSN 2071-1050) (available at: www.mdpi.com/journal/sustainability/special_issues/Waste_Strategies).

For citation purposes, cite each article independently as indicated on the article page online and as indicated below:

LastName, A.A.; LastName, B.B.; LastName, C.C. Article Title. *Journal Name* **Year**, *Volume Number*, Page Range.

ISBN 978-3-0365-8233-7 (Hbk)
ISBN 978-3-0365-8232-0 (PDF)

Contents

Preface to "Waste Treatment and Environmental Sustainability: Current Trends, Challenges and Management Strategies"

This volume of "Waste Treatment and Environmental Sustainability: Current Trends, Challenges, and Management Strategies" addresses the critical issues surrounding waste management and presents innovative strategies for achieving environmental sustainability. Through a collection of research and review articles, this reprint explores sustainable waste management approaches and offers valuable insights into the future of waste treatment.

The improper management of solid waste poses significant risks to the environment and human health. Traditional waste treatment methods have proven inadequate in mitigating these risks, necessitating the exploration of alternative approaches. This reprint focuses on techniques such as bioremediation, phytoremediation, and green technologies that promote sustainable waste management by reducing waste generation and optimizing resource utilization.

The Special Issue presented in this reprint comprises 11 articles that delve into various aspects of waste treatment and provide potential directions for future research. The topics covered include predicting O_3 levels, the impact of COVID-19 on household waste, the environmental profiles of university campuses, the relationship between municipal waste and reverse logistics, data analytics in waste management, water evaporation efficiency in bioretention, reverse supply chain strategies, efficient treatment approaches, cyanide degradation, the long-term storage of microorganisms, and the impact of biochar on phytoremediation.

We express our gratitude to the authors, reviewers, and Editors-in-Chief for their invaluable contributions to this journal volume. Special thanks are extended to the Special Issue Editor and the Editorial team for their unwavering support in successfully bringing this Special Issue to publication.

We hope that this reprint serves as a valuable resource for researchers, practitioners, policymakers, and anyone interested in waste management and environmental sustainability. May the knowledge shared within these pages inspire and guide future endeavors in addressing the challenges of waste treatment and fostering sustainable practices.

Sunil Kumar, Pooja Sharma, and Deblina Dutta
Editors

sustainability

MDPI

Article

Forecasting Daytime Ground-Level Ozone Concentration in Urbanized Areas of Malaysia Using Predictive Models

NurIzzah M. Hashim [1], Norazian Mohamed Noor [1,2,*], Ahmad Zia Ul-Saufie [3], Andrei Victor Sandu [4,5,6,*], Petrica Vizureanu [4], György Deák [6] and Marwan Kheimi [7]

1 Faculty of Civil Engineering Technology, Universiti Malaysia Perlis, d/a Pejabat Pos Besar, P.O. Box 77, Kangar 01007, Malaysia; nurizzah_hashim91@yahoo.com
2 Sustainable Environment Research Group (SERG), Centre of Excellence Geopolymer and Green Technology (CEGeoGTech), Universiti Malaysia Perlis, d/a Pejabat Pos Besar, P.O. Box 77, Kangar 01007, Malaysia
3 Faculty of Computer and Mathematical Sciences, Universiti Teknologi Mara (UiTM), Shah Alam 40450, Malaysia; ahmadzia101@uitm.edu.my
4 Faculty of Materials Science and Engineering, Gheorghe Asachi Technical University of Iasi, 61 D. Mangeron Blvd., 700050 Iasi, Romania; peviz@tuiasi.ro
5 Romanian Inventors Forum, St. P. Movila 3, 700089 Iasi, Romania
6 National Institute for Research and Development in Environmental Protection INCDPM, Splaiul Independentei 294, 060031 Bucharest, Romania; dkrcontrol@yahoo.com
7 Department of Civil Engineering, Faculty of Engineering—Rabigh Branch, King Abdulaziz University, Jeddah 21589, Saudi Arabia; mmkheimi@kau.edu.sa
* Correspondence: norazian@unimap.edu.my (N.M.N.); sav@tuiasi.ro (A.V.S.)

Abstract: Ground-level ozone (O_3) is one of the most significant forms of air pollution around the world due to its ability to cause adverse effects on human health and environment. Understanding the variation and association of O_3 level with its precursors and weather parameters is important for developing precise forecasting models that are needed for mitigation planning and early warning purposes. In this study, hourly air pollution data (O_3, CO, NO_2, PM_{10}, NmHC, SO_2) and weather parameters (relative humidity, temperature, UVB, wind speed and wind direction) covering a ten year period (2003–2012) in the selected urban areas in Malaysia were analyzed. The main aim of this research was to model O_3 level in the band of greatest solar radiation with its precursors and meteorology parameters using the proposed predictive models. Six predictive models were developed which are Multiple Linear Regression (MLR), Feed-Forward Neural Network (FFANN), Radial Basis Function (RBFANN), and the three modified models, namely Principal Component Regression (PCR), PCA-FFANN, and PCA-RBFANN. The performances of the models were evaluated using four performance measures, i.e., Mean Absolute Error (MAE), Root Mean Squared Error (RMSE), Index of Agreement (IA), and Coefficient of Determination (R^2). Surface O_3 level was best described using linear regression model (MLR) with the smallest calculated error (MAE = 6.06; RMSE = 7.77) and the highest value of IA and R^2 (0.85 and 0.91 respectively). The non-linear models (FFANN and RBFANN) fitted the observed O_3 level well, but were slightly less accurate compared to MLR. Nonetheless, all the unmodified models (MLR, ANN, and RBF) outperformed the modified-version models (PCR, PCA-FFANN, and PCA-RBFANN). Verification of the best model (MLR) was done using air pollutant data in 2018. The MLR model fitted the dataset of 2018 very well in predicting the daily O_3 level in the specified selected areas with the range of R^2 values of 0.85 to 0.95. These indicate that MLR can be used as one of the reliable methods to predict daytime O_3 level in Malaysia. Thus, it can be used as a predictive tool by the authority to forecast high ozone concentration in providing early warning to the population.

Keywords: air quality modeling; ozone; multiple linear regression; artificial neural network

Citation: Hashim, N.M.; Noor, N.M.; Ul-Saufie, A.Z.; Sandu, A.V.; Vizureanu, P.; Deák, G.; Kheimi, M. Forecasting Daytime Ground-Level Ozone Concentration in Urbanized Areas of Malaysia Using Predictive Models. *Sustainability* **2022**, *14*, 7936. https://doi.org/10.3390/su14137936

Academic Editors: Vincenzo Torretta, Sunil Kumar, Pooja Sharma and Deblina Dutta

Received: 16 March 2022
Accepted: 23 June 2022
Published: 29 June 2022

1. Introduction

Ground-level ozone (O_3) is an important component of the atmosphere because it is a major oxidant and a greenhouse gas [1,2]. At ground-level, O_3 is seen in the form of a secondary atmospheric pollutant created by a number of chemical reactions that are typically linked to degradation of air quality in the air [3], which leads to adverse effects on the health of human beings, crop production, material quality, and ecosystems. High concentration of ground-level ozone can affect human health via short-term and long-term impacts. Short-term impacts include mortality and breathing morbidity and are likely to lead to eye irritation and can also influence the airway [4], while lung damage and inflammatory reactions can be caused over the long term [5].

Ground-level O_3 is one of the global air pollution problems. In Malaysia, since 1997, ground-level O_3 has been recognized as one of the significant contaminants of air due to the growing ozone precursors [6]. Rapid economic development and high emissions of pollutants in nearby urban and industrialized areas were detected as the main contributors to the increase in O_3 precursors such as NOx, VOCs, and CO. The main sources of O_3 precursors were reported to be industrial and vehicle emission [6]. Vehicle emission can lead to high emission of NO due to higher titration processes between NO and O_3. VOCs, that often found in urban and industrial areas, in the other hand, lead to the formation of peroxy radical (RO_2) that later undergoes photoreaction to produce O_3 [7].

Thus, due to its long- and short-term impacts on human health, the variation and relations of ground-level ozone and its precursors require much investigation [8]. Nowadays, the number of studies reported on O_3 concentration in Asia has increased, particularly in Malaysia. The monitoring data in several large cities demonstrated that O_3 level are increasing and are not always at acceptable concentration in accordance with the Malaysia Ambient Air Quality Standard (MAAQS). Thus, it is very important to understand the behavior of ground-level ozone in order to explain the association of O_3 level with its precursors and weather parameters [9].

Forecasting high ozone concentration events using mathematical tools is very useful in providing early warning to the population. However, the prediction of ground-level O_3 is more complicated due to its origin as a secondary pollutant if compared to modeling primary pollutants such as particulate matter (PM_{10}) [10]. Thus, statistical approaches had been widely used by the researchers to study the variation of O_3 concentration with its precursors and weather parameters. Multiple linear regression is one of the most common techniques used in the prediction of ground level O_3 level. The objective is to model a linear relationship between the explanatory (independent) and the answer (dependent) variables, thus the relationship of O_3 level and other variables (including other air pollutants, its precursors and meteorology parameters) can be observed [10]. In several studies conducted by Hassanzadeh et al. [11], Barrero et al. [12], Banja et al. [13], and Allu et al. [14], the connection between weather parameters and ozone concentration in Portugal, Spain, Albania, and India has been described respectively. Even though many studies have been carried out in the world investigating the association between the weather and the ozone concentration using MLR, in southeast Asia in particular there is still a shortage of work. While certain studies have been conducted by Azmi et al. [15] and Awang et al. [16], their study only focused on the trend or variation of ozone concentration in Klang Valley, Malaysia. However, there are a few studies on O_3 level prediction in Malaysia. Abdullah et al. [17] studied the high night-time O_3 concentrations in Kemaman, Terengganu, while Ghazali et al. [18] related the nitrogen dioxide transformation into the ozone and predict the ozone concentration using the multiple linear regression techniques.

Besides giving a simple linear relationship of ozone concentration with its precursors and weather parameters, linear regression may not provide accurate predictions in some complex situations such as non-linear data and extreme values data. Machine learning is an effective technique for understanding the inter-dependence of climatic data and air pollution since it supports exploratory analysis of data without using an empirical model [19,20]. Further, machine learning addresses the non-linearity problem, enhancing the model's predictive performance [21,22]. Artificial neural networks (ANN), one of the most common machine learning techniques, can be a useful tool to extract information from imprecise and non-linear data such as air quality and meteorology. Currently, the applications of machine learning neural networks have become more popular for predicting ground-level O_3 concentration. A lot of researchers have effectively adopted ANN as a predictive tool to model O_3 concentration [23–26].

Modification of MLR and ANN has been conducted by many researchers to increase the accuracy of the predicted model. One of the main disturbances that will cause a reduction in the performance of the model and reduce the efficiency of the model is multicollinearity [27,28]. To deal with multidimensional issues and overcome feature redundancy, many researchers suggested various techniques for dimension reduction and feature extractions. One of the widely employed methods for these purposes is the principal component analysis (PCA) [29,30]. Basically, modification by substituting the input into principal components was accomplished. Most of the research had successfully increased the accuracy of the predictive models for particulate matter or O_3 level by modifying the input of regression models using principal components [9,31–33]. However, there were also some studies that reported the opposite results, where MLR predicts the air pollutant concentration better than PCR [1,34]. A modified model of ANN (using PCs as input to train and validate FFANN model) was implemented to increase the accuracy of the model. A few studies have applied the modified FFANN model with PCA and successfully increased the accuracy of the model in predicting PM_{10} level [31,35] and ground-level O_3 concentration [25,36,37].

Recently, despite the superiority of ANN algorithm, other machine learning algorithms such support vector regression (SVR) and support vector machine (SVM) have become popular options among researchers due to their architectural simplicity and precision [38,39]. Balogun and Tella [40] applied four machine learning algorithms (Random Forest, Decision Tree Regression, Linear Regression, and Support Vector Regression) to predict O_3 level limited to the west coast region of peninsular Malaysia. Ayman et al. [41] applied six machine learning algorithms, namely Linear Regression (LR), Tree Regression (TR), Support Vector Regression (SVR), Ensemble Regression (ER), Gaussian Process Regression (GPR), and Artificial Neural Networks models (ANN) to model only an urban area in Malaysia, i.e., Lembah Kelang. They reported that the proposed models were capable of predicting the concentrations with higher accuracy level.

Despite these sophisticated methods, there is a lack of comprehensive studies on O_3 level prediction that involve most of the urban areas in Malaysia. Hence, thorough study on suitability of using established linear and non-linear model in predicting O_3 level in Malaysia is much needed to investigate the best method that can be used as a reliable predictive tool to estimate O_3 level. In this research, linear and non-linear models with their modified models were developed and evaluated using performance indicators. The best model selected from this study is ready to be used by the authorities as the predictive strategy of Malaysia and will be very helpful in understanding how these elements interact with O_3 content.

2. Materials and Methods

2.1. Study Area

In this study, five urban areas in Malaysia were selected. Four out of five locations are located in Malaysia's peninsula and one station is situated in East Malaysia (Kuching). The four locations in peninsular Malaysia are distributed in the north (Perai) the center (Shah Alam), the south (Melaka), and the east (Kuala Terengganu) of peninsular Malaysia.

The selected air quality monitoring stations are displayed in Figure 1, while Table 1 provides a description and locations of all stations covered in this study. All the sampling sites are located in schools that are closed to residential areas. However, all of these areas are surrounded by the urban center and residential areas.

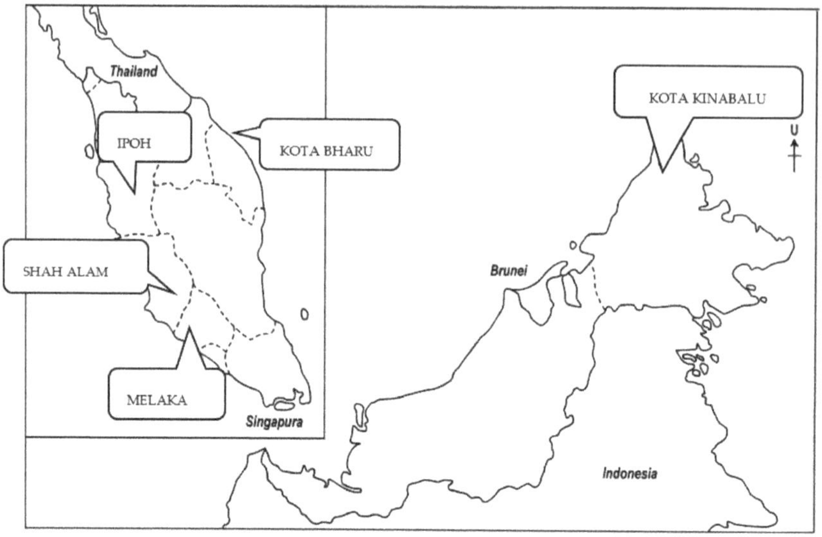

Figure 1. Location of the five selected air quality monitoring stations in Malaysia.

Table 1. Locations and the description of monitoring stations.

Region	Monitoring Site	Latitude, Longitude	Area Description
North	Ipoh	N 4.6305, E 101.1178	Urban area Residential area
Center	Shah Alam	N 3.1066, E 101.5573	Urban area Residential area Near industrial area
South	Melaka	N 2.1919, E 102.2545	Urban area Residential area Near industrial area
East Peninsular	Kota Bharu	N 6.1464, E 102.2481	Urban area Residential area
East Malaysia	Kota Kinabalu	N 5.9532, E 116.0551	Urban area Residential area

2.2. Air Pollutant Dataset

The air quality data were gathered from the Air Quality Division of the Department of Environment (DoE), Malaysia. The data were collected and monitored by Alam Sekitar Malaysia Sdn. Bhd. (ASMA), the authorized agency for DoE. The equipment used by ASMA to monitor the air quality data is from Teledyne Technologies Inc. USA (Thousand Oaks, CA, USA), and Met One Instrument Inc. USA (Grants Pass, OR, USA). Based on the Standard Operating Procedures for Continuous Air Quality Monitoring (2007), the analyzer used by ASMA to monitor PM_{10} was a BAM-1020 Beta Attenuation Mas Monitor from Met One Instrument, Inc. USA. This instrument has a high resolution of 0.1 $\mu g\ m^{-3}$ at a 16.7 L min^{-1} flow rate, with lower detection limits of <4.8 $\mu g\ m^{-3}$ and <1.0 $\mu g\ m^{-3}$ for 1 h and 24 h, respectively. The instruments used by ASMA to monitor SO_2, CO, and O_3 were the Teledyne API Model 100A/100E, Teledyne API Model 200A/200E, Teledyne API Model 300/300E, and Teledyne API Model 400/400E, respectively, from Teledyne Technologies Inc., USA [1], while SO_2 measurement was based on the UV fluorescence method, where the lowest level of detection is at 0.4 ppb. CO was measured using the non-dispersive, infrared absorption (Beer Lambert) method with 0.5% precision and the lowest detection of

0.04 ppm. Ozone concentration was measured through the UV absorption (Beer Lambert) method with a detection limit of 0.4 ppb. The measurements of SO_2, CO, and O_3 were at a precision level of 0.5%. For NmHC, the analyzer used by ASMA measured using a Teledyne API M4020 from Teledyne Technologies Inc., USA, which is equipped with aflame-ionization detector (FID) and a measurement accuracy of 1%. These instruments were used due to well-proven accuracy, reliability, and robustness.

In this study, two sets of air pollutants data were used. The first set of data was identified based on the availability of data from 1 January 2003 to 31 December 2012. These long-term datasets were used to develop the prediction models of the daytime O_3 level. The predicted models were then validated using several performance measurements. The air pollutants (O_3; PM_{10}, NO_2, SO_2, NmHC, and CO) and weather parameters (WS, WD, H, T, and UVB) used in this research are tabulated in Table 2.

Table 2. The units of air pollutants and weather parameters.

Air Pollutant/Weather Parameters	Unit
Ground-level ozone (O_3)	ppb
Nitrogen dioxide (NO_2)	ppm
Carbon monoxide (CO)	ppm
Sulphur dioxide (SO_2)	ppm
Particulate matter (PM_{10})	($\mu g/m^3$)
Non-methane Hydrocarbon (NmHc)	ppm
Ambient temperature (T)	°C
Humidity (H)	%
Wind speed (WS)	km/h
Wind direction (WD)	degree (°)
Ultraviolet radiation (UVB)	W/m^2

Table 3 shows the summary of the mean and standard deviation for hourly dataset of the air pollutants and weather parameters at the five study areas from 2003 to 2012. These long-term data were used to develop the prediction models to estimate the hourly O_3 level in the band of great solar intensity.

Table 3. Descriptive statistics (mean ± standard deviation) of air pollutants and weather parameters from 2003 to 2012. O_3: ozone; PM_{10}: particulate matter; CO: carbon monoxide; SO_2: sulphur dioxide; NO_2: nitrogen dioxide; and NmHC: non-methane hydrocarbons.

Area/Parameter	Ipoh	Shah Alam	Melaka	Kota Bharu	Kota Kinabalu
Wind Speed (km/h)	9.18 ± 2.71	8.75 ± 2.26	8.70 ± 2.77	8.47 ± 3.33	8.73 ± 2.26
Temperature (°C)	33.39 ± 2.36	32.81 ± 2.47	31.60 ± 1.74	30.78 ± 79.55	31.59 ± 2.31
Solar Radiation (W/m^2)	677.27 ± 183.81	533.16 ± 191.70	Not Available	553.42 ± 215.86	668.93 ± 7.98
Humidity (%)	56.88 ± 8.66	59.37 ± 9.74	61.87 ± 8.50	63.26 ± 10.24	68.93 ± 7.98
NmHC (ppm)	0.13 ± 0.056	0.22 ± 0.12	Not Available	0.20 ± 0.11	Not Available
SO_2 (ppm)	0.0018 ± 0.0012	0.0038 ± 0.0037	0.0022 ± 0.0021	0.00064 ± 0.0010	0.00055 ± 0.00070
NO_2 (ppm)	0.0093 ± 0.0039	0.012 ± 0.0074	0.0043 ± 0.0021	0.0054 ± 0.0035	0.0022 ± 0.0019
CO (ppm)	0.43 ± 0.19	0.51 ± 0.34	0.32 ± 0.19	0.46 ± 0.22	0.23 ± 0.12
PM_{10} ($\mu g/m^3$)	43.96 ± 19.17	47.23 ± 32.21	34.98 ± 21.25	35.62 ± 13.55	29.55 ± 12.56
O_3 (ppb)	27 ± 6.1	31 ± 7.6	20 ± 6.5	18 ± 5.6	15 ± 3.9

As for the model deployment, the prediction models (developed from previous dataset) were later deployed on the 2018 dataset. Table 4 shows the mean and standard deviation for hourly dataset at the five selected areas in 2018. Statistically, no significant differences were detected among the variables in 2018 when comparisons were made with the identical variables from the previously presented dataset. Verification of the best prediction model aimed to prove that the model was still reliable in predicting daytime O_3 level in different years.

Table 4. The mean and standard error of all the parameters at the five stations in 2018.

Area/Parameter	Ipoh	Shah Alam	Melaka	Kota Bharu	Kota Kinabalu
Wind Speed (km/h)	8.38 ± 3.19	8.14 ± 2.81	8.62 ± 2.94	7.98 ± 3.35	8.82 ± 2.37
Temperature (°C)	33.37 ± 2.62	32.81 ± 2.63	31.46 ± 1.89	30.58 ± 2.54	31.97 ± 2.38
Solar Radiation (W/m^2)	742.25 ± 195.66	592.80 ± 187.38	Not Available	596.79 ± 197.86	619.79 ± 192.54
Humidity (%)	57.67 ± 8.94	59.35 ± 9.99	62.31 ± 9.10	63.87 ± 11.03	67.64 ± 7.96
NmHC (ppm)	0.13 ± 0.06	0.24 ±0.12	Not Available	0.19 ± 0.09	Not Available
SO_2 (ppm)	0.0019 ± 0.0015	0.0037 ± 0.004	0.0021 ± 0.0026	0.0057 ± 0.001	0.005 ± 0.007
NO_2 (ppm)	0.0094 ± 0.0051	0.012 ± 0.0089	0.0044 ± 0.0026	0.005 ± 0.0038	0.0021 ± 0.019
CO (ppm)	0.416 ± 0.202	0. 535 ± 0.348	0.328 ± 0.193	0.443 ± 0.233	0.234 ± 0.128
PM_{10} ($\mu g/m^3$)	45.12 ± 22.66	47.34 ± 32.27	35.02 ± 22.10	36.35 ± 17.18	29.90 ± 15.37
O_3 (ppb)	39 ± 16.0	52 ± 32.0	34 ± 11.0	24 ± 9.0	22 ± 6.0

In this study, the multivariate air pollutant data were treated as cross sectional data which focus on observing information of air pollutants concentration at a particular time, in various locations, and depend on the information sought. In order to predict the daytime O_3 level in the band of greatest solar radiation, the hourly dataset (2003–2012) for model development was chosen to be during noon (12.00 p.m. to 4.00 p.m.) as O_3 level was observed to be highest once it received the greatest amount of solar radiation [15]. A total number of 14,124 datasets were used to develop and validate the prediction model. Out of the total data, random partition of the dataset was conducted using SPSS where 80% of the data were used for model development and the remaining data (20%) were used for model validation. Table 5 shows the results of Kolmogorov–Smirnov test of normality for hourly O_3 measurement record (12.00 p.m. to 4.00 p.m.) from the first dataset (2003 to 2012) for all study areas. It indicates that the datasets used were normally distributed as the p-values > 0.05.

Table 5. Kolmogorov–Smirnov Test of Normality.

Station	Kolmogorov-Smirnov [a]		
	Statistics	df	p-Value
Ipoh	0.163	14,124	0.200
Shah Alam	0.142	14,124	0.200
Melaka	0.154	14,124	0.200
Kota Bharu	0.170	14,124	0.200
Kota Kinabalu	0.168	14,124	0.200

[a] Lilliefors Significance Correction.

2.3. Principle Component Analysis (PCA)

Prior to conducting Principal Component Analysis (PCA), the Kaiser–Meyer–Olkin (KMO) and Bartlett's test of sphericity tests needed to be performed. KMO test was used to measure sampling adequacy for each variable in the model. The value of KMO must be greater than 0.5, showing that the data are adequate [31]. In addition, Bartlett's test of sphericity was applied to show a high degree of relationship between the parameters and that the data are suitable for factor analysis ($p < 0.001$). These requirements had been completed before the Principal Component Analysis.

In this study, PCA was used to group the large amounts of air pollutants data and weather parameters into a few sets of groups named as principal components. These principal components were later used as the input to the modified model. PCA is generally written as below [31]:

$$PC_i = A_{1i}X_{1j} + A_{2i}X_{2j} + \cdots + A_{ni}X_{nj} \tag{1}$$

where PC_i is ith principal component, A_{ji} is the loading of the observed variable, X is the measured value of variables, i is the component number, j is the sample number, and n is the total number of variables.

The principal components (PCs) generated by PCA are sometimes not readily available for interpretation; therefore, it is advisable to rotate them by varimax rotation with the eigenvalues greater than 1 [42–44]. Varimax factors (VFs) coefficient with a correlation from 0.75 are considered as a strong significant factor loading; those that range from 0.50–0.74 are moderate, while 0.30–0.49 are classified as weak significant factor loading [31]. The equation is expressed as below:

$$Z_{ij} = af_1X_{1i} + af_2X_{2i} + \cdots + af_mX_{mi} + e_{fi} \quad (2)$$

where Z is the measured value of a variables, a is the factor loading, f is the factor score, e is the residual term accounting for errors or other sources variation, i is the sample number, j is the variable number and m is the total number of factors.

2.4. Prediction Model

Overall, six models were developed, i.e., linear (Multiple Linear Regression) and non-linear model (Artificial Neural Network) including their modified models. Three models were developed from Multiple Linear Regression (MLR), Feed-Forward Neural Network (FFANN), and Radial Basis Function Neural Network (RBFANN); the remaining three models were their modified model, i.e., combination of MLR, FFANN, and RBFANN with PCA, namely PCR, PCA-FFANN, and PCA-RBFANN.

For MLR, FFANN, and RBFANN models, the measured records of hourly ground level ozone (O_3) concentration, weather parameters (wind speed (WS) ambient temperature (T), humidity (H), and other pollutants (NmHC PM_{10}, SO_2, NO_2, and CO) were used as input. In the modified model, the principal components were used as input. The output for this study is the prediction value of maximum hour of ozone concentration for the next day, known as $O_{3(t+1)}$.

2.4.1. Multiple Linear Regression (MLR)

A random response Y relating to a set of independent variables $x_1, x_2, \ldots, x_k$ based on the multiple regression model is as shown below [26,45]:

$$Y = \gamma + \beta_1x_1 + \beta_2x_2 \cdots + \beta_kx_k + \varepsilon \quad (3)$$

where γ, β_1, β_2, and β_k are unknown parameters and ε is an error term factor.

Multicollinearity occurs when there are high correlations between two or more predictor variables. Multicollinearity is a problem because it weakens the statistical significance of an independent variable; thus, it causes larger standard error of a regression coefficient. As a result, this coefficient will be less likely to be significant statistically [45].

Multicollinearity assumption was verified by Variance Inflation Factor (VIF) accompanied with the regression output. The average value of VIF under 10 is acceptable, signifying multicollinearity does not exist among independent variables [46]. The VIF is given by:

$$\mathrm{VIF} = \frac{1}{1 - R_i^2} \quad (4)$$

where VIF_i is the variance inflation factor associated with the ith predictor, and R_i^2 is the multiple coefficients of determination in a regression of the ith predictor on all other predictors. In this study, the VIF was calculated for the prediction calculated by MLR and PCR models to evaluate whether multicollinearity existed in the models.

2.4.2. Feed-Forward Artificial Neural Network Model (FFANN)

The most common and popular neural network architecture is the Feed-Forward Artificial Neural Network (FFANN), which typically contains three layers such as the input layer, hidden layer, and output layer. This study uses FFANN for its simplicity as one of the predictive models and it was built using Matlab Script.

In this study, the tansig-purelin was used as the transfer function; tansig from hidden node to output layer; and purelin as the transfer function from input to hidden node. The number of hidden layers and hidden neurons (nodes) were tried and increased systematically, checking each time if the prepared neural network obtained the stable performance error with the fixed number of neurons. A three-layer neural network with two hidden layers was used in this study. The tested number of neurons used were from 2 until 10 by incremental of two units. In FFANN, the final layer output is the function of the linear combination of the unit's activation function and the non-linear input weighted sum function. Assuming that the pattern of concentration does not change significantly from day to day, the model proposed can be used by providing values of new predictive variables to predict concentrations for a consecutive hour. The performance of the model was monitored to make sure that the model stopped training and chose the best number of neurons.

2.4.3. Radial Basis Function Artificial Neural Network (RBFANN)

The basic idea of RBFANN network is to fit a curve of the data into a high dimensional space. RBFANN networks represent another type of ANN with an input layer, an output layer, and a hidden layer of radial units, each actually modelling a Gaussian response surface. The main important advantage of the RBF approach is that the RBF network can yield the minimum approximating error of any function; thus, it is suitable for modelling complex input–output mappings [47]. RBF is also one of the unusual but extremely fast and effective methods which has smoothness function (σ) that relies from 0.1 to 0.9 [48]. In RBFANN, the critical step for good prediction is selection of the smoothness parameter (σ) [48]. Hence, this study uses σ value from 0.1 until 0.9 by incremental of 0.1. Ten variables (O_3, PM_{10}, CO, NO_2, SO_2, NmHC, UVB, humidity, wind speed, and temperature) were used as inputs for RBF model.

2.4.4. Modified Models

Hybrid models are combination of MLR, FFANN, and RBFANN models with the principal components analysis (PCA). The aim is to reduce the complexity of the model and to determine the relevant independent variables to predict the future O_3 concentrations. The differences between modified models and the models of MLR, FFANN, and RBFANN were the input variables.

In the modified models, the selected principal components from the output of Principal Component Analysis (PCA) were used as input. The scores of high loadings components with an eigenvalue greater than or equal to 1 explain most of the variation in all datasets, which is ideal to use as in regression equations as independent or predictor variables; thus, PCR, PCA-FFANN, and PCA-RBFANN establish the relationship between the dependent or response variable and the selected PCs of the independent variables [49]. The sub-model for every principal component (PC) according to the study areas are given in Section 3.1.

2.5. *Performance Indicators*

Performance indicators were used to evaluate the goodness of fit of the predicted models in the sample locations. According to Ahmat et al. [50] and Ghazali et al. [51], in order to describe the fitness for each of the selected distribution, there are at least four types of performance indicators which need to be used. There are many performance measures that have been used by researchers to describe the performances of their developed models. They are usually divided into two types, i.e., error measurement and performances measurement. For error measurement, the smaller the value to zero, the smaller the differences between the predicted and the observed values. For performances measures, it describes the linear relationship between the observed and predicted values. Hence, the closer the predicted values to the straight line, the better the agreement between observed and predicted values or the closer the value to 1, the better the prediction model. Thus, the best model is selected based on the highest accuracy measures and the smallest error measures between the predicted and their corresponding observed values [52].

In this research, the performance measures selected were the Mean Absolute Error (MAE), Root Mean Squared Error (RMSE), Coefficient of Determination (R^2), and Index of Agreement (IA). Table 6 shows the equations for each of the performance measures.

Table 6. The Performance Indicators [52].

Performance Index	Equation		Description
Mean Absolute Error (MAE)	$\mathrm{MAE} = \frac{1}{n}\sum_{i=1}^{n}\lvert Pi - Oi\rvert$	(5)	Value close to zero indicates better method.
Root Mean Squared Error (RMSE)	$\mathrm{RMSE} = \left(\frac{1}{n}\sum_{i=1}^{n}[Oi - Pi]^2\right)^{1/2}$	(6)	Value closer to zero indicates better method.
Coefficient of determination (R^2)	$R^2 = 1 - \frac{\sum_{i=1}^{n}(Pi-\eth)^2}{\sum_{i=1}^{n}(\lvert Oi-\eth\rvert^2)}$	(7)	Value closer to one indicates better method.
Index of Agreement	$\mathrm{IA} = 1 - \frac{\sum_{i=1}^{n}(P-Oi)^2}{\sum_{i=1}^{n}\lvert Pi-\eth\rvert+\lvert Oi-\eth\rvert^2}$	(8)	Value close to one indicates better method.

3. Results

3.1. Principle Components Analysis (PCA)

Table 7 shows the results of Keiser–Meyer–Olkin (KMO) and Bartlett's Test. The KMO values were greater than 0.5 and the significant p-value for Bartlett's Test were smaller than 0.001 for all stations. Hence, these datasets were suitable for PCA.

Table 7. KMO and Bartlett's Test.

Station	KMO Measure of Sampling Adequacy	Bartlett's Test of Sphericity	
		Approximate Chi-Square	*p*-Value
Ipoh	0.700	54,319	<0.000
Shah Alam	0.716	68,026	<0.000
Melaka	0.575	42,357	<0.000
Kota Bharu	0.709	73,029	<0.000
Kota Kinabalu	0.664	37,185	<0.000

After the extraction of PCA was applied, factors were considered as the principal component based on eigenvalues of more than 1 (>1.0) and varimax rotation was used as a criterion. Due to excessive factors with more significant variables, the eigenvalues with less than one (<1.0) were overlooked due to multicollinearity being present among original variables [31]. The eigenvalues for all linear components before extraction, after extraction, and after rotation are shown in Table 8. Based on the percentages of the eigenvalues, the most significant principal component in explaining the amount of variance is the first, followed by the second and third principal components.

The scores of high loadings components with an eigenvalue greater than or equal to 1 were selected as an input to the modified models. The sub-models of each principal component according to the study areas is given in Table 9. Only the strong factor loading of the Varimax factors (VFs) and coefficient ($\geq$0.75) are considered as the components of each principal component (PCs).

Table 8. Total Variance Explained.

Component	Station	Initial Eigenvalues		
		Total	Variance (%)	Cumulative (%)
1	Ipoh	2.752	27.520	27.520
2		2.665	26.646	54.166
3		1.248	12.483	66.649
1	Shah Alam	3.221	32.209	32.209
2		2.607	26.074	58.283
3		1.058	10.579	68.862
1	Melaka	2.352	29.395	29.395
2		2.028	25.350	54.746
3		1.212	15.146	69.891
1	Kota Bharu	3.587	35.866	35.866
2		1.980	19.800	55.666
3		1.105	11.051	66.717
4		1.025	10.251	76.969
1	Kota Kinabalu	2.775	30.838	30.838
2		1.800	20.004	50.843
3		1.232	13.692	64.535

Table 9. Sub model for PCR.

Area	Principle Components (PCs)	Sub-Model
Ipoh	PC1	$0.781PM_{10} + 0.760CO + 0.739NO_2 + 0.713NmHC$
	PC2	$-0.934H + 0.871T + 0.772UVB$
	PC3	0.819 WS
Shah Alam	PC1	$0.928T - 0.923H + 0.735UVB + 0.717O_3$
	PC2	$0.824PM_{10} + 0.812CO$
	PC3	−0.883WS
Melaka	PC1	$0.924T - 0.896H$
	PC2	$0.880CO + 0.875PM_{10}$
	PC3	$0.907SO_2$
Kota Bharu	PC1	$0.929T - 0.923H + 0.852NmHC$
	PC2	$0.815NmHC + 0.806NO_2 + 0.774CO$
	PC3	$0.855O_3 + 0.735PM_{10}$
	PC4	0.903WS
Kota Kinabalu	PC1	$0.900T + 0.855UVB - 0.824H$
	PC2	$0.809PM_{10} + 0.758CO$

The descriptions of principal components for each study areas are explained according to Tables 8 and 9. Each principal component with the specific significant variables was used as the input to the hybrid models.

For Ipoh, the first component described 27.520% with four significant variables which were PM_{10}, CO, NO_2, and NmHC. The second component explained 26.646% which consists of three significant variables (H, T, and UVB) and the remaining 12.483% was explained as the third component (wind speed) which made the cumulative variance 68.89%. Subsequently in Shah Alam, there were three principal components with the first component being 32.209%, which was made up of four significant factor loadings (T, H, UVB, and O_3); the second component was 26.074% with two significant variables (PM_{10} and CO), and the remaining 10.579% (WS).

However, for Melaka, the first component explained 29.395% with two significant factor loadings which were T (0.924) and H (−0.896); 25.350% for the second component which were significantly contributed to by CO (0.880 and PM_{10} (0.875) and the remaining 15.146% (PC3) was strong explained by SO_2 (0.907).

Kota Bharu showed that the two principal components are formed with cumulative of variance at 55.666% where first factor is higher than second factor with 35.866% (T, H, UVB) and 19.800% (NmHC, NO_2, CO) of variability. The third and fourth factors were 11.051% (O_3 and PM_{10}) and 10.251% (WS) respectively. For Kota Kinabalu, the first component explained 30.838% of the total variance, 20.004% for PC2, and the remaining 13.693% explained the third component. The weather parameters (T, UVB, and H) were the strong loading factor for the first component, while, for the second component, PM_{10} and CO were the important factors and for PC3, NO_2 was the only strong factor.

3.2. Development of Ground-Level O_3 Prediction Models and Their Performances

3.2.1. Multiple Linear Regression (MLR) and Its Modification (Principal Component Regression (PCR))

The summary of the developed model (MLR and PCR) and range of Variance of Inflation Factor (VIF) are given in Table 10. The VIF values for MLR and PCR models were lower than 10, which proved that multi-collinearity issue does not exists in the model. Hence, in this case, the developed MLR and PCR models had minimal relationship between independent variables that resulted in good-fitted model.

Table 10. Summary of the Multiple Linear Regression (MLR) models and Principal Component Regression (PCR) models for O_3 concentration forecasting. VIF: Variance of Inflation Factor.

Location	Method	Models	Range of VIF
Ipoh	MLR	O_{3+1} = 61.914 + (0.001 CO) − (0.387 Humidity) − (1.923 NmHC) + (0.341 NO_2) + (0.41 O_3) − (0.003 PM_{10}) − (0.454 SO_2) − (0.657 Temperature) − (0.002 UVB) + (0.568 Wind Speed)	1.147–4.170
	PCR	O_{3+1} = 12.564 + (0.067 PC1) + (0.072 PC2) + (1.021 PC3)	1.027–1.062
Shah Alam	MLR	O_{3+1} = 109.995 + (0.002 CO) − (0.404 Humidity) − (0.00001392 NmHC) + (0.07 NO_2) + (0.351 O_3) − (0.001 PM_{10}) − (0.048 SO_2) − (1.727 Temperature) + (0.002 UVB) + (0.21 Wind Speed)	1.227–4.373
	PCR	O_{3+1} = 52.582 + (0.002 PC1) + (0.000 PC2) − (0.012 PC3)	1.062–1.151
Melaka	MLR	O_{3+1} = 11.902 − (0.001 CO) + (0.033 Humidity) + (0.35 NO_2) + (0.337 O_3) + (0.022 PM_{10}) − (0.148 SO_2) + (0.252 Temperature) − (0.07 Wind Speed)	1.066–4.364
	PCR	O_{3+1} = 23.715 + (0.105 PC1) + (0.003 PC2) − (1.031 PC3)	1.032–1.061
Kota Bharu	MLR	O_{3+1} = 14.267 − (0.002 CO) − (0.027 Humidity) − (0.004 NmHC) + (0.127 NO_2) + (0.617 O_3) + (0.1 PM_{10}) + (0.188 SO_2) − (0.146 Temperature) − (0.022 Wind Speed)	1.141–5.751
	PCR	O_{3+1} = 10.296 + (0.001 PC1) − (0.005 PC2) + (0.464 PC3) − (0.130 PC4)	1.047–1.259
Kota Kinabalu	MLR	O_{3+1} = 14.267 + (0.002 CO) − (0.027 Humidity) − (0.004 NmHC) + (0.127 NO_2) + (0.617 O_3) + (0.1 PM_{10}) + (0.188 SO_2) − (0.146 Temperature) − (0.022 Wind Speed)	1.153–2.799
	PCR	O_{3+1} = 16.655 − (0.013 PC1) + (0.028 PC2) − (0.333 PC3)11	1.052–1.205

Table 11 shows the performance measurements of the predicted daytime O_3 level by MLR and PCR. In terms of predictive model by MLR, generally, for all study areas, MLR model gave very good predictions compared to its modified version model, PCR. The predicted values by MLR gave lower error compared to PCR with the range of error (MAE) within 2.684 to 11.59 and 3.597 to 13.92 for MLR and PCR, respectively. The predicted values of O_3 level in Kota Bharu and Kota Kinabalu gave smaller value of error compared to other places. For goodness of fit test (PA, IA, and R^2), MLR fit the observed data better than PCR with the range of 0.757 to 0.952 and 0.531 to 0.870.

Table 11. Model validation based on all parameters and PCA as inputs.

Location	Method	MAE	RMSE	IA	R^2
Ipoh	MLR	7.055	8.901	0.874	0.887
	PCR	8.355	10.692	0.806	0.694
Shah Alam	MLR	11.59	15.053	0.757	0.903
	PCR	13.92	18.482	0.563	0.531
Melaka	MLR	5.855	7.737	0.772	0.952
	PCR	6.969	9.17	0.636	0.672
Kota Bharu	MLR	2.684	3.373	0.949	0.944
	PCR	3.731	4.885	0.870	0.800
Kota Kinabalu	MLR	3.119	3.779	0.884	0.866
	PCR	3.597	4.794	0.658	0.531

The performances of MLR and PCR in predicting the daytime O_3 concentration can further be observed using graphical presentation. Figure 2 shows the observed and predicted value of O_3 level for the five study areas. From the plot, the MLR model fits the data very well compared to PCR for all the stations. The high R^2 values of MLR model were due to small and unbiased differences between the observed values and the model's predicted values. This can be observed as the distance between the fitted line and all the data points was minimized. The more variance that is accounted for by the regression model, the closer the data points will fall to the fitted regression line. Contrarily, for PCR model, wider distance between the regression line and all the points can be seen. Hence, reduced R^2 values were observed for the predicted values of PCR model.

3.2.2. Artificial Neural Network (ANN) and Its Modification (PCA-FFANN)

The best performance indicated by the different number of neurons for the neural network analysis and the hybrid is shown in Table 12. Overall, FFANN performed very well compared to it modified-version model, i.e., PCA-FFANN. Basically, the predicted O_3 level using FFANN had low percentage of measured error (RMSE) compared to PCA-FFANN by around 20.3 percent. Furthermore, very good agreement between observed and predicted O_3 level was detected with FFFANN model due to very close value of the performance measures (PA, IA, and R^2) to 1. This indicates that the prediction of maximum hour O_3 level were very close to the observed concentration of O_3. A number of researchers have been applying ANN for prediction of ambient air pollutants concentration. ANN was identified as one of the best models for PM_{10} level prediction [31,53] and ground-level O_3 [24,25].

Table 12. Model validation based on all parameters and PCA as inputs.

Location	Method	No. of Neuron	MAE	RMSE	IA	R^2
Ipoh	FFANN	2	6.937	9.071	0.871	0.839
	PCA-FFANN	2	8.402	10.693	0.804	0.706
Shah Alam	FFANN	2	12.090	15.677	0.729	0.846
	PCA-FFANN	6	13.233	17.534	0.638	0.576
Melaka	FFANN	2	5.599	7.850	0.769	0.853
	PCA-FFANN	6	6.438	8.816	0.684	0.647
Kota Bharu	FFANN	2	2.449	3.519	0.940	0.949
	PCA-FFANN	4	3.708	4.918	0.870	0.771
Kota Kinabalu	FFANN	8	2.619	3.579	0.841	0.691
	PCA-FFANN	4	3.583	4.779	0.658	0.540

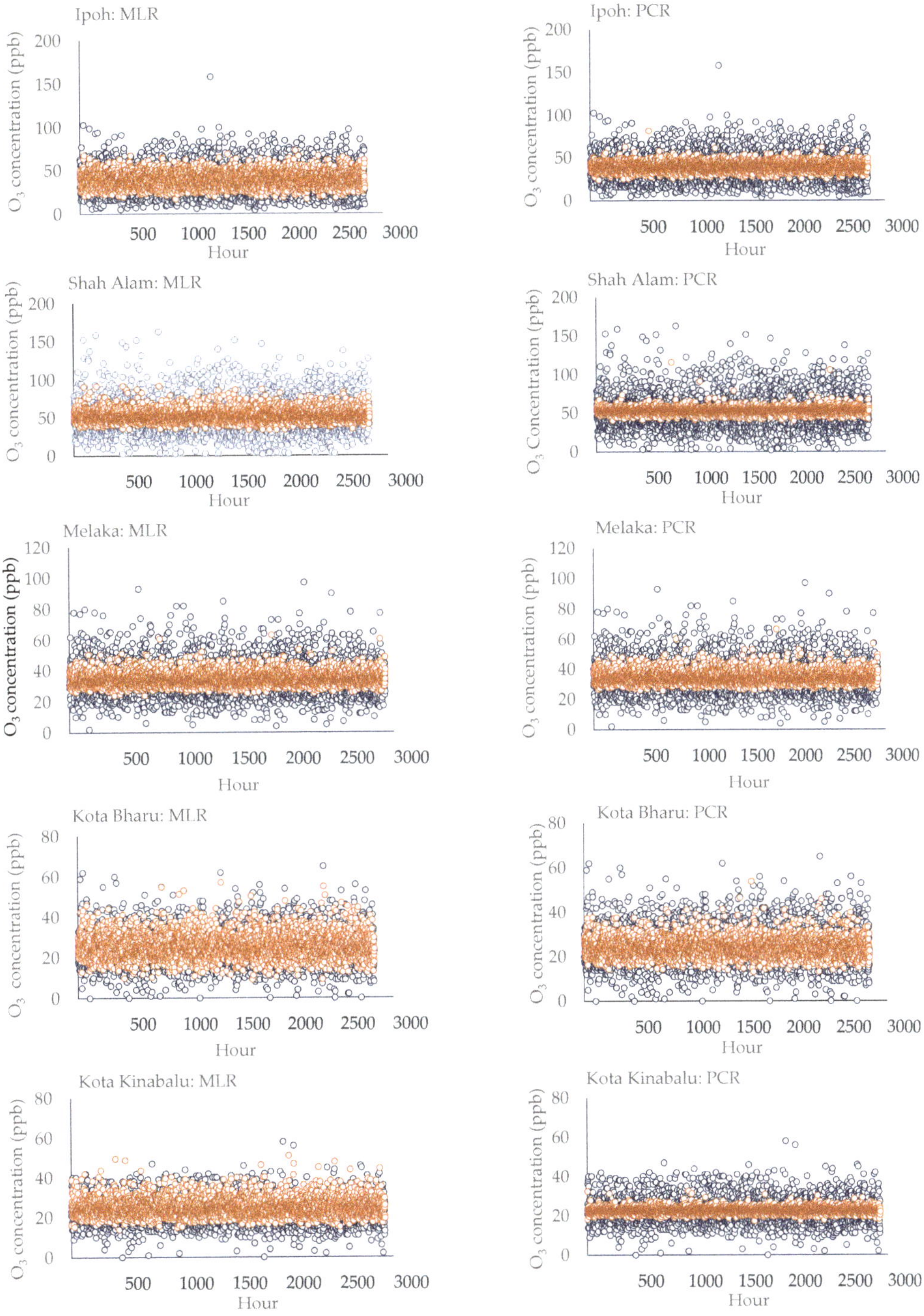

Figure 2. Observed versus predicted values of ground-level O_3 concentration using MLR and PCR. The blue marker is the observed value and the brown marker shows the predicted values.

Comparatively, PCA-FFANN model performed moderately compared to FFANN in predicting O_3 level for all the areas. Principal components (PCs) were used as input to FFANN to reduce the dimension of a given data set, making the data set more approachable and computationally easier to handle, while preserving most patterns and trends. Modified model of FFANN (by using PCs as input to train and validate FFANN model) was expected to increase the accuracy of the model. A few studies have applied the modified FFANN model with PCA and had successfully increased the accuracy of the model in predicting PM_{10} level [31,35] and ground-level O_3 concentration [25,36].

Graphical presentation of the predicted and observed O_3 level is presented in Figure 3. Generally, it can be seen that the predicted O_3 level using FFANN and PCA-FFANN was fitted with the range of the best fitted values by the model, which in this case were more distributed at the center of the observed data points. In addition, the range of O_3 level predicted by PCA-FFANN was observed to be smaller than the value predicted by FFANN, or, in other words, the range of the best fitted values was more narrowed compared to its non-modified model. However, better variation of the predicted values (FFANN was better than PCA-FFANN) was observed in Kota Bahru and Kota Kinabalu where the error was significantly small (Table 12) compared to other areas.

3.2.3. Radial Basis Functions (RBFANN) and Its Modification (PCA-RBFANN)

Table 13 shows the validation of models according to the best spread number for RBFANN and its modified model (PCA-RBFANN).

Table 13. Model validation based on all parameters and PCA as inputs.

Location	Method	Smoothness Function (σ)	MAE	RMSE	IA	R^2
Ipoh	RBFANN	0.2	8.675	11.118	0.770	0.558
	PCA-RBFANN	0.1	9.148	11.640	0.746	0.587
Shah Alam	RBFANN	0.1	12.049	16.418	0.741	0.531
	PCA-RBFANN	0.1	13.941	18.422	0.567	0.539
Melaka	RBFANN	0.1	6.247	8.620	0.710	0.852
	PCA-RBFANN	0.1	7.577	9.977	0.506	0.649
Kota Bharu	RBFANN	0.1	3.173	4.579	0.899	0.775
	PCA-RBFANN	0.1	4.316	5.690	0.791	0.771
Kota Kinabalu	RBFANN	0.1	3.036	4.292	0.783	0.379
	PCA-RBFANN	0.1	3.774	4.990	0.587	0.483

Generally, prediction of maximum O_3 level made by RBFANN was found out to be moderately good for all the study areas except for Melaka and Kota Bharu with the range of R^2 value from 0.531 to 0.852. Predicted O_3 levels using RBF neural network at these two cities were quite well-correlated with the value of R^2 of 0.852 and 0.775 for Melaka and Kota Bharu, respectively. Comparable findings were identified by a study conducted by Abdullah et al. [40], where RBFANN was used to predict PM_{10} concentration in Pasir Gudang, Malaysia. The results showed that RBFANN model was able to explain 65.2% and 84.9% variance in the data during training and testing, respectively. Hence, it is proven that RBFANN is a promising nonlinear model which has high ability in representing the complexity and nonlinearity of ambient air pollutant concentration in the atmosphere.

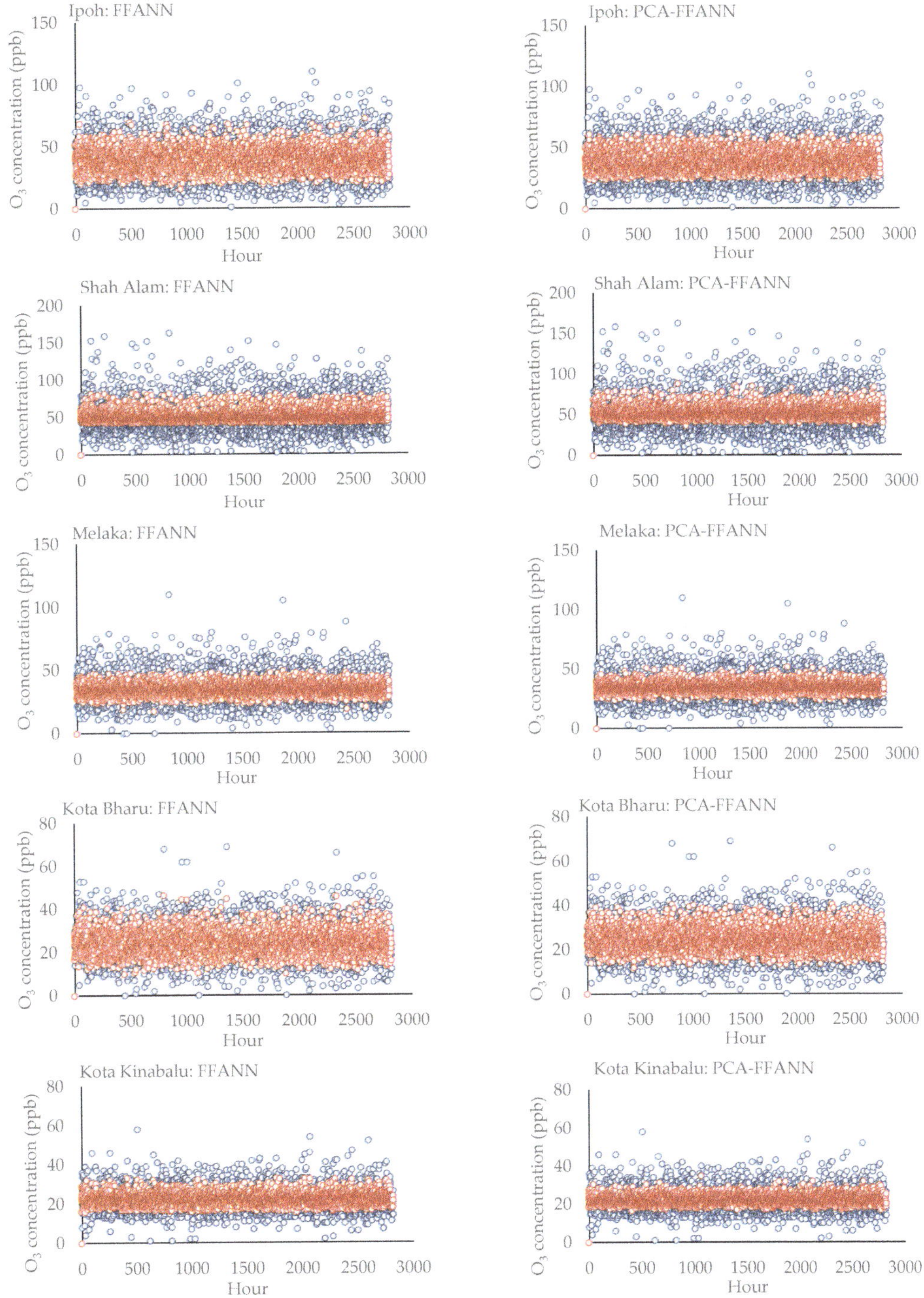

Figure 3. Observed versus predicted ground-level O_3 concentration using FFANN and PCA-FFANN. The blue marker is the observed values and the red marker shows the predicted values.

The performance of RBFANN in predicting air pollutant level, if compared to another well-known neural network such as multi-layer perception (MLP), is known to be less accurate than other neural networks. This was supported by the findings from the study conducted by Kumar et al. [54] that compared RBF with MLP neural network for prediction of O_3 level in India. The results suggested that MLP had slightly better prediction of O_3 level with the range of RMSE value of 5.4 to 15.4 compared to RBF, with the range of 5.2 to 18.6.

Its modified version, PCA-RBFANN, performed less accurately than its basis model. However, noticeable improvement was observed on R^2 value for the Ipoh, Shah Alam, and Kota Kinabalu, where better goodness-of-fit measure of the predicted data to the regression line was detected. When a regression model accounts for more of the variance, the data points are closer to the regression line; hence, a better fitted model was witnessed.

Figure 4 shows the graphical exhibition of the predicted and observed O_3 level at all study areas. As a whole, inconsistent performances of the predicted values using RBFANN or PCA-RBFANN can be observed. For prediction using RBF neural network (RBFANN), all of the cities except for Melaka were detected to have wider range of the best fitted values compared to the predicted values by FFANN (Section 3.2.2). Predicted data points of O_3 level by PCA-RBFANN were observed to have narrower range of the best fitted values, especially in Ipoh and Kota Kinabalu. Oppositely, in Melaka, the predicted data points using RBFANN had a very constricted range of the best fitted values compared to its modified version (PCA-RBFANN).

3.3. Summary

Table 14 summarizes the performance of the six models used to predict the ground-level O_3 in Malaysia. Overall, MLR gives small error (6.061 and 7.769 for MAE and RMSE respectively) and offer most fitted data to the regression line, with the value of R^2 and IA close to 1. FFANN and RBFANN fitted the observed O_3 data points well but were slightly less accurate compared to MLR. Interestingly, all the unmodified models (MLR, FFANN, and RBFANN) significantly outperformed their modified version model (PCR, PCA-FFANN, and PCA-RBFANN). The sequence of model from the best fitted model to the least is as follows:

$$\text{MLR} > \text{FFANN} > \text{RBFANN} > \text{PCR} > \text{PCA-FFANN} > \text{PCA-RBFANN}$$

Table 14. Summary of performance measures for the six prediction models.

Model	Performance Indicators			
	MAE	RMSE	IA	R^2
MLR	6.061	7.769	0.847	0.905
PCR	7.314	9.605	0.707	0.648
FFANN	5.939	7.937	0.830	0.877
PCA-FFANN	7.073	9.348	0.731	0.641
RBFANN	6.636	9.009	0.781	0.619
PCA-RBFANN	7.751	10.144	0.639	0.606

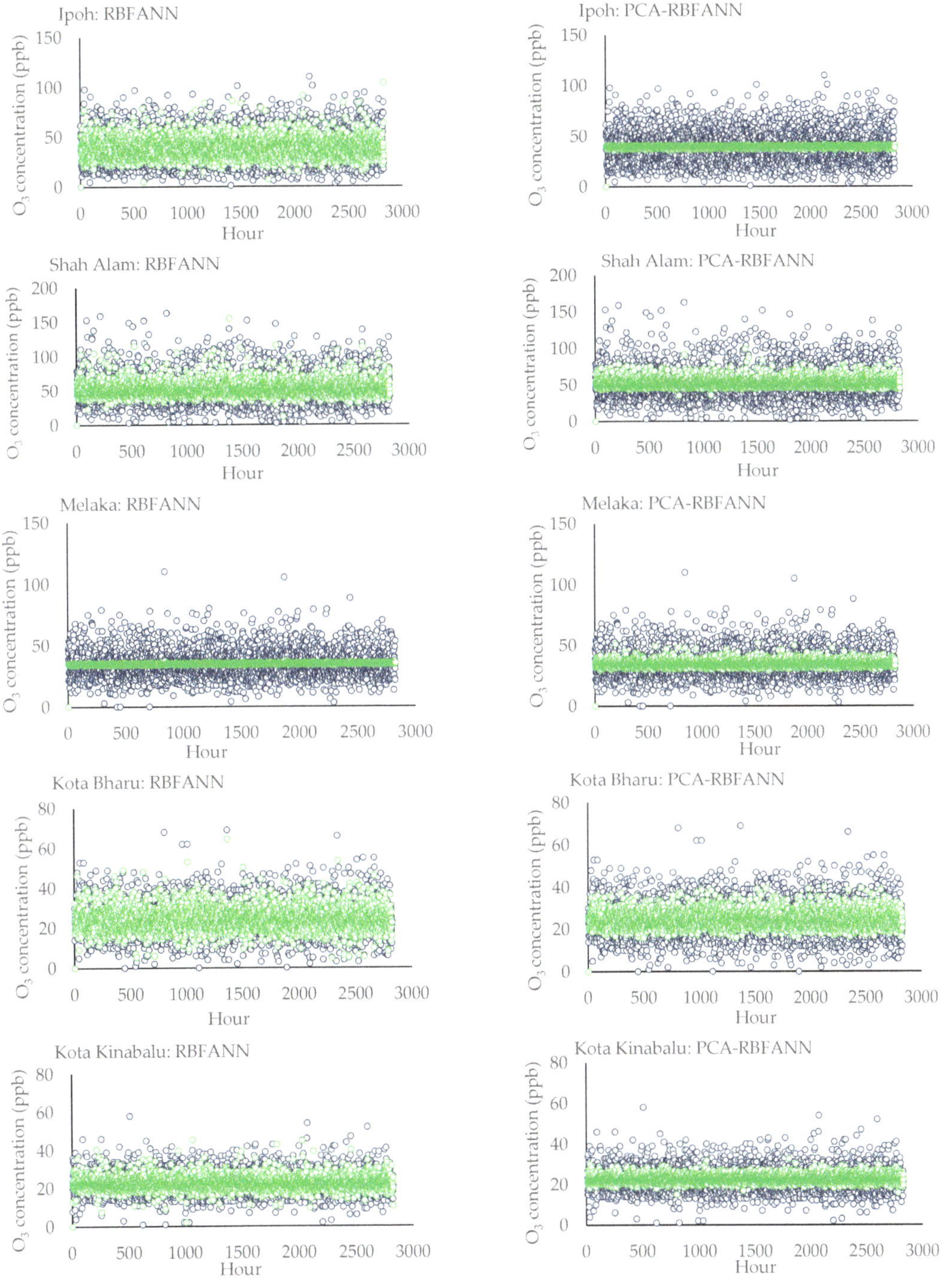

Figure 4. Observed versus predicted ground-level O_3 using RBFANN and PCA-RBFANN. The blue marker is the observed values and the green marker shows the predicted values.

3.4. *Deployment of the Best Selected Prediction Model of Ground-Level* O_3

Deployment of the best-chosen model (MLR) was done in order to prove that the model was able to predict the maximum hour of O_3 concentration using different years of dataset.

Table 15 shows the results of performance measures for the predicted O_3 level using MLR for the dataset of 2018. Small error measurement was detected, with ranges from 2.3 to 14.7 that resulted in small differences between the predicted and observed values of O_3 level. High accuracy of performance measures (IA and R^2) indicates high agreement between the observed and predicted data points. Therefore, with this high agreement between the predicted data and the observed data, it was proven that the linear regression models can be used to predict the O_3 level at any year provided no to small significant change on the dataset variability. Good prediction model was able to be developed due to the long-term period (2003 to 2012), which was taken into account during development of model where 80% of the dataset was used for model development and the remaining was used to validate the performances of the model. Thus, deployment of MLR as the best selected model for predicting daytime O_3 concentration was considered effective.

Table 15. Performance measures of model verification for the five study areas.

Area/Performance	MAE	RMSE	IA	R^2
Ipoh	8.030	10.815	0.759	0.887
Shah Alam	11.470	14.678	0.736	0.903
Melaka	9.263	12.331	0.744	0.952
Kota Bharu	2.363	3.208	0.951	0.944
Kota Kinabalu	2.447	3.197	0.928	0.866

4. Discussion

4.1. *Performances of the Predictive Models (Basis Model)*

Multiple Linear Regression predicted the maximum O_3 concentration better than other predictive models including the hybrid methods. High agreement between the observed and predicted values was witnessed with the calculated R^2 value > 0.8 for each study areas. MLR successfully modeled the relationship between the independent variables (previous O_3, NmHC, PM_{10}, SO_2, NO_2, and CO, wind speed, ambient temperature, humidity) and a dependent variable ($O_{3(t+1)}$), by fitting a linear equation to the observed data.

Multiple linear regression is one of the most widely used methods for predicting ozone concentrations with weather parameters and different atmospheric pollutants. In several studies conducted by Hassanzadeh et al. [11] and Barrero et al. [12], the connection between weather status and ozone concentration has been observed using this method. The best prediction equation for ozone and weather variables is found in Hassanzadeh et al. [11] using a multiple regression procedure. Barrero et al. [12] also show that the MLR allows maximum O_3 concentration to be predicted in city areas within several hours in advance. Banja et al. [13] applied multiple linear regression to predict the next day's maximum ozone concentration for the first time in Tirana, Albania. The relationship between daily maximum ozone values and weather variables was investigated. MLR analysis has been performed to establish the relationship between the weather parameters and peak ozone concentration. It was found out that MLR performed well with the value of $R^2 = 0.87$. Abdullah et al. [55] investigated the variation of O_3 concentrations in Klang, Malaysia from 2012 to 2015. MLR model was developed and signifies that nitrogen oxides (NO), relative humidity (RH), NO_2, CO, wind speed, temperature, and sulphur dioxide (SO_2) are the significant predictors for O_3 concentration. The calculated value of R^2 for MLR is 0.810. Since MLR is a simple linear regression method that can easily be used to correlate other pollutants and weather parameters, it was abundantly used to model O_3 concentration. Hence, from the above mentioned studies, it can be proven that the maximum O_3 concentration was best explained by the simple linear regression.

FFANN gives good prediction for the maximum concentration of O_3; however, it was less accurate than MLR. The reduction percentage of R^2 for prediction model using FFANN and MLR was 8.2%, indicating that FFANN performed slightly less well than MLR in predicting maximum O_3 concentration in Malaysia. The main purpose of using neural artificial networks to model ozone is to capture the non-linear characteristics of the relationship overlooked by a conventional statistical technique (e.g., regression model) [54]. Even though ANN was known as a powerful predictive model, the main factor that influences the accuracy of the model was the associations of air pollutants and weather parameters. It was proven by the research conducted by Pawlak and Jaroslawski [25] that developed artificial neural network models for the prediction of the daily maximum hourly mean of surface ozone concentration for the next day at rural and urban locations in central Poland. The models were generated with six input variables: forecasted basic meteorological parameters and the maximum O_3 concentration recorded on the previous day and number of the month. The mean error (ME) value indicates a tendency to overestimate the predicted values by 4.8 $\mu g/m^3$ for Belsk station and to underestimate the predicted values by 0.9 $\mu g/m^3$ for Warsaw station. The analysis of days when the relative error value was >50% revealed that all predictions with extremely high relative error value were associated.

RBF gives the worst performance in predicting maximum O_3 concentration in Malaysia if compared to MLR and FFANN. Implementation of RBF model to predict air pollutant is still very recent. Abdullah et al. [47] trained and tested the nonlinear model, namely Radial Basis Function (RBF), to predict particulate matter (PM_{10}) concentration in an industrial area of Pasir Gudang, Johor, Malaysia. Daily observations of PM_{10} concentration, meteorological factors (wind speed, ambient temperature, and relative humidity), and gaseous pollutants (SO_2, NO_2, and CO) from 2010 to 2014 were used. Results showed that RBF model was able to explain 65.2% ($R^2 = 0.652$) and 84.9% ($R^2 = 0.849$) variance in the data during training and testing, respectively. This finding was found to be similar to this study, where the prediction of maximum O_3 concentration using RBFANN in Malaysia was in the range of 0.38 to 0.85 (R^2). Thus, it is proven that a nonlinear model has high potential in virtually representing the complexity and nonlinearity of O_3 in the atmosphere without any prior assumptions.

4.2. *Performances of the Modified Models*

Overall, the hybrid models of PCR, PCA-FFANN, and PCA-RBFANN performed worse than their basis models, i.e., MLR, FFANN, and RBFANN, respectively. The lesser accuracy of the models was mainly due to application of principal components (PCs) as an input to the modified models. Hair et al. [56] outlined that the variables needed to be included in the analysis of PCA should be ideally derived from past research studies or based on the judgement of other researchers. However, in Malaysia, there were very limited studies focusing on modeling of O_3 concentration for all regions of Malaysia. Most of the studies were performed at the Lembah Klang that was the most populous area in Malaysia [1,23,45].

As stated previously, the main propose of using principal components as input to the modified models was to reduce the dimension of the dataset, though, PCA as a dimension reduction methodology is applied without considering the association between the dependent variable (O_3 concentration) and independent variables (PM_{10}, CO, NO_2, SO_2, NmHC, UVB, humidity, wind speed, and temperature). Thus, PCA is termed as an unsupervised dimension reduction methodology [57]. The performance of the modified models was not as good as compared to its basis model alone because the principal components that were used as the input to this hybrid model is governed by cumulative of variance during grouping the factors. For example, the percentage of cumulative variances are 69% for Shah Alam, and 66% for Ipoh which is lower than 70% (from Table 7). This means that only 69% and 66% of the total variance is explained. Lower percentage of reliability will affect the performance of hybrid models that used the principal components as the input parameters. In detail, specifically take Shah Alam as an example. The first principal component (PC) contributed 32% (refer Table 8) of the total variance explained that the group of parameters were correlated to the O_3 concentration. However, for the second PC that was calculated 26% of the total variance was less explanatory for the target compared to the first factor. The third factor contributed only 10.6% of the total variance and it can be related to the target.

A few studies obtained similar findings as reported in this study. Ozbay et al. [58] also reported comparable findings with PCR showing significantly lower R^2 than MLR when studying the variation in O_3 at Dilovasi, Turkey. Elbayoumi et al. [57] also reported that the use of PCR does not increase the accuracy in predicting indoor PM_{10} and $PM_{2.5}$ in the Gaza Strip (Palestine) compared with the use of MLR. For both of the studies, significant reductions of ranging from 20% to 30% have been reported [57,58]. Elbayoumi et al. [57] related the poor performance of PCR due to the fact that PCA is an unsupervised dimension reduction methodology.

Furthermore, PCs were usually not understandable. Most of supervised learning algorithms (for example logistic regression, tree-based algorithm, or neural network) can evaluate the importance of input features. These features are important as they help users to distinguish which data were further needed for exploration and features that might be beneficial or worth more [58]. These features, when combined with the machine learning algorithm, are expected to enhance the accuracy of the model [40,41]. Balogun and Tella [40] reported that the input feature of Random Forest when combined with regression contributed to higher accuracy model with high of R^2 (0.97). However, when applying principal components as the input, reduction of predictive tools can be expected since it is done in such a way that the principal components are orthogonal and have the largest possible variances which did not truly interpret the actual situation.

All the issues in applying PCA that were highlighted above might be the reasons that lessen the performances of the modified models in predicting the daytime concentration of O_3 in Malaysia.

5. Conclusions

Six models (MLR, FFANN, RBFANN, and their modified models, namely PCR, PCA-FFANN, and PCA-RBFANN) were developed to predict daytime O_3 level at five specified study areas. Out of six models, MLR outperformed other methods with highest accuracy prediction for all study areas. This indicates that the daytime O_3 level in major urban areas in Malaysia is best described using linear regression. This might be due to very limited or less extreme concentration observed in the O_3 dataset; hence, a linear regression model is applicable to predict daytime O_3 concentration for most urban areas in Malaysia.

For nonlinear models, FFANN gives better prediction compared to RBFANN. The differences in the basis function of the two machine algorithms might be the reason for the worse predictions made by RBFANN. RBFANN has localized basis functions (e.g., Gaussian) whereas FFANN has global basis functions (sigmoid). Since this study predicted O_3 level at the greatest band of solar intensity, sigmoid basis function was more relevant in fitting the dataset.

On the other hand, all the modified models show underperform prediction compared to the unmodified models. The main reason for this reduced efficiency was the input to the modified models. In the modified models, the selected principal components from the output of Principal Component Analysis (PCA) were used as input. These inputs were selected based on the high factor loading and eigenvalues (>1) for each principal component. However, this selection of inputs leads to multidimensional issues and feature redundancy that lead to a less effective model.

Author Contributions: Conceptualization, investigation, project administration, supervision, funding acquisition, and writing—original draft preparation N.M.H. and N.M.N.; software and data curation A.Z.U.-S.; formal analysis, validation, writing—review and editing N.M.N.; data curation and visualization N.M.N. and A.Z.U.-S.; validation, writing—review and editing and funding acquisition A.V.S. and P.V.; investigation, writing—review and editing, G.D.; resources and data curation M.K. All authors have read and agreed to the published version of the manuscript.

Funding: This research was funded by Malaysian Ministry of Higher Education, grant number FRGS/1/2015/TK10/UNIMAP/02/1 (FRGS 9003-00508) and CNFIS Romania, Grant no CNFIS-FDI-2021-0354.

Institutional Review Board Statement: Not applicable.

Informed Consent Statement: Not applicable.

Data Availability Statement: No new data were created or analyzed in this study. Data sharing is not applicable to this article.

Acknowledgments: The author would like to thank the Department of Environment Malaysia for the air pollutant dataset.

Conflicts of Interest: The authors declare no conflict of interest.

References

1. Awang, N.R.; Elbayoumi, M.; Ramli, N.A.; Yahaya, A.S. Diurnal variations of ground-level ozone in three port cities in Malaysia. *Air Qual. Atmos. Health* **2015**, *9*, 25–39. [CrossRef]
2. Yin, Y.; Fook, S.; Glasow, R.V. The influence of meteorological factors and biomass burning on surface ozone concentrations at Tanah Rata, Malaysia. *Atmos. Environ.* **2013**, *70*, 435–446.
3. Tan, K.C.; Lim, H.S.; Zubir, M.; Jafri, M. Prediction of column ozone concentrations using multiple regression analysis and principal component analysis techniques: A case study in peninsular Malaysia. *Atmos. Pollut. Res.* **2016**, *7*, 533–546. [CrossRef]
4. Faris, H.; Alkasassbeh, M.; Rodan, A. Artificial neural networks for surface ozone prediction: Models and analysis. *Pol. J. Environ. Stud.* **2014**, *23*, 341–348.
5. Eum, J.; Kim, H. Effects on Air Pollution in Assaults: Finding from South Korea. *Sustainability* **2021**, *13*, 11545. [CrossRef]
6. Department of Environment Malaysia. *Environmental Quality Report 2018*; Department of Environment Malaysia: Selangor, Malaysia, 2019; pp. 142–156.
7. Teixeira, E.C.; de Santana, E.R.; Wiegand, F.; Fachel, J. Measurement of surface ozone and its precursors in an urban area in South Brazil. *Atmos. Environ.* **2009**, *43*, 2213–2220. [CrossRef]
8. Al-Shammari, E.T. Towards an accurate ground-level ozone prediction. *Int. J. Electr. Comput. Eng.* **2018**, *8*, 1131–1139.
9. Verma, N.; Kumari, S.; Lakhani, A.; Kumari, K.M. 24 Hour Advance Forecast of Surface Ozone Using Linear and Non-Linear Models at a Semi-Urban Site of Indo-Gangetic Plain. *Int. J. Environ. Sci. Nat. Res.* **2019**, *18*, 555982.
10. Verma, N.; Satsangi, A.; Lakhani, A.; Kumari, K.M. Prediction of Ground level Ozone concentration in Ambient Air using Multiple Regression Analysis. *J. Chem. Biol. Phys. Sci.* **2015**, *5*, 3685–3696.
11. Hassanzadeh, S.; Hosseinibalam, F.; Omidvari, M. Statistical methods and regression analysis of stratospheric ozone and meteorological variables in Isfahan. *Phys. A Stat. Mech. Appl.* **2008**, *387*, 2317–2327. [CrossRef]
12. Barrero, M.A.; Grimalt, J.O.; Canto'n, L.M. Prediction of daily ozone concentration maxima in the urban atmosphere. *Chemometr. Intell. Lab. Syst.* **2006**, *80*, 67–76. [CrossRef]
13. Banja, M.; Papanastasiou, D.K.; Poupkou, A.; Melas, D. Atmospheric Pollution Research Development of a short–term ozone prediction tool in Tirana area based on meteorological variables. *Atmos. Pollut. Res.* **2012**, *3*, 32–38. [CrossRef]
14. Allu, S.K.; Srinivasan, S.; Maddala, R.K.; Reddy, A.; Anupoju, G.R. Seasonal ground level ozone prediction using multiple linear regression (MLR) model. *Model. Earth Syst. Environ.* **2020**, *6*, 1981–1989. [CrossRef]
15. Azmi, S.T.; Latif, M.T.; Jemain, A.A. Trend and status of air quality at three different monitoring stations in the Klang Valley, Malaysia. *Air Qual. Atmos. Health* **2010**, *3*, 53–64. [CrossRef]
16. Awang, M.B.; Jaafar, A.B.; Abdullah, A.M.; Ismail, M.B.; Hassan, M.N.; Abdullah, R.; Johan, S.; Noor, H. Air quality in Malaysia: Impacts, management issues and future challenges. *Respirology* **2000**, *5*, 183–196. [CrossRef] [PubMed]
17. Ismail, M.; Abdullah, S.; Yuen, S.F.; Ghazali, N.A. A ten-year investigation on ozone and it precursors at Kemaman, Terengganu, Malaysia. *EnvironmentAsia* **2016**, *9*, 1–8.
18. Ghazali, N.A.; Ramli, N.A.; Yahaya, A.S.; Yusof, N.F.F.M.; Sansuddin, N.; Al Madhoun, W.A. Transformation of nitrogen dioxide into ozone and prediction of ozone concentrations using multiple linear regression techniques. *Environ. Monit. Assess.* **2010**, *165*, 475–489. [CrossRef]
19. Tong, W. Chapter 5-machine learning for spatiotemporal big data in air pollution. In *Spatiotemporal Analysis of Air Pollution and its Application in Public Health*; Li, L., Zhou, X., Tong, W., Eds.; Elsevier: Amsterdam, The Netherlands, 2020.
20. Dou, J.; Yunus, A.P.; Tien Bui, D.; Merghadi, A.; Sahana, M.; Zhu, Z.; Chen, C.-W.; Khosravi, K.; Yang, Y.; Pham, B.T. Assessment of advanced random forest and decision tree algorithms for modeling rainfall-induced landslide susceptibility in the Izu-Oshima Volcanic Island, Japan. *Sci. Total Environ.* **2019**, *662*, 332–346. [CrossRef]
21. Ma, J.; Ding, Y.; Cheng, J.C.P.; Jiang, F.; Tan, Y.; Gan, V.J.L.; Wan, Z. Identification of high impact factors of air quality on a national scale using big data and machine learning techniques. *J. Clean. Prod.* **2020**, *244*, 118955. [CrossRef]
22. Li, R.; Cui, L.; Meng, Y.; Zhao, Y.; Fu, H. Satellite-based prediction of daily SO_2 exposure across China using a high-quality random forest-spatiotemporal Kriging (RF-STK) model for health risk assessment. *Atmos. Environ.* **2019**, *208*, 10–19. [CrossRef]

23. Al-Alawi, S.M.; Abdul-Wahab, S.A.; Bakheit, C.S. Combining principal component regression and artificial neural networks for more accurate predictions of ground-level ozone. *Environ. Model. Softw.* **2008**, *23*, 396–403. [CrossRef]
24. Padma, K.; Samuel Selvaraj, R.; Arputharaj, S.; Milton Boaz, B. Improved Artificial Neural Network Performance on Surface Ozone Prediction Using Principal Component Analysis. *Int. J. Curr. Res. Rev.* **2018**, *6*, 1–6.
25. Pawlak, I.; Jarosławski, J. Forecasting of surface ozone concentration by using artificial neural networks in rural and urban areas in central Poland. *Atmosphere* **2019**, *10*, 52. [CrossRef]
26. Aljanabi, M.; Shkoukani, M.; Hijjawi, M. Ground-level Ozone Prediction Using Machine Learning Techniques: A Case Ground-level Ozone Prediction Using Machine Learning Techniques: A Case Study in Amman, Jordan. *Int. J. Autom. Comput.* **2020**, *17*, 667–677. [CrossRef]
27. Castro, M.; Pires, J.C.M. Decision support tool to improve the spatial distribution of air quality monitoring sites. *Atmos. Pollut. Res.* **2019**, *10*, 827–834. [CrossRef]
28. Zhang, Y.-F.; Fitch, P.; Thorburn, P.J. Predicting the Trend of Dissolved Oxygen Based on the kPCA-RNN Model. *Water* **2020**, *12*, 585. [CrossRef]
29. Banadkooki, F.B.; Ehteram, M.; Ahmed, A.N.; Fai, C.M.; Afan, H.A.; Ridwam, W.M.; Sefelnasr, A.; El-Shafie, A. Precipitation forecasting using multilayer neural Network and support vector machine optimization based on flow regime algorithm taking into Account uncertainties of soft computing models. *Sustainability* **2019**, *11*, 6681. [CrossRef]
30. Ehteram, M.; Ahmed, A.N.; Ling, L.; Fai, C.M.; Latif, S.D.; Afan, H.A.; Banadkooki, F.B.; El-Shafie, A. Pipeline scour rates prediction-based model utilizing a multilayer perceptron colliding body algorithm. *Water* **2020**, *12*, 902. [CrossRef]
31. Ul-Saufie, A.Z.; Yahaya, A.S.; Ramli, N.A.; Rosaida, N.; Hamid, H.A. Future daily PM_{10} concentrations prediction by combining regression models and feedforward backpropagation models with principle component analysis (PCA). *Atmos. Environ.* **2013**, *77*, 621–630. [CrossRef]
32. Hashim, N.I.M.; Noor, N.M.; Annas, S. Influence of meteorological factors on variations of particulate matter (PM_{10}) concentration during haze episodes in Malaysia. In *AIP Conf. Proc.*; AIP Publishing LLC: New York, NY, USA, 2018; Volume 2045.
33. Thupeng, W.M.; Mothupi, T.; Mokgweetsi, B.; Mashabe, B.; Sediadie, T. A Principal Component Regression Model, For Forecasting Daily Peak Ambient Ground Level Ozone Concentrations, in The Presence Of Multicollinearity Amongst Precursor Air Pollutants And Local Meteorological Conditions: A Case Study Of Maun. *Int. J. Appl. Math. Stat. Sci.* **2018**, *7*, 1–12.
34. Ismail, M.; Abdullah, S.; Jaafar, A.D.; Ibrahim, T.A.E.; Shukor, M.S.M. Statistical modeling approaches for PM_{10} forecasting at industrial areas of Malaysia. *AIP Conf. Proc.* **2018**, *2020*, 020044.
35. Taspinar, F. Improving artificial neural network model predictions of daily average PM_{10} concentrations by applying principle component analysis and implementing seasonal models. *J. Air Waste Manag. Assoc.* **2015**, *65*, 800–809. [CrossRef] [PubMed]
36. Bekesiene, S.; Meidute-kavaliauskiene, I. Accurate Prediction of Concentration Changes in Ozone as an Air Pollutant by Multiple Linear Regression and Artificial Neural Networks. *Mathematics* **2021**, *9*, 356. [CrossRef]
37. Lu, W.-Z.; Wang, W.-J.; Wang, X.-K.; Yan, S.-H.; Lam, J.C. Potential assessment of a neural model PCA/RBF approach for forecasting pollution trends in Mongkok urban air, Hong Kong. *Environ. Res.* **2004**, *96*, 79–87. [CrossRef] [PubMed]
38. Tikhamarine; Yazid; Souag-Gamane, D.; Najah Ahmed, A.; Kisi, O.; El-Shafie, A. Improving artificial intelligence models accuracy for monthly streamflow forecasting using grey Wolf optimization (GWO) algorithm. *J. Hydrol.* **2020**, *582*, 124435. [CrossRef]
39. Abobakr Yahya, A.S.; Ahmed, A.N.; Othman, F.B.; Ibrahim, R.K.; Afan, H.A.; El-Shafie, A.; Fai, C.M.; Hossain, M.S.; Ehteram, M.; Elshafie, A. Water quality prediction model based support vector machine model for ungauged river catchment under dual scenarios. *Water* **2019**, *11*, 1231. [CrossRef]
40. Balogun, A.-L.; Tella, A. Modelling and investigating the impacts of climatic variables on ozone concentration in Malaysia using correlation analysis with random forest, decision tree regression, linear regression, and support vector regression. *Chemosphere* **2022**, *299*, 134250. [CrossRef]
41. Ayman, Y.; AlDahoul, N.; Birima, A.H.; Ahmed, A.N.; Sherif, M.; Sefelnasr, A.; Allawi, M.F.; Elshafie, A. Comprehensive comparison of various machine learning algorithms for short-term ozone concentration prediction. *Alex. Eng. J.* **2022**, *61*, 4607–4622.
42. Kaiser, H.F. An index of factorial simplicity. *Psychometrika* **1974**, *39*, 31–36. [CrossRef]
43. Brūmelis, G.; Brown, D.H.; Nikodemus, O.; Tjarve, D. The monitoring and risk assessment of Zn deposition around metal smelter in Latvia. *Environ. Monit. Assess.* **1999**, *58*, 201–212. [CrossRef]
44. Juahir, H.; Zain, S.M.; Yusoff, M.K.; Tengku Hanidza, T.I.; Mohd Armi, A.S.; Toriman, M.E.; Mokhtar, M. Spatial water quality assessment of Langat River Basin (Malaysia) using environmetric techniques. *Environ. Monit. Assess.* **2011**, *173*, 625–641. [CrossRef]
45. Azid, A.; Juahir, H.; Latif, M.T.; Zain, S.M. Feed-Forward Artificial Neural Network Model for Air Pollutant Index Prediction in the Southern Region of Peninsular Malaysia. *J. Environ. Prot. Sci.* **2013**, *4*, 40509. [CrossRef]
46. Azid, A.; Juahir, H.; Toriman, M.E.; Kamarudin, M.K.A.; Saudi, A.S.M.; Hasnam, C.N.C.; Aziz, N.A.A.; Azaman, F.; Latif, M.T.; Zainuddin, S.F.M.; et al. Prediction of the Level of Air Pollution Using Principal Component Analysis and Artificial Neural Network Techniques: A Case Study in Malaysia. *Water Air Soil. Pollut.* **2014**, *225*, 2063. [CrossRef]
47. Abdullah, S.; Mohd Napi, N.N.L.; Ahmed, A.N.; Wan Mansor, W.N.; Abu Mansor, A.; Ismail, M.; Abdullah, A.M.; Ramly, Z.T.A. Development of Multiple Linear Regression for Particulate Matter (PM_{10}) Forecasting during Episodic Transboundary Haze Event in Malaysia. *Atmosphere* **2020**, *11*, 289. [CrossRef]

48. Sun, G.; Hoff, S.J.; Zelle, B.C.; Nelson, M.A. Development and Comparison of Backpropagation and Generalized Regression Neural Network Models to Predict Diurnal and Seasonal Gas and PM_{10} Concentrations and Emissions from Swine Buildings. *Trans. Am. Soc. Agric. Biol. Eng.* **2008**, *51*, 685–694.
49. Gvozdic, V.; Kovac-Andric, E.; Brana, J. Influence of meteorological factors NO_2, SO_2, CO and PM_{10} on the concentration of O_3 in the urban atmosphere of Eastern Croatia. *Environ. Model. Assess.* **2011**, *16*, 491–501. [CrossRef]
50. Ahmat, H.; Yahaya, A.S.; Ramli, N.A. PM_{10} Analysis for Three Industrialized Areas using Extreme Value. *Sains Malays.* **2015**, *44*, 175–185. [CrossRef]
51. Ghazali, N.A.; Yahaya, A.S.; Mokhtar, M.I.Z. Predicting Ozone Concentrations Levels Using Probability Distributions. *ARPN J. Eng. Appl. Sci.* **2014**, *9*, 2089–2094.
52. Ul-Saufie, A.Z.; Yahaya, A.S.; Ramli, N.A.; Hamid, H.A. Performance of Multiple Linear Regression Model for Longterm PM_{10} Concentration Prediction based on Gasesous and Meteorological Parameters. *J. Appl. Sci.* **2012**, *12*, 1488–1494. [CrossRef]
53. Abdullah, S.; Ismail, M.; Ahmed, A.N. Multi-layer perceptron model for air quality prediction. *Malays. J. Math. Sci.* **2019**, *13*, 85–95.
54. Kumar, N.; Middey, A.; Rao, P.S. Prediction and examination of seasonal variation of ozone with meteorological parameter through artificial neural network at NEERI, Nagpur, India. *Urban Clim.* **2017**, *20*, 148–167. [CrossRef]
55. Abdullah, A.; Ismail, M.; Fong, S.Y. Multiple Linear Regression (MLR) Models for Long Term PM_{10} Concentration Forecasting During Different Monsoon Seasons. *J. Sustain. Sci. Manag.* **2017**, *12*, 60–69.
56. Hair, J.F.; Anderson, R.E.; Tatham, R.L.; Black, W.C. *Multivariate Data Analysis with Reading*, 4th ed.; Prentice-Hall: Englewood Cliffs, NJ, USA, 1995.
57. Elbayoumi, M.; Yahaya, A.S.; Ramli, N.A.; Noor Md Yusof, N.F.F.; Al Madhoun, W.; Ul-Saufie, A.Z. Multivariate methods for indoor PM_{10} and $PM_{2.5}$ modelling in naturally ventilated schools buildings. *Atmos. Environ.* **2014**, *94*, 11–21. [CrossRef]
58. Ozbay, B.; Keskin, G.A.; Dogruparmak, S.C.; Ayberk, S. Multivariate methodsforground level ozone modeling. *Atmos. Res.* **2011**, *102*, 57–65. [CrossRef]

sustainability MDPI

Article

COVID-19 and Households Waste in Hispanic America: An Assessment of Trends

Walter Leal Filho [1], Amanda Lange Salvia [2], Javier Sierra [1,3,*], Carly A. Fletcher [4], Craig E. Banks [4], Luis Velazquez [5], Rosley Anholon [6], Izabela Simon Rampasso [7], Claudia Maclean [8], Jelena Barbir [1] and Samara Neiva [9]

1 Faculty of Life Sciences, Hamburg University of Applied Sciences, 21033 Hamburg, Germany
2 Graduate Program in Civil and Environmental Engineering, University of Passo Fundo, Passo Fundo 99052-900, Brazil
3 Research Centre on Global Governance, Educational Research Institute, Department of Applied Economics, Faculty of Law, University of Salamanca, 37007 Salamanca, Spain
4 Faculty of Science and Engineering, Manchester Metropolitan University, Manchester M1 5GD, UK
5 Industrial Engineering Department, University of Sonora, Hermosillo 83100, Mexico
6 School of Mechanical Engineering, University of Campinas, São Paulo 13083-970, Brazil
7 Departamento de Ingeniería Industrial, Universidad Católica del Norte, Antofagasta 1270709, Chile
8 Centro de Investigación GAIA Antártica (CIGA), Universidad de Magallanes, Avenida Bulnes 0185, Punta Arenas 6210427, Chile
9 Graduate Program in University Management, Federal University of Santa Catarina, Florianópolis 88040-900, Brazil
* Correspondence: jsierra@usal.es

Citation: Leal Filho, W.; Lange Salvia, A.; Sierra, J.; Fletcher, C.A.; Banks, C.E.; Velazquez, L.; Anholon, R.; Rampasso, I.S.; Maclean, C.; Barbir, J.; et al. COVID-19 and Households Waste in Hispanic America: An Assessment of Trends. *Sustainability* **2022**, *14*, 16552. https://doi.org/10.3390/su142416552

Academic Editor: Ming-Lang Tseng

Received: 2 November 2022
Accepted: 6 December 2022
Published: 9 December 2022

Abstract: The COVID-19 pandemic has caused many social and economic problems in Hispanic America, a region with fragile health and economic systems and many inequalities. The pandemic has negatively influenced various aspects of life and led to changes in various habits and behaviours, including consumption. However, the extent to which the pandemic has influenced households, and waste production, in particular, is not well known. In this context, this paper reports on a study aimed at identifying changes in waste production across households under the special conditions created by the pandemic in Hispanic America. The majority of the respondents stated that their level of satisfaction with waste management policies in their countries did not change much during the pandemic. Only a few stated that they were more satisfied than before. Overall, the results suggest that, like previous crises, the COVID-19 outbreak may generate changes regarding household consumption and waste management in Hispanic America. At the same time, these findings stress the need to improve waste management practices. Some measures that may be adopted to allow Hispanic American countries to better cope with increases in waste production in times of pandemics are listed.

Keywords: COVID-19; sustainability; consumption; waste management; behaviour; Latin America

1. The COVID-19 Pandemic in Latin America

The COVID-19 pandemic has significantly impacted the lives of many [1,2] and has certainly changed the way people behave, act, and face the future [3,4]. As a result of the relevant decrease in greenhouse gas emissions and the subsequent air pollution, some positive outcomes were observed for ecosystems worldwide during a short period at the beginning of the pandemic. Nevertheless, there were also inevitable negative effects generated by the rapid increase in demand for plastic products such as food and grocery delivery packaging [5] or medical and personal protective equipment [6]. In this regard, plastic pollution is crucial because of the extensive footprints of plastics on natural ecosystems and public health [7].

In some regions, changes are likely to be more extreme and their impact felt for an extended period, exacerbating existing problems associated with poverty and social inequity. Indeed, this is the case in Latin America, where the pandemic has instigated a humanitarian crisis [8,9], already identified as the worst in recent history [10].

Regarding health care, most Latin American countries present fragile and segmented systems that make their populations vulnerable [8,11,12]. Notwithstanding differences between countries, common problems occurring across Latin America include the inability of healthcare systems to deal with the considerable increase in infection resulting from the second wave of COVID-19 and the appearance of new virus variants [13].

In addition, by aggravating the problems associated with the informal economy, instigating high levels of unemployment, and increasing gender inequalities, the pandemic has led to an increase in the number of people experiencing poverty or extreme poverty across Latin America [9,14]. Indeed, much of the social progress achieved over the last 20 years has practically been reversed in the last twelve months [8]. Projections from the Inter-American Development Bank (IDB) highlight increased poverty and public debt in the short term when the actions by governments in the region are taken into account [15].

A survey performed by the United Nations Economic Commission for Latin America and the Caribbean (ECLAC) across 18 Latin American countries found that in 2020, 33.7% of the population was living in poverty conditions (circa. 209 million people) and 12.5% in extreme poverty conditions (circa. 78 million people) [16]. As well as the acute impacts felt by individuals, high levels of poverty also have implications for broader society. Poverty and other social inequalities have grave implications for waste management. Indeed, they have been shown to encourage waste generation, limit access to waste infrastructure, and hinder policy implementation [17]. In addition to presenting a global health crisis and causing an economic crisis in many countries, the COVID-19 pandemic has driven a distinct shift in household consumption behaviours and exacerbated environmental problems due to the increased consumption of single-use goods, take away products, and pre-packed food [5].

Regarding the environmental aspects of the region, Latin America has presented serious deficiencies regarding the lack of wastewater treatment and waste management of plastic materials. The pandemic aggravated the scenario due to the increased utilization of single-use plastic materials, in particular, the use of plastic personal protective equipment. The situation can be more critical when the ocean coast is considered [18]. Previous studies have addressed this issue from a local or national perspective, but there seems to be a gap in understanding the challenges generated by the pandemic regarding waste management from a regional perspective.

Against this background, this research aims to explore the effects generated by the COVID-19 pandemic on household waste in Hispanic America. For this purpose, we applied a survey used by Leal Filho et al. (2021) [5]. This research presents several contributions to the literature. First, it offers preliminary results regarding changes experienced by households' waste management habits because of the COVID-19 pandemic. Second, it shows some behavioural trends common to several countries in the region. It also presents some changes in consumption habits associated with the COVID-19 pandemic in Hispanic American countries, as well as some interesting lessons regarding some of the areas where attention is needed in terms of responses to pandemic situations.

The paper is structured as follows. It starts by reviewing the impact of the COVID-19 pandemic in Latin America, with a special focus on Hispanic America. It then continues by presenting the research methodology applied in this study. This is followed by a detailed presentation and discussion of the main results. The last section is dedicated to a conclusion, where the main contributions are presented, as well as some limitations and further lines of research.

2. Impact of COVID-19 on Household Consumption and Waste Generation

The relevance of proper solid waste management is commonly underestimated as regards its effects on public health [19]. When focused on household consumption and subsequent waste generation, the various impacts of COVID-19 can be categorized across three interlinking themes: structural constraints, economic circumstances, and social-psychological factors, which are noted variably within the literature, for example, [20–23], and are explored in more detail in the ensuing text.

Structural constraints are caused by direct measures, such as national lockdowns and explicit stay-at-home instructions [23], as well as indirect consequences from the closure of extradomestic consumption channels, for example, restaurants/cafes, schools, and workplace canteens [24], the cancellation of mass events, and mobility restrictions [23]. Globally, these structural constraints changed consumption habits by forcing households to prepare and consume all their meals at home [25]. In turn, this meant that the generation of waste changed location from a place(s) of work (often serviced via commercial waste management contracts) to generation within the household (serviced via domestic services or local authorities) [26]. Indeed, [24] found that the partial reallocation from extra-domestic (i.e., restaurants, cafes, canteens, etc.) consumption to home preparation caused a 12% increase in the amount of food waste generated. Likewise, a 5–30% increase in the amount of residential solid waste generated was reported in New York City [27], complemented by a 50% reduction in the amount of waste generated by the commercial and industrial sectors.

The onset of mobility restrictions led to the expansion of online shopping platforms, where consumers can purchase products and have them delivered to their households with ease [28]. Indeed, [29] noted that the number of consumers using grocery pickup services and other delivery services increased significantly from the onset of the pandemic, driven by fears over COVID-19 and concerns about safety. The increased use of online shopping outlets and home delivery services had consequences for energy use (e.g., in transportation) and waste production [30]. Concerning the latter, the increased purchasing of online food and groceries led to the increased generation of common packaging wastes such as single-use plastics and cardboard [26].

Changes in economic circumstances also had implications for consumer behaviour. The increased prevalence of working from home, redundancies, and furlough schemes has led to households having more spare time and sustained cohabitation among families [25]. While this may be a positive consequence for some, other negative impacts have become apparent, particularly where reduced pay and redundancies jeopardize household food security [31]. Furthermore, the subsequent negative influence on disposable income affects the net volume and patterns of household consumption. A shift has been noted from the consumption of durable and non-essential goods towards food staples and other essentials [32].

The perception of risk and the subsequent behaviours employed to minimize it have impacted household consumption through the increased prevalence of panic purchasing [24], stockpiling, and overbuying [29]. The reduction in shopping frequency is also attributed to anxiety and a need to minimize perceived risks, coinciding with a decline in the purchase of fresh produce, where consumers have favoured non-perishable items instead [23]. These behaviours, driven by socio-psychological factors, have been exacerbated by supply chain issues and food shortages within stores [29]. The implications of increased panic buying, stockpiling, and overbuying on waste generation include the increased disposal of perishable products and leftovers [33]. Here, the panic-buying behaviours noted at the beginning of the pandemic led to more food waste being generated due to a lack of foresight, bad cooking habits, limited storage, and the hoarding/overbuying of food items with low-shelf lives [24,26].

Furthermore, increased levels of stress and anxiety experienced during the pandemic have led to the greater consumption of "comfort foods" [23]. While the fondness for "comfort foods" (sensorily appealing foods high in fat, sugar, and carbohydrates) has been universal during the pandemic, specific food groups have differed among nations. For example, European studies have shown that Spanish consumers ate more indulgent foods and snacks and drank more alcohol throughout the day [24]. In comparison, the Danish consumed more pastries and alcohol, the Norwegians consumed more high-sugar food and beverages [23], and the French consumed more processed meats, sugary foods and beverages, and alcohol [34]. In contrast, Italian residents, as well as consuming more chocolate, chips, and snacks [23], increased their consumption of homemade pizza, bread, and desserts [25], items that have symbolic family and cultural values and are linked with pre-existing socialization habits [35].

A further point of note is the implication of these changing behaviours on attitudes towards sustainability. Indeed, Escursell, Llorach-Massana, and Roncero (2021) highlight that while consumers are becoming more environmentally conscious regarding their purchases from brick-and-mortar stores, when purchasing goods from online outlets, other variables such as price, volume, and time of delivery are prioritized over sustainability. This trend has been mirrored during the COVID-19 pandemic with a returned preference for single-use plastics, where the action to reduce the risks of contamination/infection outweighs any ambitions to behave more sustainably [20].

In summary, the COVID-19 pandemic has accentuated existing problems across Latin America and cancelled out the social gains achieved over the past two decades [8]. As Rivarola Puntigliano [10] suggests, the current situation is not simply a pandemic crisis, but a "multiple crises" scenario. Sharma et al. [20] note that the advent of the pandemic has resulted in an unprecedented socio-economic-ecological crisis that has, and will continue to, threaten the lives and livelihoods of people while rewinding decades of sustainable development. Literature from around the world has also indicated that the ongoing pandemic has implications for waste management through increased levels of poverty [27] and/or changes to household consumption [36]. Indeed, Liang et al. [36] conclude that changes to consumption habits during the pandemic have "without exception" led to difficulties in plastic waste management and the reversal of policies that focus on the reduction in single-use plastic consumption. While studies have investigated the impact of the pandemic on consumption and waste within prominent nations such as China, Japan, Brazil, Canada, and Germany [23,36,37], impacts across Latin America are less well known. At the same time, in line with the effects that the pandemic generated in other parts of the world, it poses some challenges for household waste management, but at the same time, it contributed to the stress of some opportunities for improving the provision of this public service at the local level [38]. In this context, this research aims to shed light on the effects of the pandemic on household waste production across Hispanic America.

3. Methodology

To explore the implications of the COVID-19 pandemic on waste generation across Hispanic America, this study employed a quantitative approach. The objective was to evaluate the opinion of households across Hispanic America concerning changes in household consumption and in waste production as a consequence of the pandemic. Data were collected using an online survey developed by the research team, and the results were analysed using descriptive statistics. The survey was based on Leal Filho et al. (2021) [5], who performed an international study on the increased consumption and subsequent changes in the amounts of waste produced since the COVID-19 pandemic.

Consequent to the purpose of this research, the novel aspect considered in the present study was the inclusion of a section dedicated to the formulation of public policies, and the perception of respondents regarding public waste management during the pandemic. For this reason, the survey contained four sections: Section 1: Background: to collect information on the respondents' country, gender, age group, level of education, occupation, household income, housing, living place, and stages of lockdown. Section 2: Level of Consumption: to assess the respondents' perception of the changes in the consumption of packed, fresh, and delivered food. Section 3: Waste generation: to assess the respondents' perception of the changes in waste generation and the reasons for these changes. Section 4: Waste Management and Formulation of Public Policies: to assess the respondents' perception of the changes in household waste management and in public policies related to waste management. Items in each of the four sections were identified to capture the major structural issues regarding the perception of households about the problems addressed in this research. The survey included 31 questions in total: 11 questions regarding the general background (Section 1); 7 questions about the level of consumption (Section 2); 4 questions concerning waste generation (Section 3); and 9 questions focused on waste management and the formulation of public policies (Section 4).

The questionnaire was validated by a group of international and regional experts in the fields of environment and sustainability, waste management and recycling, sustainable consumption and production, and public health. Some of the comments from the expert validation included adjustments in the number of sections and items within each section and adaptation of some questions/options for clarity. Upon addressing the feedback from the expert validation process, a pilot application (i.e., a pre-test) was performed with six respondents to ensure completeness and clarity. The findings from this pilot application revealed that the survey instrument was satisfactory, with minor changes. The final survey instrument was created in Spanish using Google Forms (Google LLC, Mountain View, CA, USA) and is presented in Supplementary Materials.

To recruit participants, a purposive sampling approach was employed, in combination with techniques such as chain referrals and snowballing. This consisted of an invitation to participate alongside a link to the online survey, which was shared with Hispanic American audiences via the research team, the networks of the European School of Sustainability Science and Research (ESSSR), and the Inter-University Sustainable Development Research Programme (IUSDRP). Data collection was carried out between March and April 2021. In line with research ethics protocols, the participants were informed that participation was entirely voluntary, that all responses would be treated in the strictest confidence, and that data protection rules would always be upheld. The results of the survey are presented in the subsequent section and all figures show information gathered using that questionnaire.

4. Results and Discussion

4.1. Demographics

As mentioned earlier, we applied an adapted version of the survey used by Leal Filho et al. [5], and the study received a total of 102 responses from consumers in the following 10 Hispanic American countries: Argentina (17%), Bolivia (2%), Chile (13%), Colombia (15%), Costa Rica (2%), Dominican Republic (1%), Ecuador (5%), Guatemala (8%), Mexico (34%), and Peru (4%). Their distribution across the region is presented in Figure 1. Responses were not solicited from Brazilian consumers in this study as detailed research has already been performed on Brazilian consumers in the Portuguese language. Therefore, the results represent a preliminary analysis of the situation in the region that shows trends in several Hispanic-American countries.

Descriptive statistics from participants reveal that the majority of the respondents are female (71%) and have achieved higher educational levels (31% undergraduate and 51% postgraduate level). Over half of the respondents are within the youngest age group (18–29 years, 54%); correspondingly, 40% belong to the category "Student" in the occupation assessment.

Nevertheless, as the survey focused on households, the background section also collected information on net monthly household income, type of household, and household composition to help better comprehend the socio-economic characteristics of the participants and allow the identification of trends in different countries within the region. Almost 50% of respondents have a net monthly income between USD501 and USD1500, and the majority live in detached houses (63%). As per household characteristics, in 44%, the number of people is two adult residents in total (44%), and the number of households with no children is 66%. Complete demographic categories and distributions can be seen in Table 1.

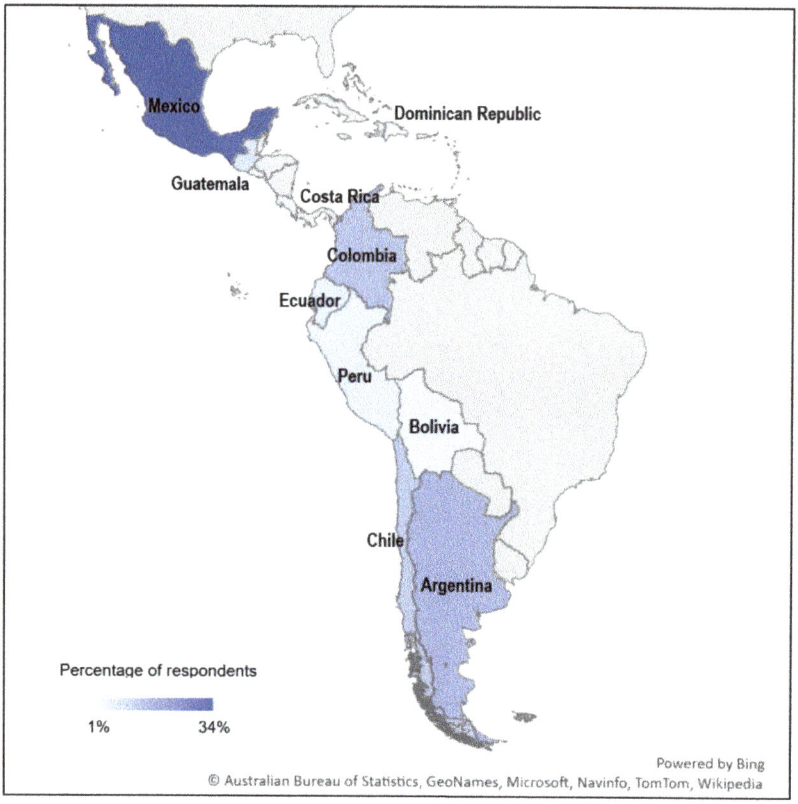

Figure 1. Survey participating countries.

With respect to the national pandemic response, the use of widespread lockdowns was particularly common in all countries included in this research. The results show that respondents experienced the trend of full lockdown during the worst phase of the pandemic (67%) and partial lockdown during survey participation (March–May 2021; 65%).

4.2. Consumption

Regarding household consumption during the pandemic, 30% of respondent households reported an increase in the consumption of industrially processed foods, while 26% said that the consumption of such products had been reduced. It is important to mention that 43% of the respondents did not report a change in this aspect of their food consumption. Overall, this indicates that the trend of change was minimal, with those that increased their consumption of industrially processed food balancing out those that decreased consumption.

On the other hand, trends in the consumption of organic and natural food paint a different picture. Indeed, 51% of respondents stated an increase in the consumption of organic and natural foods such as fruits and vegetables during the COVID-19 pandemic. Among the respondents that increased their consumption of organic and natural food, less than a quarter (23%) increased their intake by 30%, while 40% increased their intake by only 10–20%. Only 10% of respondents reduced their consumption of organic and natural foods. Figure 2 compares the results of these categories.

Table 1. Sample demographic characteristics.

Age Group	Distribution	Net Monthly Household Income	Distribution
18–29	54%	Below USD500	6%
30–39	26%	USD500 to USD1000	25%
40–49	13%	USD1001 to USD1500	21%
50–59	5%	USD1501 to USD2000	11%
60+	2%	USD2001 to USD2500	8%
Education Level	Distribution	USD2501 to USD3000	4%
Postgraduate	51%	Above USD3000	14%
University	31%	Prefer not to say	13%
High School	14%	Type of housing	Distribution
Primary school/secondary school	4%	Detached house	63%
Gender	Distribution	Flat	34%
Female	71%	Semi-detached house	2%
Male	29%	Rural	1%
Occupation	Distribution	Number of adults	Distribution
Student	40%	1	14%
Skilled Labourer	15%	2	44%
Consultant	10%	3	21%
Administrative Staff	10%	4 or more	21%
Self-employed/Partner	7%	Number of children (<18 yo)	Distribution
Middle Management	7%	none	66%
Temporary Employee	5%	1	19%
Upper Management	4%	2	11%
Junior Management	2%	3	4%
Trained Professional	1%	4 or more	0%

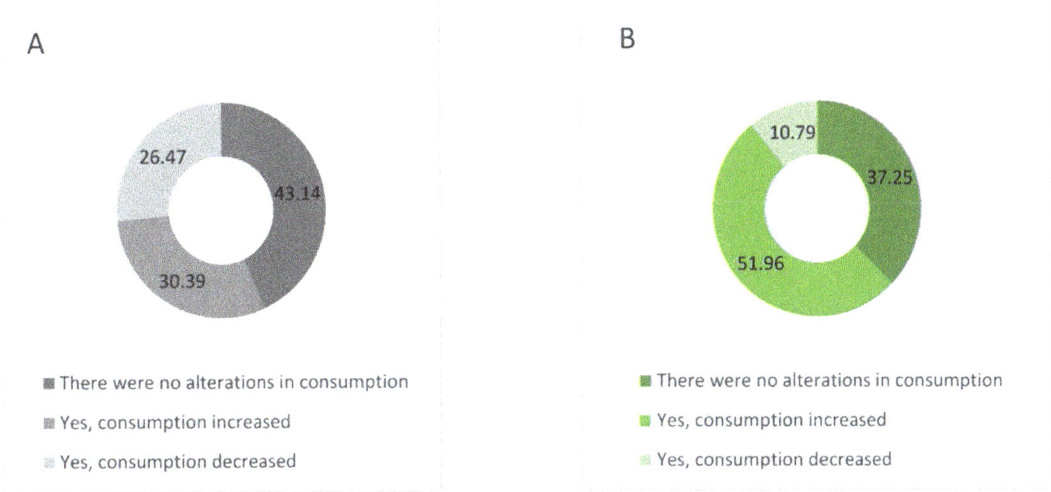

Figure 2. Variation in the consumption of industrially processed foods (**A**) and of fresh/organic products (e.g., fruits and vegetables) (**B**) during confinement in 2020.

These results indicate a trend towards the increased consumption of organic and natural foods, which is aligned with Ouhsine et al. [39]. This study, based in Morocco, observed a slight increase in the consumption of fruits and vegetables during the lockdown alongside a corresponding drop in the consumption of meat products. As detailed in that article, this trend can be explained by consumer intentions to strengthen the immune system through the increased intake of fibre and vitamins, often found in fruit and vegetables. As well as a change in the types of products consumed, the survey respondents indicated a change in how products were purchased. In this research, 63% of the households acknowledged an increase in the use of online shopping and home delivery channels, while less than a third (29%) stated no change, and only 8% reported a decrease.

Overall, these results indicate a trend towards the increased consumption of organic and natural foods, which is aligned with Ouhsine et al. [39]. This study, focused in Morocco, observed a slight increase in the consumption of fruits and vegetables during the lockdown alongside a corresponding drop in the consumption of meat products. Likewise, a study by Di Renzo et al. [40] in Italy also demonstrated that 15% of the interviewed respondents changed to the organic consumption of fruits and vegetables during the lockdown and a reduction in the consumption of savoury snacks and processed meat. A similar trend is seen in another study conducted in Morocco, Algeria, and Tunisia during the COVID-19 pandemic [41]. This study also shows the increased consumption of healthy foods such as organic and biobased foods to improve people's immunity to prevent infection. These trends can be explained by consumer intentions to strengthen the immune system through the increased intake of fibre and vitamins, often found in fruit and vegetables.

However, the findings of Janssen et al. [23] show that there is less consumption of fresh fruits and vegetables in Denmark, Germany, and Slovenia and more consumption of canned food, frozen food, and cake and biscuits during the COVID-19 pandemic. This can be explained by people's intention to reduce shopping frequency due to the high risk and anxiety of COVID-19, which led to an increase in the consumption of food with longer shelf life during the pandemic. Another study in Italy [35] revealed that people tended to take more pasta, flour, and long-life frozen fruits than fresh fruits and vegetables during the lockdown, which is consistent with the findings of Janssen et al. [23]. Concerning changes in the types of products consumed, the survey respondents indicated a change in how products were purchased. Here, 63% acknowledged an increase in the use of online shopping and home delivery channels. Less than a third (29%) stated no change, and only 8% reported a decrease.

Overall, the results indicate that the pandemic (specifically during lockdowns) had a marked impact on the way households were purchasing goods, and thus consumption behaviours, with an increasing trend in online shopping and home-delivered products. On the other hand, it is important to note that this trend did not extend out to takeaway food, since consumers opted to cook more at home. Of course, the products consumed, and the channels used to consume them can have a significant impact on the amount (and composition) of waste generated.

4.3. Waste Generation

Concerning household waste generation, the US HUNTER study [42] reported that home cooking increased in 54% of USA households during the lockdown. This supports findings from this study, where 57% of the respondents reported a reduction in food waste, possibly explained by the increased usage of recipes and time to prepare the food. Clearly indicating more sustainable human behaviours during the COVID-19 pandemic.

However, when efforts to ensure the correct separation of waste are considered, a different story appears.

In this research, over half of the respondents (57%) claimed to make no changes during the pandemic, in terms of efforts to separate wastes correctly (Figure 3). That being said, about 41% of respondents did state that efforts to properly separate waste were increased.

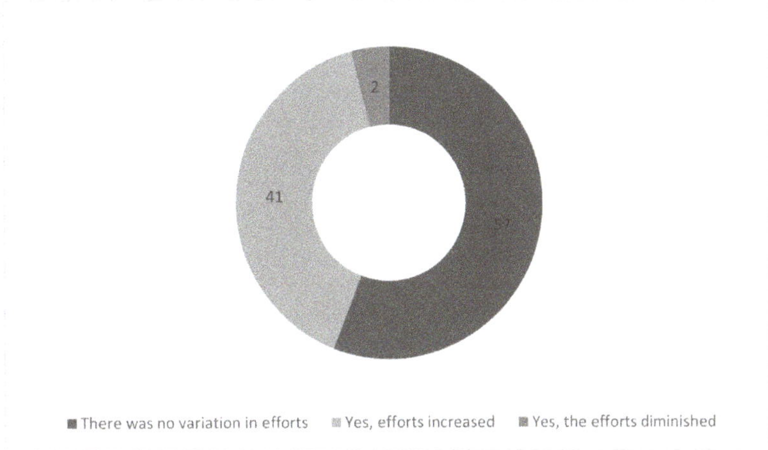

Figure 3. Variation of efforts to separate waste during confinement (in percentage).

Against this background, it is important to analyse barriers to waste separation and address the main challenges that hinder such activities within homes. When asked what the main difficulties encountered were, 49% of respondents stated the growing number (and types) of packages that are destined for waste (Figure 4). This is especially pertinent due to the many adaptations among businesses that changed the ways in which households purchase and consume food. On the other hand, 42% of the respondents affirmed that the requirements to use masks and other personal protective equipment (PPE) during the pandemic have caused a new type of health waste to emerge within their homes. In line with this, 32% of the respondents declared challenges to maintaining or increasing waste management efforts, and 28% of the participants recognized problems managing the increased amount of food waste. It is important to highlight that only 1% of the respondents declared no challenges in managing waste generated during the COVID-19 pandemic, stressing how relevant changes in households were. These results are in line with previous findings that stressed the fact that the increased waste generation experienced during the pandemic exacerbated risks for the environment and human health [7].

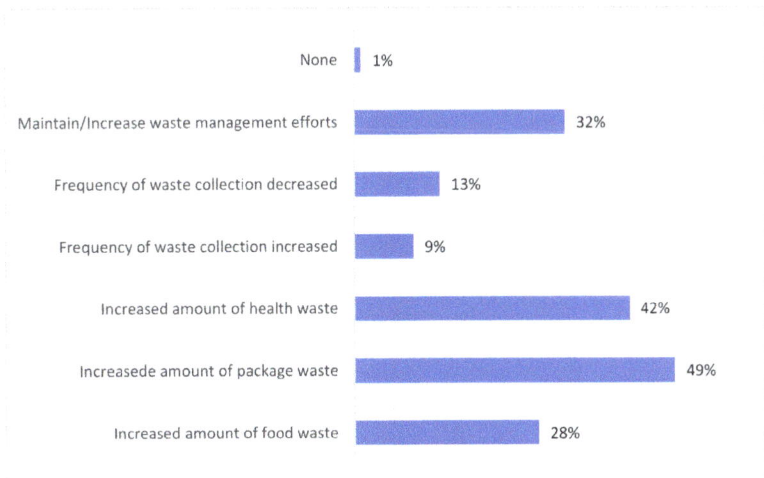

Figure 4. The main challenges regarding waste management during confinement (in percentage).

Similar findings were reported by Sarkodie et al. [33], where the use of single-use PPE such as gloves and masks led to an increase in the generation of domestic solid waste. These findings are supported by Chenarides et al. [29], where the results obtained from a survey in the US found that three-quarters of the respondents started to buy more food due to a lack of stock in supermarkets, and half of the respondents bought more food than usual. Furthermore, the number of participants using food collection services increased by about 255%, and those using food delivery services increased by 158%, thus considerably increasing domestic solid waste generation.

Finally, this study indicates the potential implications of the pandemic on consumer attitudes and behaviours, particularly concerning the management of waste. As shown in Figure 5, behavioural changes at a household level detected a growing interest in composting (32%) and recycling (40%) as well as increased efforts to eat healthier (51%). This increased awareness regarding waste generation was reported by 33% of respondents, the same share who declared increased efforts to control food waste.

Figure 5. Which items can be considered a result of the lockdown?

4.4. Waste Generation and Public Policy Formulation

Alongside consumer attitudes and behaviours concerning waste generation at a household level, this survey also sought to assess the extent to which views and policies regarding waste management, at a public level, have been affected by the pandemic. When asked if public authorities within their city made any changes to the waste collection and separation processes during the pandemic, 48% of respondents reported no change, with 27% reporting some change, and 25% stating that they did not know or preferred not to answer.

The participants were also asked which measures they thought should be introduced or intensified to improve waste management during a disaster and/or pandemic situations in their home countries and by local authorities. The key results largely referred to product design, stakeholders' consultation, and public policy. As shown in Figure 6, when presented with six proposed measures that could be introduced or intensified during subsequent pandemic or disaster situations, most respondents indicated that five out of the six proposals could help public authorities to identify citizens' priorities and thereby improve waste management services [20]. These results, comparable to Tchetchik et al. [43], suggest a trend towards increasing recycling, as well as initiatives that improve sustainable consumption and waste management.

Additionally, the participants were asked to indicate their level of agreement regarding the adequacy of current public policies and procedures for waste collection and management. As presented in Figure 7, the findings indicate a strong tendency towards dissatisfaction regarding current public policies and procedures. Across the six statements, on average, 32% of the respondents strongly disagreed, while 36% disagreed. These results suggest that citizens perceive much room for waste management improvement, confirming previous findings by Fan et al. [22].

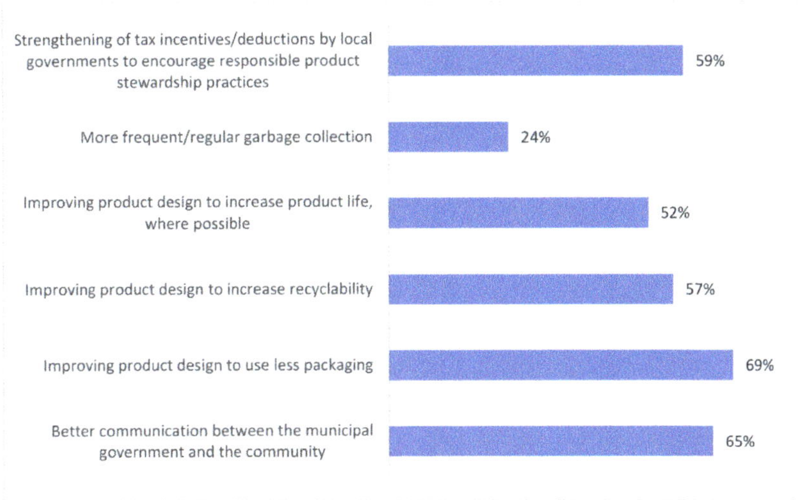

Figure 6. Measures that respondents consider should be intensified to improve waste management during a disaster and/or pandemic situation.

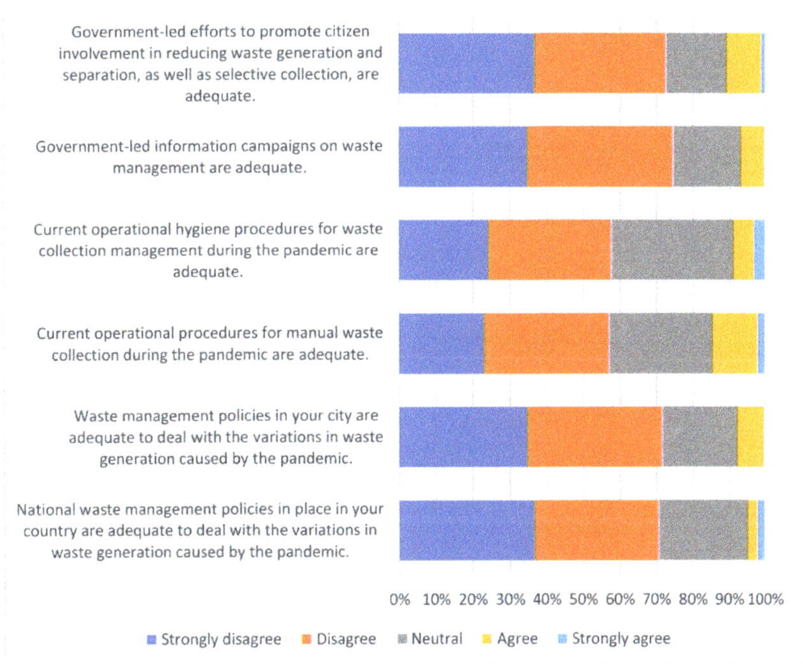

Figure 7. Participants' perception of public policies and procedures for waste collection and management.

Likewise, the respondents were asked to declare their degree of satisfaction with the political activities within their own country with respect to waste management. When considering six critical issues (as shown in Figure 8), the results highlight a relatively low level of citizen fulfilment regarding current public policies, with on average 39% of the responses corresponding to not satisfied and 41% to moderately satisfied.

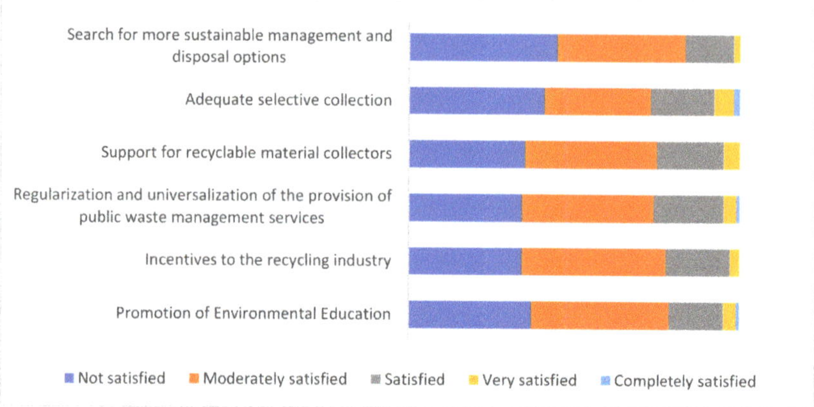

Figure 8. Participants' degree of satisfaction with the political action in waste management in their own countries.

Finally, the respondents were asked if their level of satisfaction with the waste management policy in their country had changed during the pandemic. Most of the respondent participants indicated that there has been no recognizable difference (63%), while 25% declared lower satisfaction, and only 12% stated that they were more satisfied than before.

Overall, these results suggest that, like previous crises, the COVID-19 outbreak may generate changes regarding household consumption and waste management in Hispanic America [44]. At the same time, these findings stress the need to improve waste management efficiency [45]. More precisely, this research indicated that (1) household routines changed, i.e., more working from home, more domestic consumption; (2) consumer behaviours changed, i.e., more online shopping, healthier food, and more home cooking; (3) attitudes to sustainability improved, i.e., increased interest in recycling and composting; and (4) an overall dissatisfaction towards current public policies.

5. Conclusions

This paper presents a study on the impact of the COVID-19 pandemic on household consumption and domestic waste production across Hispanic America. This research has a quantitative nature and uses descriptive statistics to analyse its results. Although the sample is non-probabilistic and follows a distribution that is not completely representative of the population of Hispanic America, the study serves to shed light on the topic and fill in the literature gap, but with no ambition of claiming to be comprehensive.

The results obtained allow some conclusions to be drawn. For instance, the sample revealed that there were no clear trends in the household consumption of industrially processed products. However, in contrast, over half of the respondents reported the increased consumption of organic and natural foods during the pandemic. In addition, the changes in demand and habits influenced better separation behaviours by about 40% of the respondents. This may be because people were at home, and either they were more aware of the waste they generated or they had more time to complete separation activities. Nevertheless, some of them have stated they have faced challenges in respect of waste separation, both in terms of the diversity of waste (e.g., food waste or packaging) and in the disposal of face masks. On the other hand, there were some positive changes associated with household waste generation. For instance, the growing interest in composting and recycling, as well as increased attention towards healthier eating, can be associated with a reduction in the amounts of waste produced, providing an even greater benefit. The fact that public authorities did not significantly change the waste management procedures during confinement, as revealed by nearly half of the respondents, suggests many were ill-prepared to handle the new situation.

This paper has some limitations. Firstly, the sample entails the participation of 102 consumers, and this cannot be regarded as comprehensive. Therefore, the findings could not be generalized to populations and are of a purely exploratory nature and show trends regarding the view of mostly young, female, educated people from middle-income households. Secondly, the respondents were the ones who could have been reached via the authors and their networks. In this context, the online questionnaire administration did not allow for clarification or follow-up on behalf of the respondents. Moreover, the online survey was open for a limited period of time, and the authors had no mechanism to validate the reliability of the answers. In these cases, the respondents might have the incentive to provide answers that present themselves as individuals with a higher commitment to sustainability than they actually are.

However, despite these constraints, which are not uncommon among similar studies, the research provides a contribution to the literature since it is one of the few studies that has reported on trends related to waste management associated with COVID-19 in Argentina, Bolivia, Chile, Colombia, Costa Rica, Dominican Republic, Ecuador, Guatemala, Mexico, and Peru.

In terms of future prospects, the pandemic represents a wake-up call in respect of the need to pay better attention to the management of waste, especially waste prevention. The contribution of this study to society is two-fold. Firstly, it shows that in times of pandemic, a new dynamic in respect of waste products may be expected. Secondly, the increased consumption was oriented towards natural and organic food, which shows a positive trend in terms of the carbon footprint of food products. This is an advancement since previous pandemics have encouraged the use of conserved food. Finally, the trend of online shopping may be a reason for concern because it may generate package waste.

This paper has two main implications: it has shown the changes in consumption habits associated with the COVID-19 pandemic in Hispanic American countries and illustrates some of the areas where attention is needed in terms of responses to pandemic situations. Moving forward, some measures that could be adopted by Hispanic American countries to better cope with changes in waste production during future pandemics include: (1) the provision of more and better infrastructure for waste collection so that household waste is not disposed of on the streets; (2) enhanced awareness of consumers in respect of ways to prevent waste generation; (3) a greater emphasis on methods and tools for waste separation since this may encourage people to segregate; and (4) more intensive efforts among producers of industrialised food to re-design their products and packaging so that they may have an appropriate shelf-life on the one hand, but may also be biodegradable on the other, hence alleviating the pressure on the environment. Finally, it seems to be necessary to encourage public authorities to prepare policies—or at least contingency plans—in the field of waste management during pandemics so that city utilities may be able to better cope with the changes in waste generation. This includes not only the collection but also the separation of the various fractions for subsequent reuse.

Supplementary Materials: The following supporting information can be downloaded at: https://www.mdpi.com/article/10.3390/su142416552/s1, Supplementary Material: Questionnaire—COVID-19 and Household Waste Production in Latina America.

Author Contributions: Conceptualization, W.L.F. and A.L.S.; Methodology, W.L.F., A.L.S., J.S., L.V. and C.M.; Formal analysis, W.L.F., A.L.S., J.S., C.A.F., C.E.B., L.V., R.A., I.S.R., C.M., J.B. and S.N.; Investigation, W.L.F., A.L.S., J.S., C.A.F., C.E.B., L.V., R.A., I.S.R., C.M., J.B. and S.N.; Resources, S.N.; Data curation, W.L.F., A.L.S., J.S., C.A.F., C.E.B., L.V., R.A., I.S.R., C.M., J.B. and S.N.; Writing—original draft, W.L.F., A.L.S., J.S., C.A.F., C.E.B., L.V., R.A., I.S.R., C.M., J.B. and S.N.; Writing—review & editing, W.L.F., A.L.S., J.S., C.A.F., C.E.B., L.V., R.A., I.S.R., C.M., J.B. and S.N.; Visualization, W.L.F., A.L.S., J.S., C.A.F., C.E.B., L.V., R.A., I.S.R., C.M., J.B. and S.N.; Supervision, W.L.F., A.L.S. and J.S.; Project administration, W.L.F., A.L.S. and J.S. All authors have read and agreed to the published version of the manuscript.

Funding: We acknowledge support for the article processing charge by the Open Access Publication Fund of Hamburg University of Applied Sciences.

Institutional Review Board Statement: The nature of the research, the methods used, and the fact that no personal data was stored or can be traced back to individuals, conforming with GDPR standards, means that the study is not subject to an ethics permit as specified by the Association of Medical Ethics Committee in Germany, the body responsible for such assessments in the country which led the study. All authors have read and agreed to the published version of the manuscript.

Informed Consent Statement: Consent was waived due to the point described above.

Data Availability Statement: Not applicable.

Acknowledgments: This paper has been prepared as part of the "100 papers to accelerate the implementation of the UN Sustainable Development Goals" initiative.

Conflicts of Interest: The authors declare no conflict of interest.

References

1. Khan, A.H.; Tirth, V.; Fawzy, M.; Mahmoud, A.E.D.; Khan, N.A.; Ahmed, S.; Ali, S.S.; Akram, M.; Hameed, L.; Islam, S.; et al. COVID-19 Transmission, Vulnerability, Persistence, and Nanotherapy: A Review. *Environ. Chem. Lett.* **2021**, *19*, 2773–2787. [CrossRef] [PubMed]
2. Khan, A.H.; Abutaleb, A.; Khan, N.A.; El Din Mahmoud, A.; Khursheed, A.; Kumar, M. Co-Occurring Indicator Pathogens for SARS-CoV-2: A Review with Emphasis on Exposure Rates and Treatment Technologies. *Case Stud. Chem. Environ. Eng.* **2021**, *4*, 100113. [CrossRef]
3. Brum, M.; De Rosa, M. Too Little but Not Too Late: Nowcasting Poverty and Cash Transfers' Incidence during COVID-19's Crisis. *World Dev.* **2021**, *140*, 105227. [CrossRef] [PubMed]
4. Maurizio, R.; Bertranou, F. The Labor Market in Latin America at the Time of the COVID-19 Pandemic: Impacts, Responses and Perspectives. *Gac. Med. Caracas* **2020**, *128*, S156–S171. [CrossRef]
5. Leal Filho, W.; Voronova, V.; Kloga, M.; Paço, A.; Minhas, A.; Salvia, A.L.; Ferreira, C.D.; Sivapalan, S. COVID-19 and Waste Production in Households: A Trend Analysis. *Sci. Total Environ.* **2021**, *777*, 145997. [CrossRef]
6. Aragaw, T.A.; Mekonnen, B.A. Understanding Disposable Plastics Effects Generated from the PCR Testing Labs during the COVID-19 Pandemic. *J. Hazard. Mater. Adv.* **2022**, *7*, 100126. [CrossRef]
7. Al Qahtani, S.; Al Wuhayb, F.; Manaa, H.; Younis, A.; Sehar, S. Environmental Impact Assessment of Plastic Waste during the Outbreak of COVID-19 and Integrated Strategies for Its Control and Mitigation. *Rev. Environ. Health* **2021**, *37*, 585–596. [CrossRef]
8. Gideon, J. Introduction to COVID-19 in Latin America and the Caribbean. *Bull. Lat. Am. Res.* **2020**, *39*, 4–6. [CrossRef]
9. Lotta, G.; Kuhlmann, E. When Informal Work and Poor Work Conditions Backfire and Fuel the COVID-19 Pandemic: Why We Should Listen to the Lessons from Latin America. *Int. J. Health Plann. Manag.* **2020**, *36*, 976–979. [CrossRef]
10. Puntigliano, A.R. Pandemics and Multiple Crises in Latin America. *Lat. Am. Policy* **2020**, *11*, 313–319. [CrossRef]
11. Pizuorno, A.; Fierro, N.A. Latin America and Chronic Diseases: A Perfect Storm during the COVID-19 Pandemic. *Ann. Hepatol.* **2021**, *22*, 100332. [CrossRef] [PubMed]
12. The Lancet COVID-19 in Latin America: A Humanitarian Crisis. *Lancet* **2020**, *396*, 1463. [CrossRef]
13. Litewka, S.G.; Heitman, E. Latin American Healthcare Systems in Times of Pandemic. *Dev. World Bioeth.* **2020**, *20*, 69–73. [CrossRef] [PubMed]
14. Caetano, G.; Pose, N. Impactos Del Covid-19 En Los Escenarios Latinoamericanos Contemporáneos. *Rev. Perfiles Latinoam.* **2021**, *29*, 2. [CrossRef]
15. IDB Opportunities for Stronger and Sustainable Postpandemic Growth: 2021 Latin American and Caribbean Macroeconomic Report. Available online: https://flagships.iadb.org/en/MacroReport2021/Opportunities-for-Stronger-and-Sustainable-Postpandemic-Growth (accessed on 14 July 2022).
16. CEPAL COVID-19 Observatory in Latin America and the Caribbean. Economic and Social Impact. Available online: https://www.cepal.org/en/topics/covid-19 (accessed on 1 May 2022).

17. Vieira, V.H.A.d.M.; Matheus, D.R. The Impact of Socioeconomic Factors on Municipal Solid Waste Generation in São Paulo, Brazil. *Waste Manag. Res.* **2018**, *36*, 79–85. [CrossRef]
18. Alfonso, M.B.; Arias, A.H.; Menéndez, M.C.; Ronda, A.C.; Harte, A.; Piccolo, M.C.; Marcovecchio, J.E. Assessing Threats, Regulations, and Strategies to Abate Plastic Pollution in LAC Beaches during COVID-19 Pandemic. *Ocean Coast. Manag.* **2021**, *208*, 105613. [CrossRef]
19. Singh, E.; Kumar, A.; Mishra, R.; Kumar, S. Solid Waste Management during COVID-19 Pandemic: Recovery Techniques and Responses. *Chemosphere* **2022**, *288*, 132451. [CrossRef]
20. Sharma, H.B.; Vanapalli, K.R.; Cheela, V.S.; Ranjan, V.P.; Jaglan, A.K.; Dubey, B.; Goel, S.; Bhattacharya, J. Challenges, Opportunities, and Innovations for Effective Solid Waste Management during and Post COVID-19 Pandemic. *Resour. Conserv. Recycl.* **2021**, *800*, 149605. [CrossRef]
21. Hantoko, D.; Li, X.; Pariatamby, A.; Yoshikawa, K.; Horttanainen, M.; Yan, M. Challenges and Practices on Waste Management and Disposal during COVID-19 Pandemic. *J. Environ. Manag.* **2021**, *286*, 112140. [CrossRef]
22. Fan, Y.V.; Jiang, P.; Hemzal, M.; Klemeš, J.J. An Update of COVID-19 Influence on Waste Management. *Sci. Total Environ.* **2021**, *754*, 142014. [CrossRef]
23. Janssen, M.; Chang, B.P.I.; Hristov, H.; Pravst, I.; Profeta, A.; Millard, J. Changes in Food Consumption During the COVID-19 Pandemic: Analysis of Consumer Survey Data From the First Lockdown Period in Denmark, Germany, and Slovenia. *Front. Nutr.* **2021**, *8*, 635859. [CrossRef] [PubMed]
24. Aldaco, R.; Hoehn, D.; Laso, J.; Margallo, M.; Ruiz-Salmón, J.; Cristobal, J.; Kahhat, R.; Villanueva-Rey, P.; Bala, A.; Batlle-Bayer, L.; et al. Food Waste Management during the COVID-19 Outbreak: A Holistic Climate, Economic and Nutritional Approach. *Sci. Total Environ.* **2020**, *742*, 140524. [CrossRef] [PubMed]
25. Caso, D.; Guidetti, M.; Capasso, M.; Cavazza, N. Finally, the Chance to Eat Healthily: Longitudinal Study about Food Consumption during and after the First COVID-19 Lockdown in Italy. *Food Qual. Prefer.* **2021**, *95*, 104275. [CrossRef] [PubMed]
26. Iivari, N.; Sharma, S.; Ventä-Olkkonen, L. Digital Transformation of Everyday Life—How COVID-19 Pandemic Transformed the Basic Education of the Young Generation and Why Information Management Research Should Care? *Int. J. Inf. Manag.* **2020**, *55*, 102183. [CrossRef]
27. Sharma, H.B.; Vanapalli, K.R.; Samal, B.; Cheela, V.R.S.; Dubey, B.K.; Bhattacharya, J. Circular Economy Approach in Solid Waste Management System to Achieve UN-SDGs: Solutions for Post-COVID Recovery. *Sci. Total Environ.* **2021**, *800*, 149605. [CrossRef] [PubMed]
28. Carolan, M. Practicing Social Change during COVID-19:Ethical Food Consumption and Activism Pre- and Post-Outbreak. *Appetite* **2021**, *163*, 105206. [CrossRef] [PubMed]
29. Chenarides, L.; Grebitus, C.; Lusk, J.L.; Printezis, I. Food Consumption Behavior during the COVID-19 Pandemic. *Agribusiness* **2021**, *37*, 44–81. [CrossRef]
30. Escursell, S.; Llorach-Massana, P.; Roncero, M.B. Sustainability in E-Commerce Packaging: A Review. *J. Clean. Prod.* **2021**, *280*, 124314. [CrossRef]
31. Arndt, C.; Davies, R.; Gabriel, S.; Harris, L.; Makrelov, K.; Robinson, S.; Levy, S.; Simbanegavi, W.; van Seventer, D.; Anderson, L. COVID-19 Lockdowns, Income Distribution, and Food Security: An Analysis for South Africa. *Glob. Food Sec.* **2020**, *26*, 100410. [CrossRef]
32. Echegaray, F. What POST-COVID-19 Lifestyles May Look like? Identifying Scenarios and Their Implications for Sustainability. *Sustain. Prod. Consum.* **2021**, *27*, 567–574. [CrossRef]
33. Sarkodie, S.A.; Owusu, P.A. Impact of COVID-19 Pandemic on Waste Management. *Environ. Dev. Sustain.* **2021**, *23*, 7951–7960. [CrossRef] [PubMed]
34. Marty, L.; de Lauzon-Guillain, B.; Labesse, M.; Nicklaus, S. Food Choice Motives and the Nutritional Quality of Diet during the COVID-19 Lockdown in France. *Appetite* **2021**, *157*, 105005. [CrossRef] [PubMed]
35. Bracale, R.; Vaccaro, C.M. Changes in Food Choice Following Restrictive Measures Due to COVID-19. *Nutr. Metab. Cardiovasc. Dis.* **2020**, *30*, 1423–1426. [CrossRef] [PubMed]
36. Liang, Y.; Song, Q.; Wu, N.; Li, J.; Zhong, Y.; Zeng, W. Repercussions of COVID-19 Pandemic on Solid Waste Generation and Management Strategies. *Front. Environ. Sci. Eng.* **2021**, *15*, 115. [CrossRef] [PubMed]
37. Ikiz, E.; Maclaren, V.W.; Alfred, E.; Sivanesan, S. Impact of COVID-19 on Household Waste Flows, Diversion and Reuse: The Case of Multi-Residential Buildings in Toronto, Canada. *Resour. Conserv. Recycl.* **2021**, *164*, 105111. [CrossRef] [PubMed]
38. Adusei-Gyamfi, J.; Boateng, K.S.; Sulemana, A.; Hogarh, J.N. Post COVID-19 Recovery: Challenges and Opportunities for Solid Waste Management in Africa. *Environ. Chall.* **2022**, *6*, 100442. [CrossRef]
39. Ouhsine, O.; Ouigmane, A.; Layati, E.; Aba, B.; Isaifan, R.J.; Berkani, M. Impact of COVID-19 on the Qualitative and Quantitative Aspect of Household Solid Waste. *Glob. J. Environ. Sci. Manag.* **2020**, *6*, 41–52. [CrossRef]
40. Di Renzo, L.; Gualtieri, P.; Pivari, F.; Soldati, L.; Attinà, A.; Cinelli, G.; Cinelli, G.; Leggeri, C.; Caparello, G.; Barrea, L.; et al. Eating Habits and Lifestyle Changes during COVID-19 Lockdown: An Italian Survey. *J. Transl. Med.* **2020**, *18*, 229. [CrossRef]
41. Ben Khadda, Z.; Ezrari, S.; Radouane, N.; Boutagayout, A.; El Housni, Z.; Lahmamsi, H.; Zahri, A.; Houssaini, T.S.; El Ghadraoui, L.; Elamine, Y.; et al. Organic Food Consumption and Eating Habit in Morocco, Algeria, and Tunisia during the COVID-19 Pandemic Lockdown. *Open Agric.* **2022**, *7*, 21–29. [CrossRef]

42. HUNTER Food Study 2020. Special Report. America Gets Cooking: The Impact of COVID-19 on America's Food Habits. Available online: https://www.slideshare.net/HUNTERNY/hunter-food-study-special-report-america-gets-cooking-231713331 (accessed on 10 July 2021).
43. Tchetchik, A.; Kaplan, S.; Blass, V. Recycling and Consumption Reduction Following the COVID-19 Lockdown: The Effect of Threat and Coping Appraisal, Past Behavior and Information. *Resour. Conserv. Recycl.* **2021**, *167*, 105370. [CrossRef]
44. Sarkis, J.; Cohen, M.J.; Dewick, P.; Schröder, P. A Brave New World: Lessons from the COVID-19 Pandemic for Transitioning to Sustainable Supply and Production. *Resour. Conserv. Recycl.* **2020**, *159*, 104894. [CrossRef] [PubMed]
45. Yousefi, M.; Oskoei, V.; Jonidi Jafari, A.; Farzadkia, M.; Hasham Firooz, M.; Abdollahinejad, B.; Torkashvand, J. Municipal Solid Waste Management during COVID-19 Pandemic: Effects and Repercussions. *Environ. Sci. Pollut. Res.* **2021**, *28*, 32200–32209. [CrossRef] [PubMed]

sustainability MDPI

Article

Environmental Impact Evaluation of University Integrated Waste Management System in India Using Life Cycle Analysis

Amit Kumar Jaglan [1], Venkata Ravi Sankar Cheela [2,*], Mansi Vinaik [3] and Brajesh Dubey [4]

1 Amity School of Architecture & Planning, Amity University, Noida 201313, India; akjaglan@amity.edu
2 Department of Civil Engineering, MVGR College of Engineering (A), Vizianagaram 535005, India
3 School of Management and Liberal Studies, The NorthCap University, Gurugram 122017, India; mansivinaik21@gmail.com
4 Department of Civil Engineering, Indian Institute of Technology Kharagpur, Kharagpur 721302, India; bkdubey@civil.iitkgp.ac.in
* Correspondence: cvrs@mvgrce.edu.in

Abstract: Decarbonization of university campuses by integrating scientific waste approaches and circular economy principles is the need-of-the-hour. Universities, the maximum energetic corporations and places for clinical studies and social activities, have a duty to assemble low-carbon campuses and play a vital function in lowering CO_2 emissions. An environmental life cycle assessment was conducted to compare proposed municipal solid waste (MSW) treatment systems with the existing system in the residential university campus (RUC) in Kharagpur, West Bengal (India). The results show the existing MSW disposal practice in RUC (baseline scenario has the highest GWP (1388 kg CO_2 eq), which can potentially be reduced by adopting integrated waste management system with source segregation as represented in futuristic scenarios (S2—50% sorting) and (S3—90% sorting)). Compared to S1, GHG emission was reduced by 50.9% in S2 and by 86.5% in S3. Adopting anaerobic digestion and engineered landfill without energy recovery offsets the environmental emissions and contributes to significant environmental benefits in terms of ecological footprints. Capital goods play a pivotal role in mitigation the environmental emissions. The shift towards S2 and S3 requires infrastructure for waste collection and sorting will contribute to reduction of associated environmental costs in the long-term.

Keywords: municipal solid waste; life cycle assessment; global warming potential; anaerobic digestion; landfill

Citation: Jaglan, A.K.; Cheela, V.R.S.; Vinaik, M.; Dubey, B. Environmental Impact Evaluation of University Integrated Waste Management System in India Using Life Cycle Analysis. *Sustainability* **2022**, *14*, 8361. https://doi.org/10.3390/su14148361

Academic Editors: Sunil Kumar, Pooja Sharma and Deblina Dutta

Received: 30 April 2022
Accepted: 4 July 2022
Published: 8 July 2022

1. Introduction

With high population density and the varied nature of domestic and scientific activities, universities could be considered as "small cities" [1–3]. These activities generate large quantities of waste, which have similar characteristics of that generated by different urban spaces such as towns and cities [4,5]. This makes them apt testing grounds for plans that can be replicated at city/town levels [6]. Therefore, it becomes imperative for universities to incorporate practices of sustainability, representing environments for practical learning, research, and operating as living labs [7–9]. Building a pervasive and holistic waste management system should be a major component of university planning and expansion [10,11]. This can create a synergetic effect and assist in spreading this practice to different urban spaces. By applying suitable technologies, an integrated solid waste management plan can reduce waste generation and propel recycling [12]. The stepping-stone of this plan is understanding the quantity, nature, and characteristics of the solid waste being generated on campus, allowing for the adoption of effective waste management strategies, carried out most efficiently by direct waste audit [13]. Cornell University, Brown University, Colorado State University, Universidad Autonoma Metropolitan, Barcelona, Asian Institute of Technology, and University of Northern British Columbia are leading examples [14].

The active role played by university campuses in reducing greenhouse gas emission, waste management, and creating awareness has been highlighted [15]. Different studies on various institutions of higher education have brought out insights into waste management. A few examples include a study on the University of Gavle, which brought out the importance of training for efficient implementation of the waste management system [16]; a study on the University of Maribor (Engineering Campus), which proposed a waste management plan with comprehensive options to manage plastic and paper waste [17,18]. Similarly, a study carried out at UK University provided a deep understanding of the classification of the WRI/WBCSD Greenhouse Gas Protocol Corporate Standard [19,20]. Universities across the world have been working on greening their campuses, establishing standards such as ISO 14001, creating an ecological campus based on an ecological technique style, ecological education, and a management style, establishing an indicator system to evaluate the performance of one green university project [21]. Life cycle assessment (LCA) is a tool to quantify or compare the environmental impacts of product(s) or service(s) throughout its life cycle, i.e., from raw material extraction to disposal. International Standard Organization (ISO) standardized the methodological framework for life cycle assessment (LCA) to evaluate impacts from an environmental perspective (ISO, 2006). The ISO 14040 (2006) and 14044 (2006) guidelines include instructions on how to perform and report LCA studies. Application of LCA in MSW provides a holistic perspective in identifying acceptable and environmentally sound solutions. Waste LCA tools have been developed since the early 1990s with an aim to evaluate the environmental performance by modelling waste management systems. Compared to product LCA tools, waste LCA evaluate environmental performance of interconnected waste treatment systems based on physico-chemical characteristics of waste from generation to final disposal. The waste LCA model has an ability to model variations in fractional waste content, operation specific emissions, substitution of energy systems, include country-specific energy mix, manufacturing of primary resources, and assessment of interconnected systems ranging from collection to final disposal. However, the models suffer lack of harmonization due to complexity in waste systems modelling and application of country-specific datasets [22–25].

Recently, the majority of higher educational institutions (HEIs) have carried out 'green' drives [26]. At present, more than 400 institutional organizations have marked a Presidents' Climate Leadership Commitment. Numerous HEIs are revealing their GHG emissions stock using the Sustainability Indicator Management and Analysis Platform (SIMAP), which was already the Clean Air-Cool Planet Campus Carbon Calculator, while others are utilizing their own custom devices and/or contracting external firms to compile their carbon footprint [27]. The current investigation aims to evaluate the environmental profiles of the use of the life cycle approach for an Indian residential university campus (RUC). This research is the maiden work in an Indian context, which can be a base model for other higher education institutes in emerging economies.

2. Materials and Methods

2.1. Study Area

The unlined landfill facility in the study area is located in the northeast corner of the university campus in Kharagpur, West Bengal state (India). The landfill is located amid a campus within a proximity of 130 km radius from the Bay of Bengal, at an elevation of 33.5 m from the mean sea level. The operations in the dumpsite began in late 2005, spreading over nine acres of land. The height of the waste heaps varies between 10 and 12 m. The average amount of MSW reaching landfills is 12.6 $\pm$ 2 tons per day. The public health and sanitation department in the Indian Institute of Technology Kharagpur is the council authority administrating the solid waste management activities. The geographic location of the study area is shown in Figure 1.

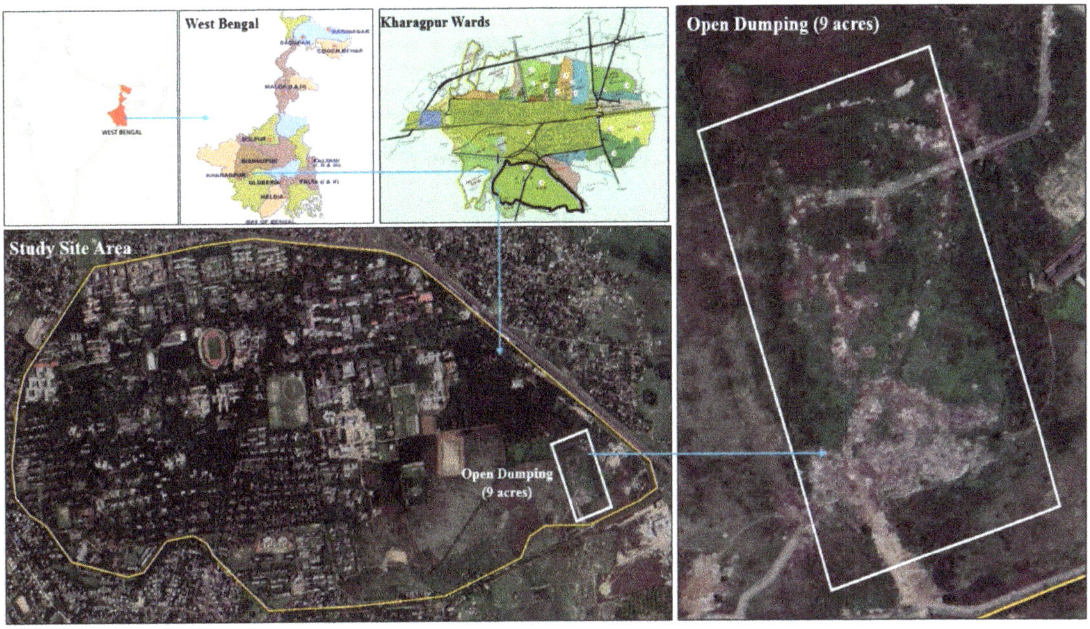

Figure 1. The geographic location of the study area.

This study aims to use the life cycle analysis (LCA) approach to compare the environmental impacts of different waste management alternatives and identify the most viable management scenario with minimal environmental impacts. The scenarios include various MSWM options, such as MRF, composting, AD, incineration, and landfilling. The impact categories, such as global warming potential (GWP), terrestrial acidification (TA), freshwater eutrophication (FEW), marine water eutrophication (ME), human toxicity (HTP), terrestrial ecotoxicity (TE), freshwater ecotoxicity (FWT), and marine ecotoxicity (MET) impacts were determined for each option.

2.2. Physical Characterization of Municipal Solid Waste

The physical characterization of MSW was performed as per ASTM 5231D: 2016 protocol. The waste sampling campaign was performed at the unlined landfill facility in the University campus. Stratified random sampling was used for the collection of waste samples. The quartering method was implemented to prepare representative samples for the determination of physicochemical characteristics. The code of practices used for the determination of moisture content was ASTM E1756-08, percentage of volatile matter [28], ash content [29], elemental analysis [30], and energy content was [31].

2.3. Waste Management Scenarios

A waste treatment strategy was developed based on the waste characteristics determined in the aforementioned section. The systems boundaries of existing and proposed waste treatment strategies are shown in Figure 2.

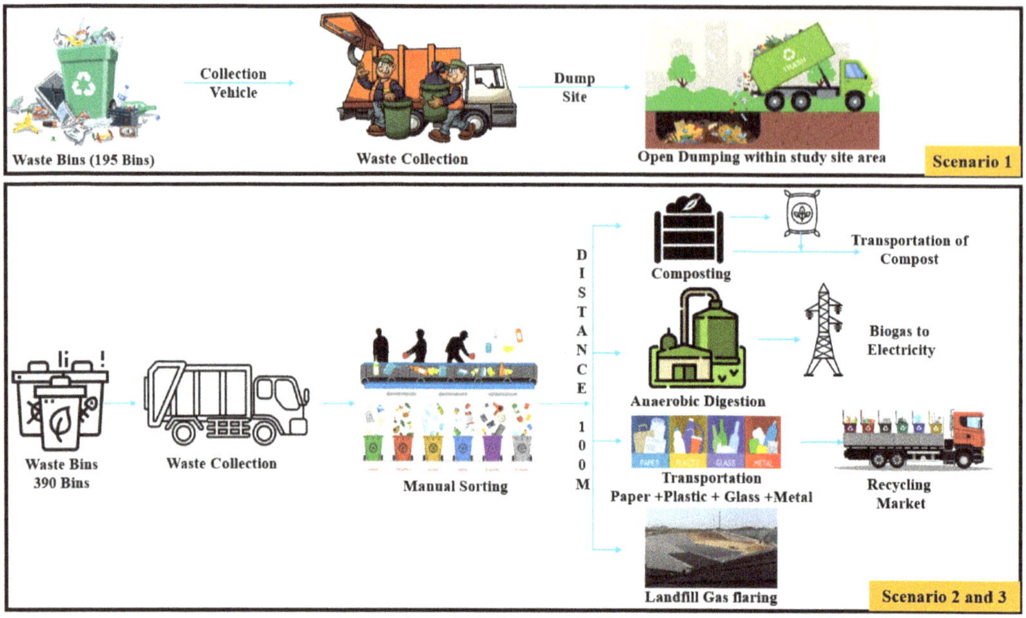

Figure 2. Scenarios for the management of municipal solid waste S1: Unlined landfill (business-as-usual); S2 and S3: Integrated waste treatment facility (waste sorting percentage is the variable parameters S2 (50%) and S3 (90%)).

2.3.1. Scenario (S1)—Business-As-Usual (BAU): Open Dumping

This scenario represents the current MSW disposal method in the RUC of Kharagpur. The waste complied by the departmental workers of sanitation is further disposed within the campus in the unlined landfill. There is a huge discharge of greenhouse gas due to the decomposition of the waste. The greenhouse gas is produced due to the emission of methane gas during the process of anaerobic decomposition of the waste. Methane (CH_4) has a global impact and methane emission is approximately 5% due to the anaerobic decomposition. The fusion of the waste with the rainwater and the moisture leads to the production of toxic liquid leachate.

2.3.2. Scenario (S2 & S3)—Integrated Waste Treatment Facility with 50 & 90% Waste Sorting

Integrated waste treatment is a systematic method used for the sustainable management of waste. In this process, waste is disposed into the landfill along with the leachate process and the gas collection system. Gas produced is further collected and flared before discharging into the environment. As suggested by the IPSS, the total landfill gas (LFG) is calculated using a first-order degradation model. The collection efficiency is estimated to be approximately 60% [32] and the surface oxidation efficiency of CH_4 from the fugitive emission to be approximately 15% to 20% [33]. The overall power consumed for the leachate treatment is about 25 to 30 kWh/ton. In this process, the organic fraction of waste is separated and transported to the anaerobic digestion unit. The leftover waste is disposed of in the engineered landfill, with no energy recovered. In this process, the leachate treatment and the gas collection processes are also considered. Before releasing the gas into the environment, it is flared. During the course of the anaerobic decomposition, about 30% of the biogas is released, which is used for the generation of electricity [34]. The collection efficiency, in this case, is found to be approximately 95%. The same process is used for the treatment of biogas slurry, which was used for the treatment of leachate. The digestive residue is applied to land as a replacement for mineral fertilizer.

2.4. Life Cycle Assessment

The LCA methodology outlined in ISO 14040-44:2006 standards was implemented for the evaluation of environmental impacts. The methodology includes four steps: (1) goal and scope, (2) life cycle inventory (LCI), (3) life cycle impact assessment (LCIA), and (4) life cycle interpretation.

2.4.1. Goal and Scope Definition

The main objective of this study is to compare various MSW management techniques using a life cycle perspective. As the study area, the RUC Kharagpur, India, was considered. The amount of MSW generated into the campus was integrated for the study. This area was considered in order to compare and analyze the various alternative processes. For the study of this scenario, an integrated waste management plan (IWMP) was implemented at the landfill site. In this process, anaerobic digestion, incineration, and engineered landfilling (with or without the energy recovery) were considered. The effect of a collection of waste and transportation of waste was presumed to be identical in all the scenarios during the process. This is due to the same process location. Emission related to the energy produced while the manufacturing of the massive types of equipment and goods that were not included in the LCA modeling process was used widely in the study [35,36]. A first-order decay model calculated the production estimate of methane and carbon dioxide on the material fraction elemental composition of waste and associated decay rates. The volume of leachate produced was calculated by the infiltration rate, waste layer height, and bulk density [37]. In the current study, an environmental assessment was performed to handle and manage 1 ton of MSW in the study area. The waste was collected at the waste facility. The complete waste was treated using a combination of treatment technologies presented in the scenarios provided. All the environmental acknowledgments corresponding to these scenarios are contemplated according to their corresponding abilities to generate the appropriate products such as slurry with the fertilizer and energy from incineration with the grid electricity. The system confines all the waste treatments and processes compiled. It helps in monitoring the performance. Thus, the allocation was not investigated.

2.4.2. Life Cycle Inventory Data

An environmental impact assessment requires the completion of a life cycle inventory (LCI). During the life cycle stages of waste management systems, LCI measures the number of inputs in terms of material and energy, as well as outputs in terms of emissions and wastes. The inventory information is used to create the flow of inputs and outputs through the various stages of the life cycle. In this study, the dataset considered is from the EASETECH database (developed by the Technical University of Denmark). This database was used for the engineered landfills. It consists of the gas flaring and energy recovery system. It was treated using EASETECH v3.7-LCA-model for the assessment of the environmental technology. There is a significant lack of Indian-specific landfill datasets for various geographical areas. Gas collection and utilization were projected to occur over 55 years, with no collection assumed for the next 45 years; however, gas oxidation was expected to occur in the top layer [38]. The dataset supplied [39] was used to examine leachate collection and treatment. Table 1 shows the input data that were utilized to simulate this system.

Table 1. Life cycle inventory data for 1 ton of MSW.

Parameter	Value	Unit	Reference
Landfill			
Diesel consumption	2	$L \cdot t^{-1}$	[40]
Methane generation	55	%	[38]
LFG Collection efficiency	90	%	[38]
LFG collected (Years 0–55)	95	%	[39]

Table 1. *Cont.*

Parameter	Value	Unit	Reference
LFG collected (Years 55–100)		%	[39]
LFG Top cover oxidation	36	% CH_4	[41]
Anaerobic digestion			
Electricity (pre-treatment)	12.6	kWh	[5]
Electricity (Reactor)	14	kWh	[39]
Methane emissions	0.5%	% of CH_4	[39]
Transport of compost	3	$L \cdot t^{-1}$	[40]
Electricityrecovery (biogas)	35	%	[40]
Land application			
N_2O-N emissions (direct)	1.25	% of N-tot	[5]
NH_3-N emissions	15	% of N-tot	[5]
NO_3^--N emissions	20	% of N-tot	[5]
Incoming N content	4.85	kg N-tot	[5]
Incoming P content	0.65	kg P-tot	[5]
Incoming K content	1.48	kg K-tot	[5]
Application of digestate	20	MJ/t digestate	[5]
Application of mineral fertilizers	0.36	MJ/kg N-tot	[5]
Incineration			
Sodium hydroxide	0.24	kg	[42]
Hydrated lime	10	kg	[42]
Activated carbon	0.25	kg	[42]
Ammonia (NH_3)	0.5	kg	[42]
Electricity consumption	0.27	MWh	[5]

LFG: landfill gas.

- Anaerobic Digestion

Anaerobic digestion (AD) is utmost the convenient method to be used for waste. Plants using a biphasic wet digester-based method produce AD as organic waste. In India, this method is in use [34]. The modelling approach makes extensive use of AD techniques found in the EASETECH databases, with inputs adjusted for Indian conditions. The volatile solid, which is food waste degradation/deterioration of around 70%, is the plant criterion utilized in the modelling process. Furthermore, the quantity of methane in biogas is approximately 63%, methane leakage from the digester is about 2%, and biogas conversion efficiency is about 35% [33].

- Windrow composting

Food waste degradation (75%), loss of volatile solids (75%), carbon (77%), and nitrogen (8%) were all used to simulate windrow composting (biological treatment technique for creating compost by stacking organic waste in long rows) [34]. Table 2 shows the input data that were utilized to simulate this system.

Table 2. Environmental impact categories evaluated in this study.

Impact Category	Unit
Global warming potential	kg CO_2 equivalent
Terrestrial acidification	kg SO_2 equivalent
Freshwater eutrophication	kg-phosphorus equivalent
Marine water eutrophication	kg-nitrogen equivalent
Human toxicity	kg 1,4-dichlorobenzene equivalent
Terrestrial ecotoxicity	kg 1,4-dichlorobenzene equivalent
Freshwater ecotoxicity	kg 1,4-dichlorobenzene equivalent
Marine ecotoxicity	kg 1,4-dichlorobenzene equivalent

2.4.3. Life Cycle Impact Assessment

Process modeling is achieved by integrating physio-chemical characteristics of the waste fractions. Using a software tool, the EASETECH v3.7-LCA-model, various scenarios of environment aspects were analyzed. For analyzing the impact study, the ReCipe 2016 midpoint world method/technique was considered. ReCipe technique is the consolidation of the CML method and Eco-indicator 99 presenting the wider view of impact categories. The LCIA impact categories were used for analyzing global warming potential, terrestrial acidification, marine eutrophication, human toxicity, terrestrial ecotoxicity, freshwater ecotoxicity, terrestrial ecotoxicity, and marine ecotoxicity. The time horizon taken for the reference was 100 years for global warming potential. This time is taken into consideration for the climate change policy [43]. Moreover, for the impact of acidification, toxicity impact levels have had a huge impact on future generations for more than 100 years. In this study, 100 years are considered as the period to find the inventory. The ReCiPE Midpoint (Heuristic) World model was used for determining the characteristics of the environmental frame/profiles of the alternatives. Impact assessment methodologies give a hierarchical view of about 100 years' time period and also include large numbers of midpoint indicators on a global scale. The impact category is subdivided into two categories, namely toxic and non-toxic. In this study, land and water use were not included due to their dependency on geographical locations [38]. The impact category was determined by analyzing LCA studies accomplished in Indian scenarios (Table 1), as indexed in Table 2.

3. Results

3.1. *Physicochemical Characteristics of MSW*

The physico-chemical characteristic of MSW collected by the first author was determined as a part of this research. The MSW is majorly composed of organic matter, which is food waste 59%, yard waste 17%, plastic 5%, cardboard 4%, polythene bags 4%, paper 2.6, glass 2.5%, inert 1.8%, leather 0.85%, other 0.80%, metal 0.60%, and e-waste 0.43%. In Figure 3, the physical characteristics of MSW are presented/demonstrated. Organic waste is mostly composed of food and yard waste. The inert and other components consist of sand, silt, dust, grit, ash, inseparable paper, and food residues, street sweeping waste, drain cleaning waste, and construction debris. As per the chemical classification of MSW, study shows that the moisture amount is about 7% to 52%, volatile solid is about 38% to 43%, carbon amount is about 33% to 47%, the oxygen content is about 36% to 65%, the hydrogen content is about 3.75% to 9.1%, sulphur and nitrogen content is about 2%, and calorific is about 4300 kcal/g to 4730 kcal/g (on dry basis).

3.2. *Life Cycle Impact Assessment Results*

3.2.1. Global Warming Potential

Global warming potential (GWP) has a huge impact on the environment, which is estimated in the terms of kg CO_2 equivalent, as demonstrated in Figure 3a. Scenario (S1) has the maximum GWP with an absolute value equivalent to the amount of 1388 kg CO_2. Fugitive emission causes approximately 99% of the total emission. The remaining 1% of the emissions is caused by the waste leveling process, compaction process, and transportation process. In comparison to scenario (S1), analysis shows that in scenario (S2), the GHG emission is reduced by 50.9%, and in scenario (S3) it is reduced by 86.5%. Landfill construction and operation, leachate collection, and management process contribute to 80% of emissions. The collection process and transportation process of MSW to the unified waste facility contribute up to 9.393 k CO_2 eq.

In scenario S2, the development of integrated waste management (IWM) systems with 50% sorting (S2) contributes to 681.8 kg CO_2 eq of GWP emissions. The LFG is the major emissions (648.6 kg CO_2 eq) contributing process. The waste technique that contributes to emission in the impact category/sector is the oxidation of gas in landfills, which is about 68.7%. The landfill construction process is about 23.8% and the gas flaring process is about 7.4%. The gases that are contributing the most in these scenarios are methane with about

76.3%, carbon dioxide with about 23.5%, and nitrous oxide with about 0.2%. In scenario S3, with 90% sorting of waste, the LFG is the major emissions (178 kg CO_2 eq) contributing process followed by AD (26.8 kg CO_2 eq) and WC (20.7 kg CO_2 eq).

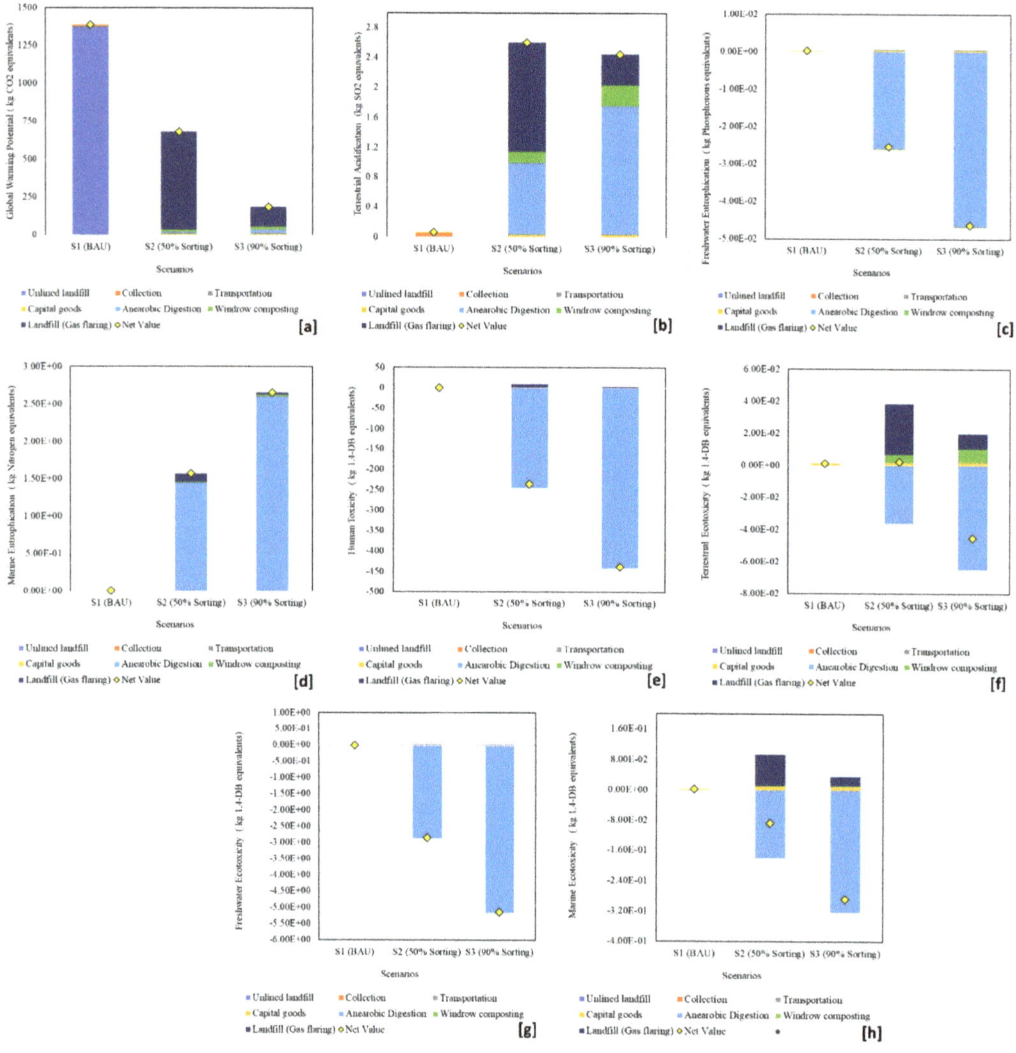

Figure 3. Characterized impact results for (**a**) global warming potential; (**b**) terrestrial acidification; (**c**) fresh water eutrophication; (**d**) marine eutrophication; (**e**) human toxicity potential; (**f**) terrestrial ecotoxicity; (**g**) fresh water ecotoxicity; and (**h**) marine ecotoxicity.

The waste technique contributing to emissions in the impact category is gas oxidation in landfills, which is about 74.3%, and the landfill construction process is about 25.7%. The gases that are contributing the most in these scenarios are methane with about 74.6%, carbon dioxide with about 25.3%, and nitrous oxide with about 0.1%. The process of disposing of treated leachate into the land leads to a reduction in emissions by 82.6%, while substituting the grid electricity to the LFG electricity reduces the emissions by 177.4%. Moreover, the usage of landfill gas (i.e., methane, which causes 28 times more global warming in comparison to CO_2) for energy substitution reduces the GHG emissions in comparison with scenario S2, which explodes the methane gas. Anaerobic digestion and landfill without energy recovery are analyzed as the best GHG reduction methods.

At the University of Leeds, UK, an Extended Environmental Input-Output Analysis (EEIO) was used to estimate the carbon emissions of purchases, documents production, waste, energy, rental properties, transportation, purchased goods, and services. Repeated semi-structured interviews were conducted with the university's procurement staff to obtain the required data. Emissions in scope 3 account for about 51% of total emissions, while scope 1 and range 2 account for 18% and 31%, respectively [44]. The University of Illinois, Chicago (UIC), in its estimate of greenhouse gas emissions from 2004 to 2008, found that in 2008, the most impactful emissions were from buildings (83%), tracked moves (16%), and waste (1%). Power plants, with cogeneration facilities, generate 63% of total GHG emissions, followed by purchased electricity (17%). UIC also used the GHG protocol defined by the World Resources Institute (WRI)/World Business Council for Sustainable Development (WBCSD) [45].

At the University of Cape Town (UCT) in South Africa, this evaluation is split into three parts: campus power emissions, shipping emissions, and goods and offerings emissions. In this study, if the facts precise to South Africa became unavailable, then Intergovernmental Panel on Climate Change (IPCC) emissions elements and applicable literature were used to estimate the carbon emissions. Electricity utilized in entire college bills for 81% of the whole carbon emissions. From the University of Montfort, UK, this study is based on the World Resources Institute (WRI)/World Business Council for Sustainable Development (WBCSD) GHG protocol. Direct and indirect emissions from fossil fuels are considered in scopes 1 and 2, and indirect emissions from shopping, tourism, commuting, and other sources are considered in scope 3 of the resolution letter designation. Scope 3 contributes 79% of the total FC, with the main contribution being supplied (48%), contributing 38% of the total emissions. The scope definition used in this study is similar to that of ISO 14064I (ISO, 2006), to quantify, report, and eliminate GHG emissions at the organizational level [46].

3.2.2. Terrestrial Acidification

Environment impacts related to terrestrial acidification are evaluated in the terms of kg SO_2 equivalent, as demonstrated in Figure 3b. In scenario (S1), the net value is 0.004 kg SO_2 eq. Techniques used for water handling (compacting, leveling, and internal transportation) at the dumping area generate environmental emissions. In scenario (S2), the net value for the terrestrial acidification category is 2.65 kg SO_2 eq. The maximum emission is produced due to the waste treatments and the construction process, and the operation of landfills is about 89.1%, leachate process is about 6.8%, and the gas-flaring process is about 4.1%. It is analyzed that in this scenario, three main gases contribute to TA, i.e., nitrogen oxide gas is about 60%, sulphur dioxide gas is about 39.8%, and ammonia gas is about 0.2%. In scenario (S3), it is analyzed that the TA impact categories net value is 0.017 PE (0.610 kg SO_2 eq.). The waste processes that contribute to the TA are the construction process and the operation of the landfill process, which is about 85.7%, energy conversation techniques are about 7.7%, and the leachate technique is about 6.6%. The most prominent gases that contribute to TA are sulphur dioxide with about 67.6% and nitrogen oxide with about 32.4%.

3.2.3. Freshwater Eutrophication

The FEW's environmental consequences are measured in kilograms of phosphorus equivalent, as shown in the Figure 3c. In scenario (S1), the effect category's net value is insignificant. In the impact category, the net value for scenario (S2) is around 2.8×10^{-4} kg-P eq. The greatest emissions created by waste treatments include leachate treatment, which accounts for about 99.5% of total emissions, and landfill and construction operations, which account for around 0.5%. The net value in scenario (S3) is 4.8×10^{-4} kg-P eq. Leachate treatment accounts for 91.8% of the FEW, whereas leachate disposal accounts for 7.7%, construction accounts for 0.5%, and landfill operation account for 0.5%. The emission offsetting unit process in this case is electricity substitution. Composting is the most effective technique for reducing emissions, whereas electricity replacement has a minor impact. In this scenario, power substitution is the primary emission-reducing unit process. The best impact categories are AD and landfills without energy recovery, and 8.1×10^{-6} units are contributed by MSW collection and transportation to the integrated waste facility.

3.2.4. Marine Eutrophication

Environment impacts related to the ME are evaluated in the terms of kg-nitrogen equivalent, as demonstrated in Figure 3d. In scenario (S2), the net value is 1.50×10^{-4} kg-N eq. in the impact category. The maximum emissions produced due to the waste treatments are leachate treatment with about 91% and the operation of landfills and construction with about 8.3%, also gas flaring with about 0.7%. In scenario (S3), the net value is 3.1×10^{-4} kg-P eq. The waste processes that contribute to the FEW are leachate treatment, which is 91.8%, leachate disposal that is 7.7%, and the construction process and the operation of landfill process, which is about 0.5%. In scenario (S3), the net value is 0.167 kg-P eq. The waste processes that contribute to the FEW are leachate treatment, which is 85.3%, the construction process and the operation of landfill process, which is about 7.8%, leachate disposal, which is about 3.8%, and electricity substitution which is about 3.0%. Landfills without energy recovery is analyzed as the best impact category. The collection and transportation of MSW to the integrated waste facility contribute 0.0036 units.

3.2.5. Human Toxicity Potential

As shown in Figure 3e, the environmental consequences of the HTP, TE, FWT, and MET are measured in kilograms of 1,4-dichlorobenzene equivalent, 5 a.m. to 5 p.m. In scenario (S2), the net value in all effect categories is zero. The net value of 3.026 kg 1,4-dichlorobenzene equivalent is found in scenario S2, HTP impact category. The greatest emissions produced by waste treatments are 44.8%from landfill operations and construction, 31.4% from leachate treatment, 12.7% from gas flaring, and 11.1% from the oxidation process of gas in landfills. The net value in scenario (S3) is 2.409 kg 1,4-dichlorobenzene equivalent. Construction and operation of the landfill process, which accounts for roughly 50.5%, and leachate treatment, which accounts for 35.5%, are the waste processes that contribute. In this case, the emission-reducing unit process is grid electricity replacement. In this scenario, the emission-reducing unit process is the land replacement, which reduces emissions by 99%. In this case, the emission-reducing unit process is electricity substitution, and 0.0182 units are contributed by MSW collection and transportation to the integrated waste facility.

3.2.6. Terrestrial Ecotoxicity Potential

The emissions associated with the TEP impact category are shown in Figure 3f. In scenario (S2), the net value is 0.051 kg 1,4 DB eq in the TEP impact category. The maximum emissions produced due to the waste treatments are leachate treatment with about 97.2% and the operation of landfills and construction with about 2.8%. In scenario (S3), the net value is 0.052 kg 1,4 DB eq. Leachate treatment, which accounts for 92.6%of the TEP, is one of the waste processes that contribute to it. In this case, the emission offsetting unit method is grid electricity replacement, and 1.5×10^{-4} units are contributed by MSW collection and transportation to the integrated waste facility.

3.2.7. Freshwater Ecotoxicity

The emissions associated with FWE impact category are shown in Figure 3g. In the FWT impact category, the net value in scenario (S2) is 0.025 kg 1,4 DB eq. The highest emission produced by waste treatments is around 98.2%by leachate treatment. The net value in scenario (S3) is 0.024 kg 1,4 DB eq. Leachate treatment, which accounts for 91.1% of the FWT, is one of the waste processes that contribute to it. In this case, the emission-reducing unit process is electricity substitution. In this case, the emission offsetting unit method is electricity substitution, and 4.1×10^{-4} units are contributed by MSW collection and transportation to the integrated waste facility.

3.2.8. Marine Ecotoxicity

The emissions associated with ME impact category is shown in Figure 3g. In scenario (S2), the net value is 0.039 kg 1,4 DB eq. in the ME impact category. The maximum emissions produced due to the waste treatments are leachate treatment with about 52.3% and the operation of landfills and construction with about 47.1%. In scenario (S3), the net value is 0. 0.029 kg 1,4 DB eq. The waste processes that contribute to the TEP are leachate treatment, which is 50.2%, and the operation of landfills and construction, which is about 45.2%. In this case, the emission-reducing unit process is grid electricity replacement. The primary emission offsetting unit process in this scenario is electricity substitution, and 1.1×10^{-3} units were provided by the collection and transportation of MSW to the integrated waste facility.

4. Discussion

4.1. Bioeconomy Fertilizers

The idea of a circular economy depends on reuse, valorization, recycling, and misuse of environmental cycles. Although this idea is broadly examined experimentally and strategically, it has just been disjointly applied practically. In the elaboration of bio-based compost innovations, the following perspectives are significant: the ecological effect that needs to be limited, resources ought to be utilized in a regenerative manner with the thought of resource shortage issue, and advancements must guarantee productivity and monetary advantages to modern attempts. Limitations of normal resources and natural security ought to be a need, however, with supporting business prerequisites for financial advantages. The commitment to discard waste is the duty of agri-food makers, for example, cultivating plants that produce organic waste. Enormous agri-food cultivating plants send business staple items to the beneficiaries, without waste, for example, chickens as carcass. Slaughter waste remains should be used sustainably. In this way, facilities delivering composts from organic waste need to be established close, so that transport is not needed. There is additionally an issue of disinfection. Organic waste in landfills causes rotting, which further causes emissions. Another obstruction to the implementation of inexhaustible raw materials in the creation of fertilizers is the changeability of the raw material. Emerging technologies should consider all of the above.

Another quite significant method is composting. Composting is an old and customary technique that works with the change of biogenic organics under controlled conditions into excrement for farming reasons. Composting measures redirect wastes from the conventional landfills and recover esteem by changing them into organic-rich manures, which consequently affects the quality and yield of farming products. At the same time, this could expand the monetary worth of waste and societal health scenarios by tending to the waste administration issues [47]. The Government of India has stated that composting is anything but a recognized type of horticultural compost and has formulated a strategy to mix it alongside inorganic manure. The use of these composts will outdate and gradually eliminate the usage of fossil-based manures [48]. A great initiative undertaken by the Government is the promotion of fertilizing the soil with Government subsidized fertilizers to encourage clean and organic farming practices. Simultaneously, the Government is setting up farming instructional hubs for giving specialized data, fundamental resources, and preparation kits. As a feature of sustainable agricultural practice, organic farming exercises are effectively increasing all over the country and the world. Organic farming uses natural fertilizers and works in agreement with nature without hurting the ecological equilibrium and at the same time accomplishes great harvest yields. Regarding developing interest in organic products, treating the soil has potential business openings in India. Because of its maintainability, compositing can likewise be stretched out to a mechanical scale using unspoiled organic MSW (municipal solid waste) as feedstock. Treating the soil can be involved with the creation of biogas as a by-product and these advances can increase the financial worth of the waste produced and set out independent job opportunities.

4.2. Bio-Methanation

Biomethanation is a technique by which organic substances are microbiologically converted into biogas under anaerobic conditions, where those microbes break down the organic substances into methane and carbon dioxide, respectively. Biogas production is presumably one of the oldest and abused biological innovations in India. The first biogas plant, otherwise called the KVIC digester, which was established in 1951, was the consequence of introductory trials led by S.V. Desai, the pioneer of anaerobic absorption [49]. The digester configuration was normalized in 1962. In the middle of 1978 and 1984, different models viz., Janata Biogas Plant and the Deenbandhu Digester were well known. Around 45 lakhs of locally designed homegrown biogas plants with 36.85% of the assessed potential were introduced with the help of the Government of India until 2013. The MNRE (Ministry of New and Renewable Energy) has financed three showing projects at Vijayawada (6 MW), Hyderabad (6.6 MW), and Lucknow (5 MW). Colossal amounts of waste produced from different areas viz. agricultural, municipal, food processing, and so on make bio-methanation measure as a first alternative [50]. Some of the most recent projects like Naturally Induced Mixing Arrangement (BIMA) Digester, the ARTI Compact Biogas Plant, Confederation of Real Estate Developers' Associations of India (CREDAI), Trash Guard, and so on, are being introduced in India. Anaerobic assimilation with a 2 MW limit was introduced by M/S Kanoria Chemicals Ltd., Kolkata, India. At present, India and China offer the anaerobic assimilation innovation similarly at a lower cost.

Now, the majority of rich organic waste that can be harnessed for utilizing in the biomethanation procedures comes from the kitchens of India. Hence, FW (food waste) management is progressively being focused upon energy and nutrient recuperation, instead of landfill, primarily because the latter includes various negative natural effects on both small and large scopes. Hence, the recuperation of energy in the form of biogases is now being conducted. In such a manner, waste management arrangements should not just depend upon a particularly traditional disposal situation, but it should be founded on coordinated procedures giving, improvement and optimization of separate municipal collection frameworks, and all the more environmentally economic disposal situations [51] to deliver new fuel sources and materials. Biogas creation from AD (anaerobic digesters) has developed quickly through the years, primarily because of the exceeding significance of environmentally friendly power organization concerning structured mitigation of GHG discharges and the requirement for practical management of organic waste. Advertisement is a grounded innovation to treat organically rich biomass, additionally as deposits and wastes that are progressively being conveyed as an environmentally friendly power generation source [52]. It is anything but a perplexing four-stage measure that includes a different array of microorganisms and methanogenic archaea that is contrasted with numerous other bioenergy innovations. AD is recognized as obliging and is a lot more extensive on the scope of substrates, even those with high moisture substance and impurities, and can be led in both enormous and limited scope digesters at all geological areas. Biogas creation yield is strictly reliant upon the nature of the input biomass, which, as a result, is greatly influenced by how biomass is gathered and overseen. Furthermore, biomass assortment and the management systems may impact the natural effects related to the later portions of the treatment chain, including biogas creation in the AD plant and energy-conversion inside the cogeneration chamber, and digestate management. The accessibility of a biogas outlet of AD plants clears the way to various opportunities for the recuperation of its energy content, beginning from the direct power creation to more refined usage arrangements, such as diverting it into the natural gas network or utilizing it to deliver biomethane for transport.

4.3. Bioethanol

Bioethanol is the derived alcohol from the fermentation of main carbohydrates that are produced in sugar or starch-bearing plants such as corn, sugarcane, sorghum, etc. The fermentation procedure requires less energy and uses a much cheaper production system than biodiesel. Around 2.2 million liters of bioethanol are created utilizing 9 MT of molasses in India ((International Standards Organization)14040, 2006). DBT-ICT Center for Energy Biosciences created bioethanol innovation and exhibited the cellulosic ethanol plant at Indian Glycol Limited, Kashipur, with the undertaking cost of $5.28 million on 22 April 2016. It was running at a pilot-scale (1 T/day) with a 750,000-L yearly liquor creation limit and was equipped for changing over to different biomasses like wheat straw, rice straw, bagasse, cottontail, bamboo, and so on, as feedstock for the liquor extraction. CSIR-NIIST, Thiruvananthapuram has introduced a pilot plant office for the lignocellulos's bioethanol program [53]. Praj Industries Ltd. (Pune, India), Pune, and Techno Sys Systems, Jaipur, have created technologies with plans of action for bioethanol creation as well. To summarize, we can say that India will have numerous bioethanol plants setting up soon because of the governmental guidelines on mixing bioethanol with gas as an obligatory necessity.

4.4. Biohydrogen

Biomass derived from plant crops, agricultural residues, woody biomass, waste inferred biomass, organic portions of MSW, and industrial wastewater (IWW) can be utilized as feedstock for the creation of biohydrogen (Bio-H2) [54]. The dark fermentation (fermentative conversion of organic substrate into biohydrogen, which is acidogenic) procedure is broadly utilized for the Bio-H2 creation as it is generally less energy consuming, operational, and financially achievable. Bio-H2 creation utilizing waste is a maintainable innovation for future energy requests and at the same time, it also adds to the development of a bio-based economy. MNRE (Ministry of New and Renewable Energy) under the mission mode category is supporting CSIR-IICT, Hyderabad, and IIT Kharagpur, for setting up a pilot plant office of 10 m^3 ability to depict the Bio-H2 creation utilizing biogenic waste as feedstock. CSIR-IICT has planned and fostered an exceptional Bio-H2 pilot plant with an amazing facility to use different sorts of biogenic wastes and assistants to specifically improve acidogenic biocatalysts and also pretreat feedstock. Aside from Bio-H2 creation, the cycle in biorefinery mode can likewise deliver side-effects, for example, biomethane and unsaturated fat-rich corrosive intermediates like acetic acid derivation, butyrate, propionate, and so on, which have direct business significance [55]. These acids have critical applications as platform compounds utilized in plastics, drugs, materials, and food additives, and are the intermediates for some major chemicals and fuels. A recycling plant or sorting plant sorts and pre-processes recyclable items. This facility only receives waste separated by source. The selected sorting process consists of a magnetic separator, a vortex separator, and a semi-automatic process with manual sorting. A selection efficiency of 95% was assumed for all materials in the sorting plant. After sorting, the collected materials are sent to a recycling facility. The latter was thought to be 95% efficient in converting the resulting material into new products [55]

The integration of biochemical wastewater treatment (especially by anaerobic digestion) and pyrolysis of sewage sludge is sustainable for urban wastewater with useful biofilter products containing pyrolytic syngas and bio, if well established. It appears as an interesting suggestion to support the basics of processing. This is a preliminary study on the recovery of Brazilian sludge from aerobic and anaerobic processes aimed at bioenergy. Further pilot-scale testing to collect evidence has not yet been conducted, but early results already indicate the possibility of using this sludge for biofuel production. Finally, the bioenergy utilization of pyrolyzed ash-rich sludge is part of the zero-landfill approach and seems to be an interesting alternative to further increasing the reading of this type of biowaste [56].

4.5. Implementation of Waste Systems across the University Campus

A zero-waste program was set up on a university campus because of grassroots students' worry over ecological administration issues. The execution technique comprised of starting conversations with the academics and neighborhood authority staff at a university environmental forum, the development of a functioning forum, the readiness of a funding proposal, and the foundation of an externally funded research system, an instructive yet limited-time program led by an academic staff member. Support from senior administration existed as a written environmental policy and a signed obligation to ecological duty in tertiary training and university financing support followed the achievement of the initial funding application. Tasks were commonly led by paid examination partners, helped by student volunteers, and administered by a program chief from the academic staff. Co-activity and support from the management office staff were acquired informally. A campus environmental forum was set up to work with correspondence on the ecological issue between the School for the Environment, senior university management, management office staff, academic staff, and students. To empower a full program advancement nonetheless, a requirement for linkages between all areas associated with the program and the presentation of a formal environmental management system was identified.

The amount of food waste creation in the Chinese catering industry is roughly 17–18 Mt each year. This area represents about 20% of the absolute food losses in China. China's National Development and Reform Commission has confirmed that 100 pilot urban areas in five bunches will be executing food waste treatment projects. Practically 80% of these ventures depend upon anaerobic processing. In this way, it is vital to see what the ecological effect of this new bioenergy or waste-to-energy chains (particularly at a limited scale) is. Therefore, a life cycle assessment contextual investigation was introduced in this work, because of an anaerobic processing plant, taken care of with the non-consumable food waste created by 29 containers, which work inside the grounds of the Huazhong University of Science and Technology (HUST). The speculated effects are climate change, acidification, eutrophication, and photochemical oxidation. The notable unit is addressed by 1 kWh of created power. This work shows that limited-scale biogas plants can be permitted to work seamlessly inside large Chinese university grounds and can efficiently diminish the ecological effect of food waste management, particularly if the pyrolysis procedure is devised to dispose of the digestate [56].

Comparing the normalized CO_2 balances of different universities, we found a clear difference. The main reason for this is that there is no unified international standardization method for calculating the carbon footprint of educational institutions with specific characteristics compared to organizations, especially those in other disciplines. It is desirable not to extrapolate the data for a specific period. Emission factors are revised and published annually, so it makes more sense to use the fiscal year (rather than the academic year) as the time base for calculating carbon footprint in educational institutions. Therefore, it is advisable to implement a mechanism for keeping historical records.

4.6. Recommendations for Future Studies

As mentioned in each section, the lack of data was a major obstacle to this study. Future studies will proactively choose between an input and output approach, process analysis, or mixed life cycle assessment to identify suitable GHG sources and recommend a more comprehensive registration page at the university. Due to the variety of activities, future carbon footprint studies are encouraged to consider all greenhouse gas emissions, discuss data assumptions, and include lifecycle stages in the assessment. This allows for more detailed comparisons and benchmarking of carbon emissions from higher education institutions. Other greenhouse gas emission sources that can be evaluated are compost production, agriculture, food, beverages, furniture, laboratory supplies, agricultural products, machinery and infrastructure, and construction activities. The GHG emission source evaluated in this study can be expanded to include additional life cycle stages upstream of scope 1 GHG emissions. Downstream impacts such as greenhouse gas emissions associated with landfill and recycling, construction, and disposal of demolition materials can also help further improve campus carbon dioxide emissions. This includes CO_2 compensation such as purchases and forest management credits. To get a complete picture of the effects of MSWM, social factors (respiratory diseases, etc.), economic factors (diesel costs, etc.), and psychological factors (abnormal noise, etc.) must be considered.

5. Conclusions

This research focuses on waste characterization and emphasizes the importance of several factors, including waste diagnosis and the need for regulatory standards on open dumping, the conversion of open dumps to sanitary landfills, and the establishment of waste treatment, material, and energy recovery units. According to the characterization research, organic waste is the most common component of MSW, accounting for 76% of the total trash. As a result of this finding, anaerobic digestion was discovered as a viable therapeutic option. It was also shown that separating MSW at the source improves the effectiveness of anaerobic digestion plants. It is also critical that residents are made aware of the necessity of trash sorting and recycling through mass awareness campaigns and education, in which universities play a key role. Universities must invest in trash collection infrastructure that allows source-segregated waste to be collected separately and transported to waste treatment facilities. For a smooth transition to structural and sustainable waste management systems, LCA is being used to analyze the present MSW management of the RUC of India. For the RUC waste management, this study evaluated a base scenario of open dumping as well as three alternative scenarios that included a combination of anaerobic digestion, composting, and landfilling. Material and energy recovery were evaluated, as well as their influence on the environment, including global warming, acidification, eutrophication, human toxicity, terrestrial and aquatic toxicity, and so on. For the GWP, FEW, HT, TE, FWT, and MET impact categories, scenario S3 (AD + LFWR) had the lowest impact. According to the study, treating MSW RUC with an integrated waste management facility that combines anaerobic digestion in small-scale units, composting, and landfill without gas recovery alternatives is the best approach for maximizing material and energy recovery while minimizing environmental impacts. The findings support the use of LCA as a valuable tool for developing integrated waste management systems because it allows council officials to compare the environmental consequences of several alternative waste treatment technologies. Furthermore, research into normalization factors and weights for mid-point and end-point effect categories relevant to the Indian context must be conducted. The goal of this study was to give a comprehensive assessment of the environmental consequences associated with existing waste management treatment options, as well as possible prospects for successful impact offsetting.

To get a complete picture of the effects of MSWM, social (respiratory diseases, etc.), economic (diesel costs, etc.), and psychological (abnormal noise, etc.) factors need to be considered for further assessment and the final decision. The foremost hassle within the evaluation is to locate the supply of facts and gather the facts to decide the greenhouse fuel line emissions. However, this looks at how the extent of consciousness amongst college and college students for feasible discount ability of GHG emissions has been raised. This can serve for example Indian universities in lessening their effects because of intake and they can also increase their carbon control plans. Therefore, following this review, there is no standardized standard for reporting university GHG emissions, primarily related to the organizational boundary aspects of indirect emissions and emission factors. The lack of common criteria, apart from the issue of comparability, results in inventories that do not clearly indicate potential opportunities for mitigation action. For this reason, and as a future research proposal, it is necessary to develop methods and simplified computational tools for the university's carbon footprint. The goal is to achieve comparable results to other universities and to be able to identify and take into account all opportunities to reduce emissions.

Author Contributions: Conceptualization, A.K.J., V.R.S.C. and B.D.; methodology, A.K.J., V.R.S.C. and B.D.; software, A.K.J. and V.R.S.C.; validation, A.K.J. and V.R.S.C.; formal analysis, A.K.J. and V.R.S.C.; investigation, A.K.J. and V.R.S.C.; resources, B.D.; data curation, M.V.; writing—original draft preparation, A.K.J.; writing—review and editing, V.R.S.C., M.V. and B.D.; supervision, B.D.; project administration, B.D. All authors have read and agreed to the published version of the manuscript.

Funding: This research received no external funding.

Institutional Review Board Statement: Not applicable.

Informed Consent Statement: Not applicable.

Data Availability Statement: Not applicable.

Acknowledgments: The first author is thankful to Ministry of Human Resources and Development, Government of India for providing scholarship. Acknowledgements to Rajesh Kola, Technical Assistant Prasad for supporting all throughout the laboratory testing and analysis.

Conflicts of Interest: The authors declare no conflict of interest.

References

1. Abdul-Azeez, I.A.; Ho, C.S. Realizing Low Carbon Emission in the University Campus towards Energy Sustainability. *Open J. Energy Effic.* **2015**, *4*, 15–27. [CrossRef]
2. Adeniran, A.E.; Nubi, A.T.; Adelopo, A.O. Solid waste generation and characterization in the University of Lagos for a sustainable waste management. *Waste Manag.* **2017**, *67*, 3–10. [CrossRef]
3. AICTE. Government of India, All India Council for Technical Education. 2016. Available online: https://facilities.aicte-india.org/dashboard/pages/dashboardaicte.php (accessed on 28 August 2021).
4. Andrianisa, H.A.; Brou, Y.O. Role and importance of informal collectors in the municipal waste pre-collection system in Abidjan, Côte d'Ivoire. *Habitat Int.* **2016**, *53*, 265–273. [CrossRef]
5. Angelo, A.C.M.; Saraiva, A.B.; Clímaco, J.C.N.; Infante, C.E.; Valle, R. Life Cycle Assessment and Multi-criteria Decision Analysis: Selection of a strategy for domestic food waste management in Rio de Janeiro. *J. Clean. Prod.* **2017**, *143*, 744–756. [CrossRef]
6. Armijo de Vega, C.; Ojeda-Benítez, S.; Ramírez-Barreto, M.E. Mexican educational institutions and waste management programmes: A University case study. *Resour. Conserv. Recycl.* **2003**, *39*, 283–296. [CrossRef]
7. Bahçelioğlu, E.; Buğdaycı, E.S.; Doğan, N.B.; Şimşek, N.; Kaya, S.; Alp, E. Integrated solid waste management strategy of a large campus: A comprehensive study on METU campus, Turkey. *J. Clean. Prod.* **2020**, *265*, 121715. [CrossRef]
8. Zen, I.S.; Subramaniam, D.; Sulaiman, H.; Saleh, A.L.; Omar, W.; Salim, M.R. Institutionalize waste minimization governance towards campus sustainability: A case study of Green Office initiatives in Universiti Teknologi Malaysia. *J. Clean. Prod.* **2016**, *135*, 1407–1422. [CrossRef]
9. Armijo de Vega, C.; Ojeda Benítez, S.; Ramírez Barreto, M.E. Solid waste characterization and recycling potential for a university campus. *Waste Manag.* **2008**, *28* (Supp. 1), S21–S26. [CrossRef]
10. Fagnani, E.; Guimarães, J.R. Waste management plan for higher education institutions in developing countries: The Continuous Improvement Cycle model. *J. Clean. Prod.* **2017**, *147*, 108–118. [CrossRef]
11. Smyth, D.P.; Fredeen, A.L.; Booth, A.L. Reducing solid waste in higher education: The first step towards 'greening' a university campus. *Resour. Conserv. Recycl.* **2010**, *54*, 1007–1016. [CrossRef]
12. Malakahmad, A.; Kutty, S.; Hasnain Isa, M. Solid Waste Characterization and Recycling Potential for University Technology PETRONAS Academic Buildings USM-MOSTI Project View project Detection Methods of Carcinogens in Estuaries View project. *Am. J. Environ. Sci.* **2014**, *6*, 422–427. [CrossRef]
13. Berchin, I.I.; Grando, V.D.S.; Marcon, G.A.; Corseuil, L.; Guerra, J.B.S.O.D.A. Strategies to promote sustainability in higher education institutions: A case study of a federal institute of higher education in Brazil. *Int. J. Sustain. High. Educ.* **2017**, *18*, 1018–1038. [CrossRef]
14. Tangwanichagapong, S.; Nitivattananon, V.; Mohanty, B.; Visvanathan, C. Greening of a campus through waste management initiatives: Experience from a higher education institution in Thailand. *Int. J. Sustain. High. Educ.* **2017**, *18*, 203–217. [CrossRef]
15. Clabeaux, R.; Carbajales-Dale, M.; Ladner, D.; Walker, T. Assessing the carbon footprint of a university campus using a life cycle assessment approach. *J. Clean. Prod.* **2020**, *273*, 122600. [CrossRef]
16. Sammalisto, K.; Brorson, T. Training and communication in the implementation of environmental management systems (ISO 14001): A case study at the University of Gävle, Sweden. *J. Clean. Prod.* **2008**, *16*, 299–309. [CrossRef]
17. Lukman, R.; Tiwary, A.; Azapagic, A. Towards greening a university campus: The case of the University of Maribor, Slovenia. *Resour. Conserv. Recycl.* **2009**, *53*, 639–644. [CrossRef]
18. Geng, Y.; Liu, K.; Xue, B.; Fujita, T. Creating a "green university" in China: A case of Shenyang University. *J. Clean. Prod.* **2013**, *61*, 13–19. [CrossRef]
19. Ozawa-Meida, L.; Brockway, P.; Letten, K.; Davies, J.; Fleming, P. Measuring carbon performance in a UK University through a consumption-based carbon footprint: De Montfort University case study. *J. Clean. Prod.* **2013**, *56*, 185–198. [CrossRef]

20. Dong, J.; Ni, M.; Chi, Y.; Zou, D.; Fu, C. Life cycle and economic assessment of source-separated MSW collection with regard to greenhouse gas emissions: A case study in China. *Environ. Sci. Pollut. Res.* **2013**, *20*, 5512–5524. [CrossRef]
21. Wang, M.Q.; Han, J.; Haq, Z.; Tyner, W.E.; Wu, M.; Elgowainy, A. Energy and greenhouse gas emission effects of corn and cellulosic ethanol with technology improvements and land use changes. *Biomass Bioenergy* **2011**, *35*, 1885–1896. [CrossRef]
22. Mandpe, A.; Bhattacharya, A.; Paliya, S.; Pratap, V.; Hussain, A.; Kumar, S. Life-cycle assessment approach for municipal solid waste management system of Delhi city. *Environ. Res.* **2022**, *212*, 113424. [CrossRef] [PubMed]
23. Cheela, V.R.S.; Dubey, B. Review of Application of Systems Engineering Approaches in Development of Integrated Solid Waste Management for a Smart City. In *Water Resources and Environmental Engineering II*; Rathinasamy, M., Chandramouli, S., Phanindra, K., Mahesh, U., Eds.; Springer: Berlin/Heidelberg, Germany, 2019; pp. 159–177.
24. Liikanen, M.; Havukainen, J.; Viana, E.; Horttanainen, M. Steps towards more environmentally sustainable municipal solid waste management—A life cycle assessment study of São Paulo, Brazil. *J. Clean. Prod.* **2018**, *196*, 150–162. [CrossRef]
25. Oliveira, L.S.; Oliveira, D.S.; Bezerra, B.S.; Pereira, B.S.; Battistelle, R.A.G. Environmental analysis of organic waste treatment focusing on composting scenarios. *J. Clean. Prod.* **2017**, *155*, 229–237. [CrossRef]
26. Sharp, L. Green campuses: The road from little victories to systemic transformation. *Int. J. Sustain. High. Educ.* **2002**, *3*, 128–145. [CrossRef]
27. Institute, U.S. Climate & Emissions Sustainability. 2018. Available online: https://www.unh.edu/sustainability/operations/air-climate (accessed on 28 August 2021).
28. *ASTM E1756-08*; Standard Test Method for Determination of Total Solids in Biomass. American Society for Testing and Materials: West Conshohocken, PA, USA, 2008. Available online: https://www.astm.org/Database.cart/historical/e1756-08.htm (accessed on 24 August 2021).
29. *ASTM E711-87*; Standard Test Method for Gross Calorific Value of Refuse-Derived Fuel by the Bomb Calorimeter (Withdrawn 2004). American Society for Testing and Materials: West Conshohocken, PA, USA, 2004. Available online: https://www.astm.org/Standards/E711.htm (accessed on 24 August 2021).
30. *ASTM E777-87*; Standard Test Method for Carbon and Hydrogen in the Analysis Sample of Refuse-Derived Fuel. American Society for Testing and Materials: West Conshohocken, PA, USA, 2004. Available online: https://www.astm.org/DATABASE.CART/HISTORICAL/E777-87R04.htm (accessed on 24 August 2021).
31. *ASTM E872-82*; Standard Test Method for Volatile Matter in the Analysis of Particulate Wood Fuels. American Society for Testing and Materials: West Conshohocken, PA, USA, 2019. Available online: https://www.astm.org/Standards/E872.htm (accessed on 24 August 2021).
32. Barlaz, M.A.; Chanton, J.P.; Green, R.B. Controls on landfill gas collection efficiency: Instantaneous and lifetime performance. *J. Air Waste Manag. Assoc.* **2009**, *59*, 1399–1404. [CrossRef] [PubMed]
33. Liu, Y.; Sun, W.; Liu, J. Greenhouse gas emissions from different municipal solid waste management scenarios in China: Based on carbon and energy flow analysis. *Waste Manag.* **2017**, *68*, 653–661. [CrossRef]
34. Bhupednra Kumar, S.; Munish, K. Life cycle assessment of potential municipal solid waste management strategies for Mumbai, India. *Waste Manag. Res. J. Sustain. Circ. Econ.* **2017**, *35*, 79–91. [CrossRef]
35. Biswas, W.K. Life cycle assessment of seawater desalinization in Western Australia. *World Acad. Sci. Eng. Technol.* **2009**, *56*, 369–375. [CrossRef]
36. Silva, F.B.; Yoshida, O.S.; Diestelkamp, E.D.; De Oliveira, L.A. Relevance of including capital goods in the life cycle assessment of construction products. *LALCA Rev. Lat.-Am. Em Avaliação Do Ciclo De Vida* **2018**, *2*, 7–22. [CrossRef]
37. CPHEEO. CPHEEO:Central Public Health & Environmental Engineering Organisation (CPHEEO), Govt of India. 2020. Available online: http://cpheeo.gov.in/cms/about-cpheeo.php (accessed on 28 August 2021).
38. Environmental Protection Agency (EPA). Report on "LFG Energy Project Development Handbook". 2020. Available online: https://www.epa.gov/sites/production/files/2016-11/documents/pdh_full.pdf (accessed on 15 April 2021).
39. Bassi, A.; Højlund, T. Environmental performance of household waste management in Europe—An example of 7 countries. *Waste Manag.* **2021**, *69*, 545–557. [CrossRef]
40. Babu, G.L.S.; Lakshmikanthan, P.; Santhosh, L.G. Life Cycle Analysis of Municipal Solid Waste (MSW) Land Disposal Options in Bangalore City. In *ICSI 2014: Creating Infrastructure for a Sustainable World—Proceedings of the 2014 International Conference on Sustainable Infrastructure*; American Society for Civil Engineers: Pomona, CA, USA, 2014; pp. 795–806. [CrossRef]
41. Bogner, J.E.; Chanton, J.P.; Blake, D.; Abichou, T.; Powelson, D. Effectiveness of a Florida Landfill Biocover for Reduction of CH_4 and NMHC Emissions. *Environ. Sci. Technol.* **2010**, *44*, 1197–1203. [CrossRef] [PubMed]
42. Turner, D.A.; Williams, I.D.; Kemp, S. Combined material flow analysis and life cycle assessment as a support tool for solid waste management decision making. *J. Clean. Prod.* **2016**, *129*, 234–248. [CrossRef]
43. Mistri, A.; Dhami, N.; Bhattacharyya, S.K.; Barai, S.V.; Mukherjee, A.; Biswas, W.K. Environmental implications of the use of bio-cement treated recycled aggregate in concrete. *Resour. Conserv. Recycl.* **2021**, *167*, 105436. [CrossRef]
44. European Commission. Committee and the Committee of the Regions a European Agenda on Migration 2. 2015. Available online: http://www.europarl.europa.eu/oeil/popups/ficheprocedure.do?lang=en&reference=2015/2660 (accessed on 25 August 2021).
45. Pooja, Y.; Sukha, R. Environmental impact assessment of municipal solid waste management options using life cycle assessment: A case study. *Environ. Sci. Pollut. Res.* **2018**, *25*, 838–854. [CrossRef]

46. Chatterjee, B.; Mazumder, D. Role of stage-separation in the ubiquitous development of Anaerobic Digestion of Organic Fraction of Municipal Solid Waste: A critical review. *Renew. Sustain. Energy Rev.* **2019**, *104*, 439–469. [CrossRef]
47. CPHEEO. Manual on Municipal Solid Waste Management—2016:Central Public Health & Environmental Engineering Organisation (CPHEEO), Govt of India. 2016. Available online: http://cpheeo.gov.in/cms/manual-on-municipal-solid-waste-management-2016.php (accessed on 28 August 2021).
48. Seruga, P.; Krzywonos, M.; Seruga, A.; Niedźwiecki, Ł.; Pawlak-Kruczek, H.; Urbanowska, A. Anaerobic Digestion Performance: Separate Collected vs. Mechanical Segregated Organic Fractions of Municipal Solid Waste as Feedstock. *Energies* **2020**, *13*, 3768. [CrossRef]
49. Goedkoop, M.; Oele, M.; Leijting, J.; Ponsioen, T.; Meijer, E. Introduction to LCA with SimaPro Title: Introduction to LCA with SimaPro. 2016. Available online: www.pre-sustainability.com (accessed on 25 August 2021).
50. El Hanandeh, A.; El-Zein, A. Life-cycle assessment of municipal solid waste management alternatives with consideration of uncertainty: SIWMS development and application. *Waste Manag.* **2010**, *30*, 902–911. [CrossRef]
51. Ingrao, C.; Faccilongo, N.; Di Gioia, L.; Messineo, A. Food waste recovery into energy in a circular economy perspective: A comprehensive review of aspects related to plant operation and environmental assessment. *J. Clean. Prod.* **2018**, *184*, 869–892. [CrossRef]
52. Styles, D.; Dominguez, E.M.; Chadwick, D. Environmental balance of the UK biogas sector: An evaluation by consequential life cycle assessment. *Sci. Total Environ.* **2016**, *560–561*, 241–253. [CrossRef]
53. National Research Council. *Waste Incineration and Public Health*; National Academies Press: Washington, DC, USA, 2000.
54. Kanhar, A.H.; Chen, S.; Wang, F. Incineration Fly Ash and Its Treatment to Possible Utilization: A Review. *Energies* **2020**, *13*, 6681. [CrossRef]
55. Cheela, V.R.S.; John, M.; Biswas, W.; Dubey, B. Environmental Impact Evaluation of Current Municipal Solid Waste Treatments in India Using Life Cycle Assessment. *Energies* **2021**, *14*, 3133. [CrossRef]
56. Zhou, H.; Yang, Q.; Gul, E.; Shi, M.; Li, J.; Yang, M.; Yang, H.; Chen, B.; Zhao, H.; Yan, Y.; et al. Decarbonizing university campuses through the production of biogas from food waste: An LCA analysis. *Renew. Energy* **2021**, *176*, 565–578. [CrossRef]

sustainability

MDPI

Article

Case Study of Municipal Waste and Its Reliance on Reverse Logistics in European Countries

Olga Lingaitienė, Aurelija Burinskienė * and Vida Davidavičienė

Faculty of Business Management, Vilnius Gediminas Technical University, LT-10223 Vilnius, Lithuania; olga.lingaitiene@vilniustech.lt (O.L.); vida.davidaviciene@vilniustech.lt (V.D.)
* Correspondence: aurelija.burinskiene@vilniustech.lt

Abstract: The authors have examined municipal waste, its components and their integration with reverse logistics processes. Background: The theoretical part begins with a definition of municipal waste. Later, the integration between municipal waste and reverse logistics is provided, including presentation of the hierarchy of qualitative methods and models. Methods: The authors constructed a correlation matrix and applied a dynamic regression model to identify that the level of municipal waste impacts recycling of biowaste which demands reverse logistics. Results: The authors provided a dynamic regression model which could be applied for forecasting the size of recycled municipal waste into biowaste indicated in European Union countries. Conclusions: The variety of components in municipal waste prevents the increase of the recycling rates and has to be changed to ones that have higher recycling rates.

Keywords: municipal waste; packaging; recycling; logistics processes; reverse logistics

Citation: Lingaitienė, O.; Burinskienė, A.; Davidavičienė, V. Case Study of Municipal Waste and Its Reliance on Reverse Logistics in European Countries. *Sustainability* **2022**, *14*, 1809. https://doi.org/10.3390/su14031809

Academic Editors: Sunil Kumar, Pooja Sharma and Deblina Dutta

Received: 2 December 2021
Accepted: 29 January 2022
Published: 5 February 2022

1. Introduction

Package design is paramount to grabbing buyers' attention. In direct relation to the creation of the packaging, the demand for the product grows [1]. Product packaging design, like branding, is critical to product positioning in the industry and can either drive sales or block sales entirely [2].

The globalisation of modern goods and transactions has presented a new perspective in the return management process [3]. Unfortunately, most companies emphasise getting the product out the door and overlook the need for returns and business management acts as if not expecting the potential return of the shipped product. This gap is often attributed to a deficiency of system automation required to manage the returned product. Indeed, they focus more on immediate reverse logistics customer issues but concentrate less on returns [4].

According to Gartner, Inc., the net profit loss due to improper handling of returns results in loss of control and loss of income and inventory can be estimated to be 35%. Therefore, it is economically prudent and makes good business sense to create an efficient and reliable reverse logistics process [5].

The reverse logistics activities occur during recycling, where packing material is used as a component of municipal waste [6]. The report *"The Future of World Packaging to 2022"* indicates that the need and demand for packaging will gradually grow by 2.9% to reach $980 billion in 2022. Global packaging sales will increase by 3% at an annual rate of 4% from 2018. In Western Europe, packaging sales accounted for 22%, in North America 23%, and in Asia even 36% of total sales [1,2].

The management of municipal waste demands reverse logistics. The physical return triggers reverse logistics processes. The authors investigate the connection points between reverse logistics and municipal waste management.

An analysis of literature (i.e., review of books published by Oxford University Press, Cambridge University Press, Harvard University Press, Springer, M.E. Sharpe, Routledge, other publishers was performed and identified that the theme of municipal waste is rarely discussed in the literature on reverse logistics. The analysis presented in Table 1 shows that only 0.49 per cent of the above publications describe investigations in that research area.

Table 1. Review of literature.

Year	Literature of Reverse Logistics	Literature of Municipal Waste	Thematic of Municipal Waste Under the Literature of Reverse Logistics
1994–1998	813	45,600	1
1999–2003	3460	46,300	5
2004–2008	8130	44,100	10
2009–2013	11800	46,600	117
2014–2018	36600	58,200	137
2019–2021	30200	61,700	173
Total	91003	302,500	443
%	100%		0.49%

Source: Constructed by authors, according to publications published by Oxford University Press, Cambridge University Press, Harvard University Press, Springer, M.E. Sharpe, Routledge, and other publishers.

Reverse logistics differs from waste management in that it focuses on the addition of value to a product to be recovered and then the outcomes will be used by forward logistics while waste management involves mainly the collection and treatment of the waste products that have got no new use.

That is why the article aims to identify the trends towards recycling and the size of the flows of municipal waste that are recycled. The recycling of municipal waste without the proper organization of waste flows, which are handled by reverse logistics and proper infrastructure, which is a necessity for reverse logistics, is hardly possible.

The article investigates the links between reverse logistics and municipal solid waste management. The article consists of seven main sections. The study starts with the introduction and the summary of the literature. In the second section, the authors discuss the research of various authors to reveal the essential elements of reverse logistics and municipal solid waste management processes. The third section presents the hierarchy of qualitative methods and models for researching reverse logistics and waste collection aspects. The fourth section analysis the role of reverse logistics; and the fifth part—packing and recycling aspects. The sixth section highlights the relationship between a three-level methodology between reverse logistics and municipal waste and describes a dynamic regression model. The seventh section presents results of the dynamic regression model, i.e., helping to forecast the recycled biowaste amounts.

2. Literature Review

Reverse logistics is the activity that includes the reverse distribution of materials, as well as reducing the number of new materials in the forward system [7–9]. According to De Brito et al., reverse logistics focuses on the recovery of products when they are no longer desired (end-of-life products such as computers or mobile phones) or can no longer be used (end-of-life products, i.e., tires and packaging) to obtain economic returns through reuse, recycling or recycling in new production [10–12].

Other authors note the importance of environmental requirements and the increasing role of the reverse supply chain in the extraction of materials [13–16]. Valenzuela et al. examined recycling models for plastics based on a reverse logistics model and a waste recycling model and also pointed out that the reverse logistics process is a process from consumption to point of origin that includes the planning, execution and effective control of the costs of raw materials, work in progress, finished goods and related information, to recover the primary value of materials or dispose of them properly [17]. Table 2 presents the main elements of reverse logistics described by different authors.

Table 2. Essential elements of reverse logistics.

	Elements	References
Essential elements of reverse logistics	Operation cost	[18–42]
	Recapture value	[18,21,24,26,28,30,35,41]
	Technical feasibility	[24,28,29,35,40,42–44]
	Recycling network	[18,19,22,25,26,28–30,33,36,40–48]
	Return cost	[25,30,32,35,42,43,49]
	Remanufacturing network	[18,21,26,30,36,42,44,48–50]
	Recovery value	[18,25,36,44,49,50]
	Reuse network	[18,20,21,27,28,31,33,41–43,45–47,50]
	Product recovery	[18,19,21,23–26,28,30–34,39,42,49,51,52]
	Environmental impacts	[18,20,21,23,24,27–31,33,35,36,38,40–43,45–47,49–53]
	Service management	[18,20,34,37,40,44,49–51,53]
	Market demand	[18,19,30,32,34,37,44,50,53]
	Sustainable development	[22,23,27–31,35–37,40,44–46,51,53,54]
	Green effect	[20,22–27,34–37,39,40,44–46,49,51,52,54]
	Product return	[18,21,23–25,30–32,34–36,39,41,42,45,48,49,54,55]
	Closed loop supply chain	[24–26,29,30,32,35,40,42,50–52,55]
	End-of-life product	[18,19,21–26,35,39,42,43,50,55]

Table 2 shows that in recent years the publications describing the processes of reverse logistics, necessarily emphasize the impact of such processes on the environment and sustainable development, green effect and others, which are disclosed in more detail in the table.

Municipal solid waste (MSW, as specified in Appendix B) includes refuse from households, non-hazardous solid waste from agriculture areas, commercial and institutional establishments (including hospitals), market waste, yard waste, and street sweepings [56]. MSW, which is defined as waste that is collected and treated by the municipality. MSW is the term applied to domestic waste and domestic-type industrial waste (paper, plastic, electronic appliance waste).

MSW includes everyday items used and then thrown away, called rubbish or trash in everyday life. According to Vergara et al. MSW is all solid or semi-solid materials disposed of by residents and businesses, excluding hazardous wastes and wastewater [57]. MSW includes food packaging, bottles, clothing, furniture, lawn clippings, scraps of foods, newspapers, household appliances, paints, and batteries [14,47]. The primary producers of solid waste are in residential areas such as apartments, houses, companies, schools, hospitals and others.

According to Ogwueleka, MSW management is the management of waste in urban areas in terms of its collection, transfer, treatment, recycling, reuse and disposal [56]. The main goals of MSW management are related to the promotion of the quality of the urban environment, the creation of jobs and income, the protection of environmental health and the maintenance of economic efficiency and productivity [56,58]. The definition of MSW does not include building rubble, waste from the demolition of buildings and structures, and wastewater from municipal sewage treatment plants and sewer networks.

Table 3 shows the trends that many authors in their scientific papers emphasize waste management, waste minimization and reduction, green images in the above-mentioned processes.

Table 3. Essential elements of MSW.

	Elements	References
Essential elements of MSW	Waste reusing	[18,26,31]
	Waste incineration	[21,23,28,32,42,47]
	Air emission	[22,31,40,47]
	Carbon footprint	[18,20,21,26,30,36,39,46,47,49]
	Waste recycling	[19,21,32,35,37,39,51]
	Waste disposal	[20–26,31,39,41,43,47,51]
	Zero Waste	[21,32,47,51]
	Responsible use	[24,26,31,35,40,41,45,46,51]
	Waste treatment	[22,26,28,29,39,46,52]
	Waste use for energy	[21,26,45,51,53]
	Green image	[18–20,22,24–26,30,34,35,40,42,43,45,46,50,52,54]
	Waste generation	[18–20,22,28,29,31,40,42,43,47,50,55]
	Waste minimization (reduction) & recovery	[20–22,25–27,29,32,35,40,43,45,47,50–52,55]
	Waste landfilling	[21,23,24,28,30,32,35,42,47,55]

The similarity of household waste shows how household waste can be classified according to the European classification. It is the European classification possibility proposed by Eurostat for the scope of municipal waste. Waste organisation is based on the principle that municipal waste includes household waste and waste from non-household sources, regardless of the division of responsibility for municipal waste collection that rests with the municipality or individuals [59]. Typical household and commercial solid waste include clothing, disposable utensils, garden waste, cans, disposable office tables, paper and boxes. In contrast, institutional and industrial solid waste includes restaurant waste, paper, school waste, wooden pallets, plastics, corrugated boxes, etc. and office documents [60,61].

3. Municipal Waste Recovery and Disposal Categories vs. Reverse Logistics

The authors identify the interface between reverse logistics and MSW. Some directions are identified in previous studies. The authors focus on such directions: (1) the comparison of urban solid waste and municipal governance practices in several European countries to identify the characteristics and main aspects of waste management and physical reverse logistics by modelling logistics for urban solid waste; (2) the improvement of decisions at the operational reverse logistics level for electronic equipment waste; (3) the design of reverse logistics network for electrical and electronic waste in Turkey; (4) the construction of routes used for the management for urban solid waste in Brazil; (5) creation of optimal reverse logistics network for determining the carbon footprint; (6) the production planning for products repair and recycling based on reverse logistics network management and other studies cited in Table 4.

Table 4. Hierarchy of qualitative methods and models for researching reverse logistics and waste collection aspects.

Model Type	Model Technique	Solution Method	References Discussing Reverse Logistics and Waste Collection
Mathematical programming method	Single objective	Linear programming	[62–66]
	Multiple objectives	Mixed-integer linear programming	[67–72]
		Mixed-integer programming	[73,74]

Table 4. *Cont.*

Model Type	Model Technique	Solution Method	References Discussing Reverse Logistics and Waste Collection
		Multiple regression	[75,76]
		Analysis of hierarchical regression	[77]
		Fuzzy-goal programming	[78]
		Stochastic dynamic programming	[79]
		Non-linear programming	[80,81]
	Time series	Dynamic regression analysis	*this study*
Causal models	Causality identification methods	Causal effect modelling	[82]
		Diagram of causal systems	[83]
Heuristic methods	Simple heuristic	Simulated annealing heuristics	[84]
	Artificial intelligence techniques	Markov chain	[85]
		Object-oriented Petri nets	[86]
		Bayesian network modelling	[87]
		Fuzzy logic	[88]
		Rough sets	[89]
		Neighbourhood rough sets	[90]
	Metaheuristic	Genetic Algorithm Multi-objective evolutionary Algorithm Multi-objective differential evolution algorithm Particle swarm optimisation Ant Colony Optimization	[91–96]
Analytical models	Multi-criteria decision making	Analytical hierarchy process	[97]
		DEMATEL	[98]
Analytical models	Systematic models	Delphi method	[55,99]
		Network model	[100]

References Discussng Reverse Logistics and Waste Collection	Description of Study
[62]	Linear model is constructed for co-collecting separated waste streams and for integrating reverse logistics delivery and collection activities.
[63]	Linear model is presented production-recycling-reuse of plastic beverage bottles in the context of reverse logistics and waste management.
[64]	A manufacturing/remanufacturing inventory model is formed with waste disposal and developed reverse logistics process for energy used.
[65]	Linear programming model presents the impact of waste pickers activities result on the reverse MSW logistics network.
[66]	Model is suggested for integrated forward-reverse logistics with carbon footprint considerations.
[67]	MILP model includes reverse logistics networks for returned medical waste seeking to improve medical waste management.
[68]	Reverse logistics network is constructed that sources end-of-life products.
[69]	Model is developed for the planning and management of the electronic waste collection system in the city of Genoa.
[70]	MILP model is formed for determining the best WEEE recycling offer price by determining reverse logistics operation planning strategies.
[71]	Optimization reverse logistics activities are described with MILP model of end-of-life vehicle recycling.
[72]	Model is developed to determine which recycling strategy is should be selected for commercial waste.
[73]	Heuristics is suggested for the configuration of reverse logisticsnetworks serving the recycling of electronic appliances and computers.
[74]	Regression method is used to investigate the relationship between food firm competitiveness and reverse logistics attributes important for waste management.

Table 4. *Cont.*

Model Type	Model Technique	Solution Method	References Discussing Reverse Logistics and Waste Collection
[75]	Major factors are examined that may influence industries toimplement reverse logistics, among which is the regulation of the waste amount.		
[76]	Four-step hierarchical regression analysis is used seeking to motivate firms to implement reverse logistics to handle e-waste.		
[77]	Fuzzy goal programming is presented for a lead/acid battery reverse logistics network design.		
[78]	Model of stochastic dynamic programming is presented for reverse logistics interaction with production planning of the re-manufacturing system, determining the amount of recycled waste.		
[79]	Non-linear model of multifunctional reverse logistics is constructed for planning and design of an optimal computer waste management system.		
[81]	Causal effect model is proposed to describe the need for the collection of agricultural waste (i.e., pesticide packaging).		
[82]	Diagram of the causal system is used to create a new WEEE handling system based on two flows a forward and reverse logistics flows.		
[83]	Simulated annealing algorithm is tailored to generate a solution using the output of an e-recycling reverse logistics network.		
[84]	Markov chain approach is used for presenting retailer reverse logistics which deals with waste handling.		
[85]	Petri net forecasting model is suggested for household waste and model reverse logistics network.		
[86]	Bayesian network is constructed for the reverse logistics that are used for product recovery and waste reduction.		
[87]	Simulation model is presented for reverse logistics network collecting end-of-life appliances.		
[88]	Rough set theory is applied to reduce the complexity of the RL for the company involved in waste management.		
[89]	Neighborhood rough set conceptual application is developed for management decisions within the context of reverse logistics and defective products.		
[90]	Genetic algorithm is presented to model the reverse logistics network for medical waste management.		
[91]	Model is constructed for the selection of solid waste transfer stations under a reverse logistics network.		
[92]	Optimization of the flow distribution of e-waste reverse logistics network is researched to obtain Pareto optimum solution with an evolutionary algorithm.		
[93]	Differential evolution algorithm is used to design product return network to balance costs and loads.		
[94]	Reverse logistics network optimization model is proposed which fully considered environment effect and the waste recycling factors.		
[95]	Model of multi-objective ant colony optimization (MACO) algorithm is suggested for reducing reverse logistics cost considering environmental factors was verified through a simulation on waste textile product reverse logistics.		
[96]	AHP technique is proposed for the decision-making process to evaluate strategies for obtaining optimal strategies in reverse logistics that collects used products as a waste.		
[97]	DEMATEL technique is suggested as a decision-making tool used to design reverse logistics for collecting textile waste.		
[55]	Delphi method is used to develop a model to select the most appropriate firms for the treatment of infectious waste objectively and efficiently.		
[98]	Solution is proposed for assessing the green practices including reverse logistics and use of waste.		
[99]	Network model is presented to rank the alternatives for implementing the process of reverse logistics, which include three major stages: waste collecting, sorting and reprocessing.		

In Table 4, the authors examined the type of qualitative methods and models dedicated to reverse logistics and waste collection are used by other authors for their research. Among the methods, the most popular is the mixed-integer linear programming method in studies dedicated to the above-mentioned topic [100]. The authors identified that the dynamic regression model is not mentioned among above listed quantitative methods. Time series analysis could help to identify the capacity that is required to handle reverse logistics at different periods to perform waste collection and recycling.

Reverse logistics principle, evaluation and recycling of organic waste is relevant for both developing and economically developed countries from the point of view of solving current problems in waste management. Reverse logistics focuses on adding value to the product to be disposed of.

The implementation of life cycle processes (LCA) is an essential tool in reverse logistics. The life cycle assessment includes alternative material concepts and various components from the extraction of raw materials through the use phase to recycling; alternative ideas are implemented over the entire product life cycle, starting with the development process [101,102].

In Table 5, the authors briefly summarize the main sources and types of waste, with a description next to each.

Table 5. Municipal waste by source and types, modified by authors following [103].

MSW Source	Types of Solid Waste	Description
Agriculture	Solid waste, which includes waste from the food and meat processing industries, industrial, agricultural, yard, garden and plant debris, and medical solid waste, as well as hazardous solid and chemical waste	Agricultural activities associated with the preparation, production, storage, processing, and consumption of agricultural products, livestock, and processed products generally generate solid agrarian waste.
Commercial	Food wastes, metals, glass, special wastes, plastics, paper, cardboard, wood, hazardous wastes	Commercial Waste or MSW is produced in businesses and includes general waste, mixed dry recyclables and organic waste.
Household	Textiles, food wastes, metals, glass, special wastes (batteries, consumer electronics, oil, tires, bulky items), plastics, paper, cardboard, wood, ashes, household hazardous wastes	Household Waste or MSW is produced in our homes and includes general waste, mixed dry recyclables and organic waste.
Institutional	Food wastes, metals, glass, special wastes, plastics, paper, cardboard, wood, hazardous wastes	Actions address waste materials originating in institutional facilities, such as government offices, schools, hospitals, nursing homes, correctional facilities, research institutions and public buildings.
Municipal service	Street-cleaning residues, trimmings landscape and trees, collecting general wastes from recreational areas (such as parks, beaches, etc.)	Household waste and waste of a similar type and composition are taken into account by municipal waste.

Commercial household waste has similar properties and composition to household waste. Such waste can be collected separately from household waste in removable containers [104–106]. The waste management pyramid defines levels of waste prevention and management. Under the waste management pyramid there are stated waste management steps, which should be named [107–109]:

1. reduced,
2. reused,
3. repaired,
4. recycled,
5. recovered,
6. composted,
7. incinerated,
8. landfilled.

Reverse logistics are the processes of planning, implementation and control of the return flow of raw materials, production stocks, packaging and finished goods from the original production, distribution or use location to the final disposal. The flow goes from the customer, more precisely from the point of consumption to the manufacturer, more precisely to the place of origin, to dispose of them properly or to return the value.

This activity is due to some aspects of environmental protection. However, most of them relate to issues respecting the correct distribution of waste. From this, we can conclude that reverse logistics is associated with garbage, especially those that are suitable for processes that restore their useful functions and for processes that are safe distance [17,110].

Based on reverse logistics, the ultimate goal of a company must be resources reduction, that is, waste and energy savings producing greener products reduction [20,21,26]. In such cases, the company should reuse the materials and make an effort to maximise waste recycling. In this case, reverse logistics at the end of the chain would focus on waste disposal [52,86]. Any predictable returns or pattern that recurs or repeats over one year is said to be a seasonal return.

Reusing waste management, proper handling and recycling, that is, activities in the return channel indicates the connection between waste management and take-back logistics [14,15,17,110]. Remove items—any waste management element that needs to be removed.

When it comes to waste management, reverse logistics plays a vital role [6,7]. As we can see in Figure in Appendix A, reverse logistics is a process that allows organisations to reuse, reduce, use energy, landfill and recycle waste generated at various points in their value chain [18,34].

The links between the supply chain, return logistics and waste management processes are also evident in the diagram [16,35]. Returning goods will always be part of the business, but it is not worth considering only an operating expense. Properly implemented reverse logistics processes can improve a company's performance [3,15]. In addition, better management of returned goods will help reduce waste and generate higher profits, as the same goods can be reused [103,104].

Reverse logistics differs from traditional waste management in that it adds value back into the chain by recovering and repurposing products, while waste management mainly focuses on disposal [109]. Fundamental to reverse logistics is offering efficient, potentially profit-generating methods of disposing of end-of-life products and waste [17,110]. A well-integrated waste management system encompasses government, business and society [10,61,107].

What is considered a waste to someone today may be a resource in the future, asserted by the authors [111,112]. They further point out that waste in one industry may be a raw material in another industry in recent industrial development.

Municipal waste and waste, in general, can be treated as redundant objects, which have lost their initial functionality, but that present value in terms of their secondary function [113,114].

Table 6 shows that the municipal waste is distributed according to recycling, landfill, incineration, composting by million tones and by kilograms per capita in the years from 2010 to 2019, when the numbers are presented as the European Union average [115]. The numbers of recycling and composting are growing. However, the volume of landfills is decreasing, and waste incineration is increasing. The environmental impact of incinerated waste is highly dependent on the nature of the trash that is disposed of.

Table 6. Municipal waste by waste management operations, EU-27, 2010–2019 [116].

	2010	**2011**	**2012**	**2013**	**2014**	**2015**	**2016**	**2017**	**2018**	**2019**
					million tones					
Recycling	55	56	58	56	59	63	65	66	67	68
Composting	29	29	30	31	33	33	36	38	38	39
Incineration	53	55	54	56	57	57	58	59	59	60
Landfill	79	74	67	63	59	57	54	53	52	53
Other	6	6	6	5	4	4	5	6	6	4

Table 6. *Cont.*

	2010	2011	2012	2013	2014	2015	2016	2017	2018	2019
					kg per capita					
Recycling	125	128	130	128	134	141	146	148	149	152
Composting	66	66	69	71	73	75	82	85	84	87
Incineration	121	125	122	127	128	128	131	132	132	134
Landfill	178	167	153	142	134	127	121	118	116	119
Other	13	13	14	10	9	9	10	13	13	10

During recycling, waste is recycled into products or materials. Waste is sorted from the general waste stream [117]. In such cases, on-site processing in industrial plants is excluded. Other categories of waste are composition or decomposition. The processing of organic substances by aerobic or anaerobic processes and the processing into substances that can be used as fuels and for energy generation do not belong to the waste treatment category. In the case of household waste, they are recycled directly or after their pre-treatment.

Both composting and decomposition are biological processes in which biodegradable waste is decomposed aerobically or anaerobically. The products of these processes are composted or gestate. After any further processing, agriculture benefits or in the hope of using it wisely, the ecology is used as a recycled product or as a material for tillage [118]. Municipal waste can be composted directly or after pre-treatment [119]. If the product after treatment is subsequently disposed of in a landfill, incinerated or otherwise not used for the above purpose, in which case the biological treatment of the waste MBT (mechanical-biological treatment) cannot be considered as composting [118]. Wastes from composting/fermentation and incineration of debris from another recovery/disposal operation to be recycled materials as metals [115,120].

Incineration is the thermal treatment of waste in an incineration plant. Incinerate immediately or after pre-treatment, possibly municipal waste. Incineration of municipal waste makes it possible to obtain as secondary fuel [107,121,122].

The landfill is defined as waste discharge into the ground and temporary storage in permanent places for more than one year [120]. Landfills with internal (waste generator itself eliminates waste generation) and external sites are defined. Municipal waste is disposed of in landfills either directly or after pre-treatment. If a sorting step is performed in the landfill area, the sorting results are assigned to the proper recovery/disposal operations [103,110].

4. The Role of Reverse Logistics

Reverse logistics processes offer companies the opportunity to become more environmentally friendly through reuse, recycling and recovery [5,15]. A material reduction in a direct system based on minimum return quantities, possible material reuse, and simplified recycling create a more comprehensive picture of reverse logistics [109].

The traditional supply chain uses resources from the environment. Such resources are converted into valuable products, and the cycle is considered complete when the product arrives and is distributed among consumers [121–124]. It is noteworthy that secondary raw materials follow "reverse sales channels", which are more associated with a reverse logistics strategy than with a method of using traditional sales and logistics channels [125]. Reverse logistics is defined as " … *the role of logistics in product returns, serving activities such as reduction, recycling, material replacement, material reuse, waste disposal and recycling, repair and recovery* … " [126]. In reverse logistics, an integrated approach is used to be successful. By definition, reverse logistic activities take many forms. Rogers and Tib-ben-Lembke [124] classify the list of possible reverse logistics actions. The activities are classified according to the materials used in products and packages in which products are delivered (see Table 7).

Table 7. Reverse Logistics serving Activities, constructed by the authors following [123].

Materials in	Reverse Logistics Serves such Activities
Package & Products	Reduce Reuse Repair Recycle Recover Compost Incinerate Landfill

In Table 7, the authors highlight the main activities served by reverse logistics.

Reduce is an activity that minimises the quantity and use of materials that cannot be recycled.

Reuse—an activity to clean, repair products and materials in products for reuse without any pre-processing steps.

Repair—the fixing of broken items in products and materials.

Recycling is the removal of material from a recycled product or package. These are carried out to recycle a product or packaging as raw material for a new product or packaging [127,128]. During waste collection, the driver should know that will be the further step. If the waste will be recycled the driver should not mix different waste streams by putting them into the truck.

Recover activity uses technologies and methods allowing to recover materials from mixed wastes.

Composting—activity, which separately collected waste converts into biowaste.

Incineration is an activity, which helps to transform waste into energy, plastics to fuels, and other resources.

The landfill is a site for the disposal of waste materials, also known as a tip, dump, rubbish dump, garbage dump, or dumping ground.

The main problem is that all products and/or packaging are sent to landfill when recycling is not involved [129].

When forward logistics end, reverse logistics begins. In the beginning, consumers buy the product they need, such as a newspaper or a soft drink. After reading a newspaper or drinking a soft drink, the product usually runs out, or the newspaper read becomes irrelevant, the product's useful life ends [130]. And then, consumers have to decide how to dispose of waste properly. It depends on consumers decision to dispose of leftover packaging or unwanted newspapers. Their choices are potentially substantial long-term effects on the environment [131]. Several factors must be considered, such as the activities that encourage a particular disposal method or financial incentives.

The reason for this is that in many cases, the price of the container, such as an empty soda bottle or a used newspaper, is not apparent to consumers or manufacturers, so there may be an incentive to recycle used raw materials be pretty low. The value of such a secondary or residual product will only increase as a potential raw material for a new product [129]. Therefore, the supply and demand for secondary raw materials must be developed to have value from new raw materials. The focus here is on reverse logistics.

5. Packing and Recycling

The packaging protects products that are placed on the market for sale, storage, use, etc. The packing generally relates to the design, evaluation and manufacture of packaging. Packing is critical for the movement of products that are distributed in large quantities. Improper packing could cause handling problems during reverse logistics activities. Common packaging materials are boxes, cardboard boxes, cans, bottles, pouches, envelopes, packages, and containers [132].

The increased demand for packaging has led many companies to look for methods and ways to improve packaging design to increase sales of their products [133]. Attractive, durable packaging design not only protects goods from damage and/or breakage but also helps to attract the end user's attention.

Packaging types and methods are presented in Table 8.

In Table 8 the authors briefly summarize the main packaging types and methods, with a description next to each.

Decisions about the choice of primary, secondary and tertiary packaging are of great strategic importance. In this article, we describe the key concepts for choosing product packaging from a logistics perspective and the difference between primary, secondary, and tertiary packaging [134–137].

Primary packaging is designed to contain, store and protect products. It is in direct contact with the product and is designed to keep it in optimal conditions. This packaging is the least used, so it does not complicate a single sale of goods. Primary packaging forms are jars, cans, bags, bottles, sacks, etc., [138].

Secondary packaging consists of an aggregate of primary packaging. This type of packing further protects the product and makes it easier to market on a larger scale. These are mainly cardboard boxes, although they can also be plastic. For example, in the case of milk, a single carton would be the primary packaging, and the carton containing the carton would account for the secondary packaging [138,139].

We need tertiary packaging to create larger units of loads, which includes primary and secondary. The most common forms of tertiary packaging are pallets, containers, and the modular cartons they contain. A distinction is made between three packaging types, which depend on the function and purpose: individual packaging, inner packaging and outer packaging [138,140].

Individual packaging is used for packaging each particular product. The unique packaging aims to protect the product from various climatic influences such as cold, heat, moisture, light, and wind [141].

The inner packaging is a unit that is sold in retail stores. It is used to bundle or group individually packaged products into a unit, such as lollipops in pouches. Each individually packaged candy is additionally grouped by weight in a single pack. Attractive packaging design is essential to arouse the consumer's desire to buy and thus to stimulate sales. It is, therefore, necessary to pay attention to the interior design of the packaging [142].

The most extensive packaging in which products are packaged in outer packaging is the outer packaging, such as cardboard boxes or wooden boxes. The primary purpose of the outer packaging is to protect products from breakage, dirt and elements [143].

Following the pyramid of plastic waste management, it is vital to prevent plastic use, reduce unnecessary plastics, and use reusable plastics designed for long life. It is suggested to recycle low-value plastics and produce recyclable high-value materials.

Recycling is understood to mean the flow of valuable materials [7,124]. For example, the bottle-from-bottle recycling method leads the industry and results in 100% recycled plastic used in bottles for cleaning and handwashing. Thus, recycling achieves a good green triple effect: less waste is sent to landfills, 70% less energy (as primary energy) is used for resin production and beautiful bottles are made from recycled plastic [144].

- For hand washing detergents, dishwashing detergents and sprays made from 100% PCR (post-consumer recycled), 100% recycled plastic is used, i.e., all plastic bottles made from 1-PET. Compared to using pure plastic, PCR has about 70% less carbon footprint.
- 50% recycled plastic is used in 2-HDPE bottles, ranging from 25% PCR in our toilet cleaners to 50% PCR in our detergent with 8× magnification.

Recyclable design is a closed packaging solution. With this in mind, recycling systems have been carefully researched to determine which plastics and packaging materials are suitable for recycling. Bottles are designed to be compatible with this purpose whenever possible. With this method, all packaging is developed to maximise recycled content, materials, efficiency and recyclability [145].

Table 8. Packaging types and methods.

Packaging Types	Packaging Methods	Description
Anti-corrosive Packaging	• Papers and films with volatile corrosion protection (VCI) • Moisture-absorbing barrier aluminium foil • Oil or liquid coating methods	Corrosion protection packaging should protect products from corrosion and avoid lengthy operations. Various materials are used to protect goods from the effects of different climatic conditions, such as paper, oil, bubble wrap with VCI, chips and bags, which are used as part of the anti-corrosion package.
Packaging of Pharma [50]	• Primary Package • Secondary Package • Tertiary package	During packaging of drugs or pharmaceuticals, the package goes through processes from production plants via distributing companies to end consumers. The package of pharmaceutical products is designed to ensure drug safety, ease of use and product safety upon delivery. The primary purpose of pharma packaging is to equip vital medicines for surgical devices, blood and blood products, liquid and bulk dosage forms, and solid and semi-solid dosage forms while maintaining their original condition and properties. The packaging described is used for delivery, etc.
Plastics Packaging [51]	• Bottle made from bottle • Design for recyclability • Refills method	Plastic packaging is mainly used for packaging various items such as fragile or non-perishable products. In addition, plastic packaging materials are used for coating materials or plastic-related products. For reuse in their factories, most plastic packaging companies recycle waste or plastic waste and offer alternatives.
Flexible Packaging [52]	Recycling of such types *: • PET or PETE • HDPE • LDPE • PV • PP • PS • Other plastic	Flexible packaging can be easily reshaped and defined as any packaging. If you choose flexible packaging, it has some benefits: • Provides safety for food products and shelf life indications through heat seal tightness, clogging prevention, ease use solutions and high print quality. • This packaging type generally reduces landfill waste as it creates little waste in printing processes. • Improvement of the production processes, reducing greenhouse gas emissions and volatile organic compounds, energy consumption, and water. • By using lighter and more flexible bags, more respect for nature, a lower environmental impact and lower consumption of energy and fossil fuel during transport are achieved.

* PET or PETE—Polyethylene Terephthalate, HDPE—High-Density Polyethylene, LDPE—Low-Density Polyethylene, PV—Polyvinyl, PP—Polypropylene, PS—Polystyrene.

The refuelling method ensures that only the required amount of plastic and other materials are used. Most of the plastic waste consists of containers and packaging. Almost

13% of all solid household waste is made up of various plastics. When looking for ways to make packaging as green as possible, researchers always emphasise the need to find additional ways to reduce environmental impact. For example, easy-to-use bags for replenishing supplies for manual, automatic, and dishwashing detergent save around 80% of plastic, water, and energy compared to a disposable bottle [146].

Different possibilities of recycling flexible packaging and plastic materials [136,147]:

- Polyethene terephthalate is the easiest to recycle. It is widely used in beverage bottles and food packaging (PET or PETE).
- High-density polyethene usually is used recycled into plastic bottles and bags, also used for thicker bottles for motor oil, bleach, and hair products (HDPE).
- A thinner, low-density polyethene is used to make plastic freezer bags and grocery bags. It can be recycled back into plastic bags (LDPE).
- Polyvinyl chloride is difficult to recycle and is environmentally hazardous, but it is used to manufacture furniture and pipes (PVC).
- Fibre plastic—polypropylene can be recycled into fibre materials for further use in clothing, roper, and other ways (PP).
- Polystyrene is used to make packaging materials, foam cups and other lightweight products. Because of its low density, it is difficult to recycle, but it can be reused (PS).
- Other plastics include polymer fibres, acrylic, polycarbonate, nylon, and fibreglass.

In industry, it's a good idea to assess the optimisation level of the packaging in terms of the material it is made of, transportation, handling and storage, waste management, and cost. Only with the overview of the process will you be able to choose the logistics packaging best suited to the company's application [148,149].

To raise awareness among customers, help them increase sales, and make them aware of the latest trends, many companies use the packaging methods listed in Table 9.

Table 9. Types of plastics and the examples of their applications in packing.

Plastics Types	Examples of Applications
Polyethene terephthalate (PET or PETE)	Fizzy bottles, bleach, cleaners and most shampoo bottles
High-density polyethene (HDPE)	Most shampoo and cleaner bottles, bleach, milk bottles
Polyvinyl chloride (PVC)	Thermal insulation (PVC foam) and auto parts, fittings, pipes, door and window frames (hard PVC)
A thinner, low-density polyethene (LDPE)	Bin liners, packaging films, carrier bags
Fibre plastic—polypropylene (PP)	Microwave-safe food bowls, margarine barrels, carpet fibres and threads, wall coverings and upholstery for automobiles
Polystyrene (PS)	The insulating material in construction, foam boxes for eggs and hamburgers, plastic cutlery and plastic cups for yoghurt, protective packaging for electronic items and toys
Other plastics	Plastics that do not fall into any of the categories listed above, such as B. Polycarbonate, which is widely used in the aerospace industry for glazing

In Table 9, the authors provide examples of how individual types of plastics can be used in packaging.

By choosing these methods, the packing of products delivers ideal results. Using these packaging methods, the shelf life of the products is extended, and competitive advantage is created. In addition, innovative packaging design maximises product profitability [138–149].

6. Materials and Methods

The collection of municipal waste involves operations of reverse logistics, including sorting, storing, and transportation. This study aims to figure out the effective actions important for decision-making helping to achieve sustainable development.

Various stakeholders make decisions:

- Producers, which decide which materials have to be used in products and in which volume, what should be their packing materials and what should be production methods;
- Retailers, which select and provide packing materials to consumers;
- Consumers, which apply to sort and products reuse practices;
- Logistics service providers implement a reverse logistics service management system.

The authors divided the methodology into three layers which present the connections of municipal waste and reverse logistics (see Table 10).

Table 10. Three-level methodology highlighting the connection between municipal waste and reverse logistics.

Level	Relationship to Reverse Logistics	Description of Municipal Waste Generation Minimisation by Stages	The Application of Methods	Links with Sustainability
1st level Use of environmentally friendly materials	The physical system supports the production and the reduction of the use of material.	• Selection of recycling supporting products and packing materials during production and selling stage.	• Review of literature;Statistical analysis; • Statistical analysis.	• The decision helps to minimise the negative impact on the environment.
2nd level Collection of municipal waste	The physical system is used for the collection of municipal waste from end-users.	• Sorting during the collection of municipal waste.	• Panel data analysis; • Regression analysis.	• Waste reduction and long-term sustainability supporting system.
3rd level Transformation of collected municipal waste	The physical system that supports recycling.	• Selection of methods that allows prolonging the life of materials.	• Comparison; • Investigations.	• Decision helping to save natural resources.

Table 10 provides a summary highlighting the link between municipal waste and reverse logistics with the help of a three-level methodology, providing descriptions, relationships and methods specific to each level.

For the research, the authors used such indicators, such as

(1) Recycling of biowaste;
(2) The recycling rate of e-waste;
(3) The recycling rate of municipal waste;
(4) The recycling rate of packaging waste by type of packaging.

The yearly data was retrieved from Eurostat for 30 European countries (27 European Union countries, Island, Norway and United Kingdom) for the period 2000–2019 [120]. In total it was 4359 data sets with the values.

The authors revised the data, constructed a correlation matrix and selected for the regression model only elements that have a probability lower than 0.1 (Table 11). The novelty of the study is that the authors constructed a dynamic regression model, by analysing the impact in year *t* and year *t-n*. The authors of this work use the dynamic regression model first applied by Petris et al. [150]. The first step in the modelling procedure was the transformation of time series to help identify the dependent variable and its relationships to the regressors. The developed model meets the requirements important for the construction of a simple regression model but provides dynamic interrelationships.

Table 11. Correlation matrix of variables transformed into dlog.

Covariance Analysis: Ordinary					
Sample: 56 577					
Included observations: 40					
		Recycling of biowaste	Recycling rate of e-waste	Recycling rate of municipal waste	Recycling rate of packaging waste by type of packaging
Recycling of biowaste	Correlation coeficient Probability	1.0 -			
Recycling rate of e-waste	Correlation coeficient Probability	−0.08 0.59	1.0 -		
Recycling rate of e-waste (−1)	Correlation coeficient Probability	0.10 0.52	0.10 0.53		
Recycling rate of municipal waste	Correlation coeficient Probability	0.86 0,0	−0.17 0.26	1.0 -	
Recycling rate of municipal waste (−1)	Correlation coeficient Probability	0.26 0.09	0.31 0.04	0.38 0.01	
Recycling rate of packaging waste by type of packaging	Correlation coeficient Probability	0.36 0.02	−0.07 0.65	0.54 0.00	1.0 -
Recycling rate of packaging waste by type of packaging (−1)	Correlation coeficient Probability	−0.15 0.35	0.17 0.29	−0.06 0.70	−0.24 0.12

Table 11 summarizes the correlation analysis performed for this study, noting the level of correlation between the elements listed in the table. The constructed Table 11 shows the link between the recycled rate of packing waste and the recycled rate of municipal waste, as packing waste is part of municipal waste. Table 11 indicates that for the recycling of biowaste the recycling rate of municipal waste (with probability 0) the same year strongly correlates.

The authors constructed a dynamic regression model helping to identify the amount of biowaste that is recycled. The completed equation is presented below (1).

The authors use the regressors in constructing mathematical equation:

$$rec\_\mathrm{biow}_t = \beta_0 + \beta_1\, rec\_biow_{(t-n)} + \beta_2\, rec\_mu_{(t-n)} + u_t \quad (1)$$

where: rec_biow_t—dlog of recycling of biowaste in year t, which is expressed in kg per capita; β_0—intercept in the equation; $rec\_biow_{(t-n)}$—dlog of recycling of biowaste, in year $t-n$, which is expressed in kg per capita; $rec\_mu_{(t-n)}$—dlog of recycling rate of municipal waste, in year $t-n$, which is expressed in kg per capita; u_t—random error of regression model and $\beta_{1,2}$—the influence of regressors on biowaste processing reflected coefficients of elasticity.

7. Results

The results show that residuals of the equation spread following normal distribution (Figure 1).

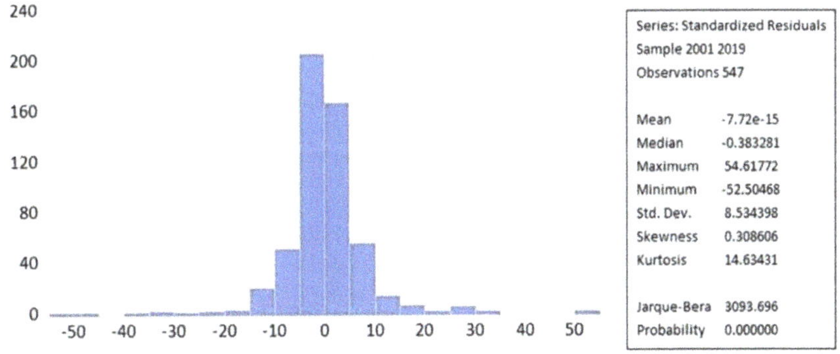

Figure 1. Normal distribution formed from equation residuals.

Figure 1 shows that the average of residuals approximates to zero. The forecasting of volumes of the recycled biowaste are presented in Figure 2.

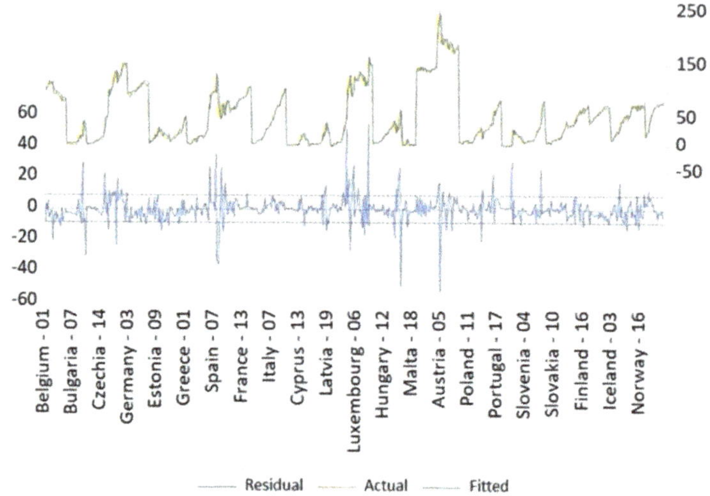

Figure 2. Forecasting the recycled biowaste level by the European Union countries.

The formed equation of the dynamic regression model is specified below (2) by defining coefficients and standard error:

$$\begin{aligned} rec\_\mathrm{biow}_t = -1.1 + 0.091\ rec\_biow_{(t-1)} + 0.23\ rec\_mu_{(t)} \\ (0.75)\quad (0.012)\rightarrow \quad\quad (0.036) \end{aligned} \quad (2)$$

Seeking to identify concrete values for dynamic regression model (2), the authors used the Panel least squares method and presented the results of method application in Figure 3. Where Durbin-Watson statistics is 2.20.

Variable	Coefficient	Std. Error	t-Statistic	Prob.
C	-1.104664	0.752624	-1.467750	0.1428
REC_BIOW(-1)	0.916663	0.012444	73.66330	0.0000
REC_MU	0.238563	0.036551	6.526858	0.0000

Root MSE	8.526593	R-squared	0.972816
Mean dependent var	53.36563	Adjusted R-squared	0.972716
S.D. dependent var	51.76258	S.E. of regression	8.550072
Akaike info criterion	7.135226	Sum squared resid	39768.43
Schwarz criterion	7.158833	Log likelihood	-1948.484
Hannan-Quinn criter.	7.144453	F-statistic	9733.870
Durbin-Watson stat	2.206447	Prob(F-statistic)	0.000000

Figure 3. Formation of equation (2): panel least squares revision method.

The application of the method shows that the adjusted *R* squared is 0.97. The statistical validity is revised by applying the Lagrange multiplier tests. The tests show the correct statistical validity. The probability for the Breausch-Pagan test is lower than 0.05. The detailed presentation of tests is presented in Appendix C.

8. Discussion

The returns management process deals with product returns from customers. These activities should be fast, controllable, visible and straightforward. Repeated returns can also occur when the seller rejects the return and returns it to the buyer without a refund. A type of reverse logistics for packaging management aims to reuse packaging materials to reduce waste and recycle. In the article, the returns management process is considered as relating primarily to the avoidance of returns and the return of products from customers. Redemption actions should be quick, simple, visible and easy to use. It is also possible that custom-made items will be returned if the seller refuses to accept the buyer's return and sends the item back to the buyer. Another type of reverse logistics discussed in the article aims to reuse packaging materials to reduce waste and recycle. Also described other types of reverse logistics management for product repair, refurbishment, and rework, including refurbishment, remanufacturing, and reconditioning activities. Repairs include the disassembly, cleaning, and reassembly of products.

To achieve sustainable development, the authors point out the need to build and expand a sustainable network in all supply chain processes, i.e., production, logistics and reverse logistics. Furthermore, particularly sustainable development aims to draw attention to an integrated approach to ecological, social and economic aspects, which leads to long term, sustainable profit growth.

The development of sustainable reverse logistics and sustainable waste management is essential for environmental protection. By reducing the overall negative impact of logistics on the environment, sustainable return logistics aims to solve environmental problems while considering the costs of recycling and disposing of waste, especially municipal waste. Therefore, the article discusses sustainable practices such as recycling, reuse, waste reduction, product return management, etc., to achieve process efficiency.

The study has some limitations: the authors do not revise the process efficiency; they identify the options for how to increase recycling rates and provide the dynamic regression model, which proves that.

9. Conclusions

The links and interdependencies between municipal waste and reverse logistics are a new topic that other authors have not explored so far. This article reveals that municipal waste is strongly and directly related to reverse logistics processes. The paper also discusses essential elements of reverse logistics and municipal solid waste. The authors constructed the hierarchy of qualitative methods and models for researching reverse logistics and waste collection aspects and figure out that most often authors apply mixed-integer linear programming method.

The authors identified aspects of materials and their recyclability opportunities. Also, highlighted reverse logistics, which is playing an important role in seeking sustainable development. The authors provided a methodology that identifies connection points between municipal waste and reverse logistics. Reverse logistics appear as supporting production processes and the collection of municipal waste from end-users. The authors identified three levels of connection points. The second level of methodology was researched in a mathematical way seeking to identify interconnections between recycling and municipal waste generation. The authors determined that the link among the above-identified components is positive.

Further research directions could evaluate the impact of specific materials and production methods on improving recycling rates. The study could be also extended to other countries and the recycling of other waste streams could be added. Also, the authors could compare the flows of reverse logistics with forwarding logistics.

Author Contributions: Conceptualisation is delivered by O.L. and A.B.; writing—original draft preparation is made by O.L.; writing—reviewing and editing are given by V.D., and methodology is constructed by A.B. All authors have read and agreed to the published version of the manuscript.

Funding: This research received no external funding.

Institutional Review Board Statement: Not applicable.

Informed Consent Statement: Not applicable.

Data Availability Statement: Not applicable.

Conflicts of Interest: The authors declare no conflict of interest.

Appendix A

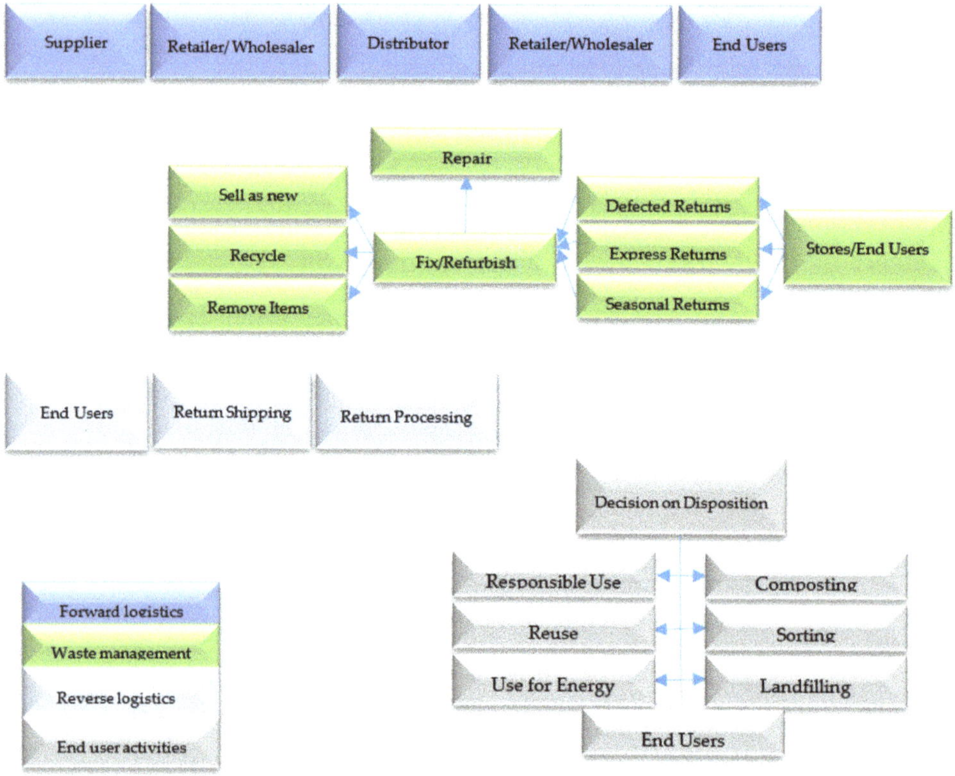

Figure A1. Reliance on reverse logistics and waste management.

Appendix B

Table A1. The abbreviations and definitions of key terms in the paper.

Full name	Abbreviation	Definition
Reverse logistics	RL	• Reverse logistics (RL) is the process of planning, implementing, and controlling the efficient and cost-effective flow of raw materials, in-process inventory, finished goods, and related information from the point of consumption to the point of origin to recapture value or proper disposal [9].
Municipal solid waste	MSW	• MSW is defined to include refuse from households, non-hazardous solid waste from industrial, commercial and institutional establishments (including hospitals), market waste, yard waste, and street sweepings [56].

Appendix C

The authors performed Residual Cross-sectional dependence test:

Residual Cross-Section Dependence Test
Null hypothesis: No cross-section dependence (correlation) in residuals
Equation: EQ01_2VAR
Periods included: 19
Cross-sections included: 30
Total panel (unbalanced) observations: 547
Test employs centered correlations computed from pairwise samples

Test	Statistic	d.f.	Prob.
Breusch-Pagan LM	615.87	435	0
Pesaran scaled LM	6.13		0
Bias-corrected scaled LM	5.29		0
Pesaran CD	-1.31		0.18

Figure A2. Residual Cross-sectional dependence test.

Figure A2 shows that Probability of Pesaran CD test is higher than 0.1. Also, the authors performed Redundant Fixed Effects tests:

Redundant Fixed Effects Tests
Equation: EQ01_2VAR
Test cross-section and period fixed effects

Effects Test	Statistic	d.f.	Prob.
Cross-section F	4.03	-29497	0
Cross-section Chi-square	115.64	29	0
Period F	1.40	-18497	0.12
Period Chi-square	27.16	18	0.07
Cross-Section/Period F	3.09	-47497	0
Cross-Section/Period Chi-square	140.34	47	0

Figure A3. Redundant Fixed Effects tests.

Figure A3 shows that Probability of Chow test is lower 0.05. Fix evaluation method is chosen properly.

References

1. Korhonen, J.; Koskivaara, A.; Toppinen, A. Riding a Trojan horse ? Future pathways of the fiber-based packaging industry in the bioeconomy. *For. Policy Econ.* **2020**, *110*, 101799. [CrossRef]
2. Boz, Z.; Korhonen, V.; Sand, C.K. Consumer Considerations for the Implementation of Sustainable Packaging: A Review. *Sustainability* **2020**, *12*, 2192. [CrossRef]
3. Hashmi, S.D.; Akram, S. Impact of green supply chain management on financial and environmental performance: Mediating role of operational performance and the moderating role of external pressures. *LogForum* **2021**, *17*, 359–371.
4. Modi, K.; Lowalekar, H.; Bhatta, N.M.K. Revolutionising supply chain management the theory of constraints way: A case study. *Int. J. Prod. Res.* **2019**, *57*, 3335–3361. [CrossRef]

5. Afum, E.; Sun, B.Z.; Kusi, C.L.Y. Reverse Logistics, Stakeholder Influence and Supply Chain Performance in Ghanaian Manufacturing Sector. *J. Supply Chain Manag. Syst.* **2019**, *8*, 13–24.
6. Vargas, M.; Alfaro, M.; Karstegl, N.; Fuertes, G.; Gracia, M.D.; Mar-Ortiz, J.; Sabattin, J.; Duran, C.; Leal, N. Reverse logistics for solid waste from the construction industry. *Adv. Civ. Eng.* **2021**, *2021*, 6654718. [CrossRef]
7. Sarkis, J.; Helms, M.M.; Hervani, A.A. Reverse logistics and social sustainability. *Corp. Soc. Responsib. Environ. Manag.* **2010**, *17*, 337–354. [CrossRef]
8. Sepúlveda, J.M.; Banguera, L.; Fuertes, G.; Carrasco, R.; Vargas, M. Reverse and inverse logistic models for solid waste management. *South Afr. J. Ind. Eng.* **2017**, *28*, 120–132.
9. Lambert, S.; Riopel, D.; Abdul-Kader, W. A reverse logistics decisions conceptual framework. *Comput. Ind. Eng.* **2011**, *61*, 561–581. [CrossRef]
10. Mahajan, J.; Vakharia, A.J. Waste Management: A Reverse Supply Chain Perspective. *Vikalpa* **2016**, *41*, 197–208. [CrossRef]
11. De Brito, M.P.; Dekker, R.; Flapper, S.D.P. Reverse logistics: A review of case studies. *Distrib. Logist.* **2005**, 243–281.
12. Turki, S.; Sauvey, C.; Rezg, N. Modelling and optimization of a manufacturing/remanufacturing system with storage facility under carbon cap and trade policy. *J. Clean. Prod.* **2018**, *193*, 441–458. [CrossRef]
13. Trochu, J.; Chaabane, A.; Ouhimmou, M. Reverse logistics network redesign under uncertainty for wood waste in the CRD industry. *Resour. Conserv. Recycl.* **2018**, *128*, 32–47. [CrossRef]
14. Mesjasz-Lech, A. Reverse logistics of municipal solid waste—Towards zero waste cities. *Transp. Res. Procedia.* **2019**, *39*, 320–332. [CrossRef]
15. Jalil, E.E.A.; Grant, D.B.; Nicholson, J.D.; Deutz, P. Reverse logistics in household recycling and waste systems: A symbiosis perspective. *Supply Chain Manag.* **2016**, *21*, 245–258. [CrossRef]
16. Alamerew, Y.A.; Brissaud, D. Modelling reverse supply chain through system dynamics for realising the transition towards the circular economy: A case study on electric vehicle batteries. *J. Clean. Prod.* **2020**, *254*, 120025. [CrossRef]
17. Valenzuela, J.; Alfaro, M.; Fuertes, G.; Vargas, M.; Sáez-Navarrete, C. Reverse logistics models for the collection of plastic waste: A literature review. *Waste Manag. Res.* **2021**, *39*, 1116–1134. [CrossRef]
18. Agrawal, S.; Singh, R.K.; Murtaza, Q. A literature review and perspectives in reverse logistics. *Resour. Conserv. Recycl.* **2015**, *97*, 76–92. [CrossRef]
19. Senthil, S.; Srirangacharyulu, B.; Ramesh, A. A robust hybrid multi-criteria decision making methodology for contractor evaluation and selection in third-party reverse logistics. *Expert Syst. Appl.* **2014**, *41*, 50–58. [CrossRef]
20. De Oliveira, U.R.; Espindola, L.S.; da Silva, I.R.; da Silva, I.N.; Rocha, H.M. A systematic literature review on green supply chain management: Research implications and future perspectives. *J. Clean. Prod.* **2018**, *187*, 537–561. [CrossRef]
21. Lai, K.H.; Wu, S.J.; Wong, C.W. Did reverse logistics practices hit the triple bottom line of Chinese manufacturers? *Int. J. Prod. Econ.* **2013**, *146*, 106–117. [CrossRef]
22. Çankaya, S.Y.; Sezen, B. Effects of green supply chain management practices on sustainability performance. *J. Manuf. Technol. Manag.* **2019**, *30*, 98–121. [CrossRef]
23. Bouzon, M.; Govindan, K.; Rodriguez, C.M.T. Evaluating barriers for reverse logistics implementation under a multiple stakeholders' perspective analysis using grey decision making approach. *Resour. Conserv. Recycl.* **2018**, *128*, 315–335. [CrossRef]
24. Govindan, K.; Bouzon, M. From a literature review to a multi-perspective framework for reverse logistics barriers and drivers. *J. Clean. Prod.* **2018**, *187*, 318–337. [CrossRef]
25. Sirisawat, P.; Kiatcharoenpol, T. Fuzzy AHP-TOPSIS approaches to prioritizing solutions for reverse logistics barriers. *Comput. Ind. Eng.* **2018**, *117*, 303–318. [CrossRef]
26. Amemba, C.S. Green supply chain best practices in hospitality industry in Kenya. *Glob. J. Commer. Manag. Perspect.* **2013**, *2*, 7–18.
27. Sharma, V.K.; Chandna, P.; Bhardwaj, A. Green supply chain management related performance indicators in agro industry: A review. *J. Clean. Prod.* **2017**, *141*, 1194–1208. [CrossRef]
28. Wijewickrama, M.K.C.S.; Chileshe, N.; Rameezdeen, R.; Ochoa, J.J. Information sharing in reverse logistics supply chain of demolition waste: A systematic literature review. *J. Clean. Prod.* **2021**, *280*, 124359. [CrossRef]
29. Lim, M.K.; Tseng, M.L.; Tan, K.H.; Bui, T.D. Knowledge management in sustainable supply chain management: Improving performance through an interpretive structural modelling approach. *J. Clean. Prod.* **2017**, *162*, 806–816. [CrossRef]
30. Lee, C.K.M.; Lam, J.S.L. Managing reverse logistics to enhance sustainability of industrial marketing. *Ind. Mark. Manag.* **2012**, *41*, 589–598. [CrossRef]
31. Schamne, A.N.; Nagalli, A. Reverse logistics in the construction sector: A literature review. *Electron. J. Geotech. Eng.* **2016**, *21*, 691–702.
32. Liao, T.Y. Reverse logistics network design for product recovery and remanufacturing. *Appl. Math. Model.* **2018**, *60*, 145–163. [CrossRef]
33. Hao, H.; Sun, Y.; Mei, X.; Zhou, Y. Reverse Logistics Network Design of Electric Vehicle Batteries considering Recall Risk. *Math. Probl. Eng.* **2021**, *2021*, 5518049. [CrossRef]
34. Huscroft, J.R.; Hazen, B.T.; Hall, D.J.; Skipper, J.B.; Hanna, J.B. Reverse logistics: Past research, current management issues, and future directions. *Int. J. Logist. Manag.* **2013**, *24*, 304–327. [CrossRef]
35. Beh, L.S.; Ghobadian, A.; He, Q.; Gallear, D.; O'Regan, N. Second-life retailing: A reverse supply chain perspective. *Supply Chain. Manag. Int. J.* **2016**, *21*, 259–272. [CrossRef]

36. Chen, L.; Duan, D.; Mishra, A.R.; Alrasheedi, M. Sustainable third-party reverse logistics provider selection to promote circular economy using new uncertain interval-valued intuitionistic fuzzy-projection model. *J. Enterp. Inf. Manag.* **2021**, 34–48. [CrossRef]
37. Magazzino, C.; Alola, A.A.; Schneider, N. The trilemma of innovation, logistics performance, and environmental quality in 25 topmost logistics countries: A quantile regression evidence. *J. Clean. Prod.* **2021**, *322*, 129050. [CrossRef]
38. Aguezzoul, A. Third-party logistics selection problem: A literature review on criteria and methods. *Omega* **2014**, *49*, 69–78. [CrossRef]
39. Li, Y.L.; Ying, C.S.; Chin, K.S.; Yang, H.T.; Xu, J. Third-party reverse logistics provider selection approach based on hybrid-information MCDM and cumulative prospect theory. *J. Clean. Prod.* **2018**, *195*, 573–584. [CrossRef]
40. Kurdve, M.; Shahbazi, S.; Wendin, M.; Bengtsson, C.; Wiktorsson, M. Waste flow mapping to improve sustainability of waste management: A case study approach. *J. Clean. Prod.* **2015**, *98*, 304–315. [CrossRef]
41. Da Silveira Guimarães, J.L.; Salomon, V.A.P. ANP applied to the evaluation of performance indicators of reverse logistics in footwear industry. *Procedia Comput. Sci.* **2015**, *55*, 139–148. [CrossRef]
42. Sangwan, K.S. Key activities, decision variables and performance indicators of reverse logistics. *Procedia Cirp.* **2017**, *61*, 257–262. [CrossRef]
43. Abdulrahman, M.D.; Gunasekaran, A.; Subramanian, N. Critical barriers in implementing reverse logistics in the Chinese manufacturing sectors. *Int. J. Prod. Econ.* **2014**, *147*, 460–471. [CrossRef]
44. Bai, C.; Sarkis, J. Flexibility in reverse logistics: A framework and evaluation approach. *J. Clean. Prod.* **2013**, *47*, 306–318. [CrossRef]
45. Hsu, C.C.; Tan, K.C.; Zailani, S.H.M. Strategic orientations, sustainable supply chain initiatives, and reverse logistics: Empirical evidence from an emerging market. *Int. J. Oper. Prod. Manag.* **2016**, *36*, 86–110. [CrossRef]
46. Dubey, R.; Gunasekaran, A.; Papadopoulos, T.; Childe, S.J.; Shibin, K.T.; Wamba, S.F. Sustainable supply chain management: Framework and further research directions. *J. Clean. Prod.* **2017**, *142*, 1119–1130. [CrossRef]
47. Căilean, D.; Teodosiu, C. An assessment of the Romanian solid waste management system based on sustainable development indicators. *Sustain. Prod. Consum.* **2016**, *8*, 45–56. [CrossRef]
48. Govindan, K.; Palaniappan, M.; Zhu, Q.; Kannan, D. Analysis of third party reverse logistics provider using interpretive structural modeling. *Int. J. Prod. Econ.* **2012**, *140*, 204–211. [CrossRef]
49. Bernon, M.; Rossi, S.; Cullen, J. Retail reverse logistics: A call and grounding framework for research. *Int. J. Phys. Distrib. Logist. Manag.* **2011**, *41*, 484–510. [CrossRef]
50. Govindan, K.; Soleimani, H.; Kannan, D. Reverse logistics and closed-loop supply chain: A comprehensive review to explore the future. *Eur. J. Oper. Res.* **2015**, *240*, 603–626. [CrossRef]
51. Sarkis, J.; Zhu, Q. Environmental sustainability and production: Taking the road less travelled. *Int. J. Prod. Res.* **2018**, *56*, 743–759. [CrossRef]
52. Vanalle, R.M.; Ganga, G.M.D.; Godinho Filho, M.; Lucato, W.C. Green supply chain management: An investigation of pressures, practices, and performance within the Brazilian automotive supply chain. *J. Clean. Prod.* **2017**, *151*, 250–259. [CrossRef]
53. Andronie, M.; Lăzăroiu, G.; Ștefănescu, R.; Uță, C.; Dijmărescu, I. Sustainable, Smart, and Sensing Technologies for Cyber-Physical Manufacturing Systems: A Systematic Literature Review. *Sustainability* **2021**, *13*, 5495. [CrossRef]
54. Jabbour, C.J.C.; de Sousa Jabbour, A.B.L. Green human resource management and green supply chain management: Linking two emerging agendas. *J. Clean. Prod.* **2016**, *112*, 1824–1833. [CrossRef]
55. Rebai, N.; Benabdelhafid, A.; Benaissa, M. A Decision Support System Proposal in a reverse logistic environment. In Proceedings of the 4th International Conference on Logistics, Hammamet, Tunisia, 31 May–3 June 2011; pp. 424–429.
56. Ogwueleka, T. Municipal solid waste characteristics and management in Nigeria. *J. Environ. Health Sci. Eng.* **2009**, *6*, 173–180.
57. Vergara, S.E.; Tchobanoglous, G. Municipal solid waste and the environment: A global perspective. *Annu. Rev. Environ. Resour.* **2012**, *37*, 277–309. [CrossRef]
58. Babatunde, B.B.; Vincent-Akpu, I.F.; Woke, G.N.; Atarhinyo, E.; Aharanwa, U.C.; Green, A.F.; Isaac-Joe, O. Comparative analysis of municipal solid waste (MSW) composition in three local government areas in Rivers State, Nigeria. *Afr. J. Environ. Sci. Technol.* **2013**, *7*, 874–881.
59. Dyczkowska, J.; Bulhakova, Y.; Łukaszczyk, Z.; Maryniak, A. Waste Management as an Element of the Creation of a Closed Loop of Supply Chains on the Example of Mining and Extractive Industry. *Manag. Syst. Prod. Eng.* **2020**, *28*, 60–69. [CrossRef]
60. Hennebert, P.; Van der Sloot, H.A.; Rebischung, F.; Weltens, R.; Geerts, L.; Hjelmar, O. Hazard property classification of waste according to the recent propositions of the EC using different methods. *Waste Manag.* **2014**, *34*, 1739–1751. [CrossRef]
61. Sztangret, I.B. The marketing value creation in the waste management sector—multi-conceptual business model. *SHS Web Conf.* **2020**, *73*, 01028. [CrossRef]
62. Beullens, P. Reverse logistics in effective recovery of products from waste materials. *Rev. Environ. Sci. Bio/Technol.* **2004**, *3*, 283–306. [CrossRef]
63. Matar, N.; Jaber, M.Y.; Searcy, C. A reverse logistics inventory model for plastic bottles. *Int. J. Logist. Manag.* **2014**, *25*, 315–333. [CrossRef]
64. Bazan, E.; Jaber, M.Y.; El Saadany, A.M. Carbon emissions and energy effects on manufacturing–remanufacturing inventory models. *Comput. Ind. Eng.* **2015**, *88*, 307–316. [CrossRef]
65. Ferri, G.L.; Chaves, G.D.L.D.; Ribeiro, G.M. Reverse logistics network for municipal solid waste management: The inclusion of waste pickers as a Brazilian legal requirement. *Waste Manag.* **2015**, *40*, 173–191. [CrossRef]

66. Choudhary, A.; Sarkar, S.; Settur, S.; Tiwari, M.K. A carbon market sensitive optimization model for integrated forward–reverse logistics. *Int. J. Prod. Econ.* **2015**, *164*, 433–444. [CrossRef]
67. Shi, L.H. A mixed integer linear programming for medical waste reverse logistics network design. In Proceedings of the International Conference on Management Science and Engineering, Moscow, Russia, 14–16 September 2009; pp. 1971–1975.
68. Gomes, M.I.; Barbosa-Povoa, A.P.; Novais, A.Q. Modelling a recovery network for WEEE: A case study in Portugal. *Waste Manag.* **2011**, *31*, 1645–1660. [CrossRef] [PubMed]
69. Siri, S.; Mendis, I.T.; Repetto, C. The facility location problem in a reverse logistic network: Weeenmodels project in the city of Genoa. In Proceedings of the 18th International Conference on Intelligent Transportation Systems, Gran Canaria, Spain, 15–18 September 2015; pp. 1581–1586.
70. Capraz, O.; Polat, O.; Gungor, A. Planning of waste electrical and electronic equipment (WEEE) recycling facilities: MILP modelling and case study investigation. *Flex. Serv. Manuf. J.* **2015**, *27*, 479–508. [CrossRef]
71. Kilic, H.S.; Cebeci, U.; Ayhan, M.B. Reverse logistics system design for the waste of electrical and electronic equipment (WEEE) in Turkey. *Resour. Conserv. Recycl.* **2015**, *95*, 120–132. [CrossRef]
72. Demirel, E.; Demirel, N.; Gökçen, H. A mixed integer linear programming model to optimize reverse logistics activities of end-of-life vehicles in Turkey. *J. Clean. Prod.* **2016**, *112*, 2101–2113. [CrossRef]
73. Tang, J.; Liu, Y.; Fung, R.Y.; Luo, X. Industrial waste recycling strategies optimization problem: Mixed integer programming model and heuristics. *Eng. Optim.* **2008**, *40*, 1085–1100. [CrossRef]
74. Wang, I.L.; Yang, W.C. Fast heuristics for designing integrated e-waste reverse logistics networks. *IEEE Trans. Electron. Packag. Manuf.* **2007**, *30*, 147–154. [CrossRef]
75. Anne, M.; Nicholas, L.; Ithinji, G.K.; Bula, H.O. Reverse logistics practices and their effect on competitiveness of food manufacturing firms in Kenya. 2016. Available online: http://41.89.227.156:8080/xmlui/handle/123456789/524 (accessed on 10 December 2021).
76. Ho, G.T.S.; Choy, K.L.; Lam, C.H.Y.; Wong, D.W. Factors influencing implementation of reverse logistics: A survey among Hong Kong businesses. *Meas. Bus. Excell.* **2012**, *16*, 29–46. [CrossRef]
77. Khor, K.S.; Udin, Z.M.; Ramayah, T.; Hazen, B.T. Reverse logistics in Malaysia: The contingent role of institutional pressure. *Int. J. Prod. Econ.* **2016**, *175*, 96–108. [CrossRef]
78. Subulan, K.; Taşan, A.S.; Baykasoğlu, A. A fuzzy goal programming model to strategic planning problem of a lead/acid battery closed-loop supply chain. *J. Manuf. Syst.* **2015**, *37*, 243–264. [CrossRef]
79. Li, C.; Liu, F.; Cao, H.; Wang, Q. A stochastic dynamic programming based model for uncertain production planning of re-manufacturing system. *Int. J. Prod. Res.* **2009**, *47*, 3657–3668. [CrossRef]
80. Ahluwalia, P.K.; Nema, A.K. Multi-objective reverse logistics model for integrated computer waste management. *Waste Manag. Res.* **2006**, *24*, 514–527. [CrossRef] [PubMed]
81. Braga Marsola, K.; Leda Ramos de Oliveira, A.; Filassi, M.; Elias, A.A.; Andrade Rodrigues, F. Reverse logistics of empty pesticide containers: Solution or a problem? *Int. J. Sustain. Eng.* **2021**, *14*, 1451–1462. [CrossRef]
82. Bonev, M. *Managing Reverse Logistics Using System Dynamics: A generic End-to-End Approach*; Diplomica Verlag: Berlin, Germany, 2012.
83. Wang, L.; Goh, M.; Ding, R.; Mishra, V.K. Improved Simulated Annealing Based Network Model for E-Recycling Reverse Logistics Decisions under Uncertainty. *Math. Probl. Eng.* **2018**, *2018*, 4390480. [CrossRef]
84. Horvath, P.A.; Autry, C.W.; Wilcox, W.E. Liquidity implications of reverse logistics for retailers: A Markov chain approach. *J. Retail.* **2005**, *81*, 191–203. [CrossRef]
85. Hanafi, J.; Kara, S.; Kaebernick, H. Reverse logistics strategies for end-of-life products. *Int. J. Logist. Manag.* **2008**, *19*, 367–388. [CrossRef]
86. Mimouni, F.; Abouabdellah, A. Proposition of a modeling and an analysis methodology of integrated reverse logistics chain in the direct chain. *J. Ind. Eng. Manag.* **2016**, *9*, 359–373. [CrossRef]
87. Olugu, E.U.; Wong, K.Y. Fuzzy logic evaluation of reverse logistics performance in the automotive industry. *Sci. Res. Essays* **2011**, *6*, 1639–1649.
88. Huang, C.C.; Liang, W.Y.; Tseng, T.L.; Chen, P.H. The rough set based approach to generic routing problems: Case of reverse logistics supplier selection. *J. Intell. Manuf.* **2016**, *27*, 781–795. [CrossRef]
89. Bai, C.; Sarkis, J. Integrating and extending data and decision tools for sustainable third-party reverse logistics provider selection. *Comput. Oper. Res.* **2019**, *110*, 188–207. [CrossRef]
90. Hejrani, S.; Ko, H.S. A reverse logistics model for medical waste management. In *IIE Annual Conference Proceedings*; Institute of Industrial and Systems Engineers (IISE): Peachtree Corners, GA, USA, 2013; pp. 97–105.
91. Jin-feng, Z. The Application of the Multi-objective Evolutionary Algorithm in the Collection and Transportation System of Solid Waste. *J. Guangdong Univ. Technol.* **2011**, *2*, 51–64.
92. Cao, S.; Zhang, K. Optimization of the flow distribution of e-waste reverse logistics network based on NSGA II and TOPSIS. In Proceedings of the International Conference on E-Business and E-Government—(ICEE), Shanghai, China, 6–8 May 2011; pp. 1–5.
93. Lo, C.C.; Chen, H.M.; Huang, H.L. A multi-objective reverse logistics network for product returns. In Proceedings of the IEEE International Conference on Industrial Engineering and Engineering Management, Singapore, 8–11 December 2008; pp. 1634–1638.

94. Lu, Y.; Li, X.; Liang, L. Multi-objective optimization of reverse logistics network based on improved particle swarm optimization. In Proceedings of the 7th World Congress on Intelligent Control and Automation, Chongqing, China, 25–27 June 2008; pp. 7476–7480.
95. Li, D.; Liu, C.; & Li, K. A remanufacturing logistics network model based on improved multi-objective ant colony optimization. *J. Eur. Des Systèmes Autom.* **2019**, *52*, 391–395. [CrossRef]
96. Hsueh, J.T.; Lin, C.Y. Integrating the AHP and TOPSIS decision processes for evaluating the optimal collection strategy in reverse logistic for the TPI. *Int. J. Green Energy* **2017**, *14*, 1209–1220. [CrossRef]
97. Güzel, B.; Taş, A. Decision analysis on criteria that affect design of reverse logistic systems in textile sector. *Tekstil ve Mühendis* **2018**, *25*, 154–168. [CrossRef]
98. Zhou, Y.; Xu, L.; Muhammad Shaikh, G. Evaluating and prioritizing the green supply chain management practices in Pakistan: Based on delphi and fuzzy AHP approach. *Symmetry* **2019**, *11*, 1346. [CrossRef]
99. Hsueh, J.T.; Lin, C.Y. Constructing a network model to rank the optimal strategy for implementing the sorting process in reverse logistics: Case study of photovoltaic industry. *Clean Technol. Environ. Policy* **2015**, *17*, 155–174. [CrossRef]
100. Kumar, S.; Kumar, R. Forecasting of municipal solid waste generation using non-linear autoregressive (NAR) neural models. *Waste Manag.* **2021**, *121*, 206–214.
101. Gorji, M.A.; Jamali, M.B.; Iranpoor, M. A game-theoretic approach for decision analysis in end-of-life vehicle reverse supply chain regarding government subsidy. *Waste Manag.* **2021**, *120*, 734–747. [CrossRef] [PubMed]
102. Da Silva, T.R.; de Azevedo AR, G.; Cecchin, D.; Marvila, M.T.; Amran, M.; Fediuk, R.; Vatin, N.; Karelina, M.; Klyuev, S.; Szelag, M. Application of plastic wastes in construction materials: A review using the concept of life-cycle assessment in the context of recent research for future perspectives. *Materials* **2021**, *14*, 3549. [CrossRef] [PubMed]
103. Abdel-Shafy, H.I.; Mansour, M.S.M. Solid waste issue: Sources, composition, disposal, recycling, and valorisation. *Egypt. J. Pet.* **2018**, *27*, 1275–1290. [CrossRef]
104. Shi, K.; Zhou, Y.; Zhang, Z. Mapping the research trends of household waste recycling: A bibliometric analysis. *Sustainability* **2021**, *13*, 6029. [CrossRef]
105. Luo, Y.; Wu, L.; Huang, D.; Zhu, J. Household food waste in rural China: A noteworthy reality and a systematic analysis. *Waste Manag. Res.* **2021**, *39*, 1389–1395. [CrossRef]
106. Adelakun, G.I.; Adigun, A.I.; Niyi, A.S.B.; Tayo, O.J.O. An Assessment of the Effectiveness of the Waste Bins Collection and Disposal in Sango-Ota, Ogun. *World Sci. News.* **2019**, *119*, 27–40.
107. Pires, A.; Martinho, G. Waste hierarchy index for circular economy in waste management. *Waste Manag.* **2019**, *95*, 298–305. [CrossRef]
108. Teigiserova, D.A.; Hamelin, L.; Thomsen, M. Towards transparent valorisation of food surplus, waste and loss: Clarifying definitions, food waste hierarchy, and role in the circular economy. *Sci. Total Environ.* **2020**, *706*, 136033. [CrossRef]
109. Alarcón, F.; Cortés-Pellicer, P.; Pérez-Perales, D.; Mengual-Recuerda, A. A reference model of reverse logistics process for improving sustainability in the supply chain. *Sustainability* **2021**, *13*, 10383. [CrossRef]
110. De Oliveira, U.R.; Aparecida Neto, L.; Abreu, P.A.F.; Fernandes, V.A. Risk management applied to the reverse logistics of solid waste. *J. Clean. Prod.* **2021**, *296*, 126517. [CrossRef]
111. Degenstein, L.M.; Mcqueen, R.H.; Krogman, N.T. What goes where ? Characterising Edmonton's municipal clothing waste stream and consumer clothing disposal. *J. Clean. Prod.* **2021**, *296*, 126516. [CrossRef]
112. Rehman, A.; Ali, K.; Hamid, T.; Nasir, H.; Ahmad, I. Effective Utilization of Municipal Solid Waste as Substitute for Natural Resources in Cement Industry. *Civ. Eng. J.* **2020**, *6*, 238–257. [CrossRef]
113. Tsai, W.T. Analysis of plastic waste reduction and recycling in Taiwan. *Waste Manag. Res.* **2021**, *39*, 713–719. [CrossRef]
114. Ghinea, C.; Drăgoi, E.N.; Comăniţă, E.D.; Gavrilescu, M.; Câmpean, T.; Curteanu, S.I.L.V.I.A.; Gavrilescu, M. Forecasting municipal solid waste generation using prognostic tools and regression analysis. *J. Environ. Manag.* **2016**, *182*, 80–93. [CrossRef] [PubMed]
115. Sofi, M.; Sabri, Y.; Zhou, Z.; Mendis, P. Transforming municipal solid waste into construction materials. *Sustainability* **2019**, *11*, 2661. [CrossRef]
116. Eurostat. 2021. Available online: https://ec.europa.eu/eurostat/statistics-explained/index.php?title=Municipal_waste_statistics#Municipal_waste_treatment (accessed on 8 August 2021).
117. Rimaitytė, I.; Ruzgas, T.; Denafas, G.; Račys, V.; Martuzevicius, D. Application and evaluation of forecasting methods for municipal solid waste generation in an eastern-European city. *Waste Manag. Res.* **2012**, *30*, 89–98. [CrossRef]
118. Purwani, A.; Hisjam, M.; Sutopo, W. Municipal solid waste logistics management: A study on reverse logistics. *AIP Conf. Proc.* **2020**, *2217*, 030181.
119. Chung, S.S. Projecting municipal solid waste: The case of Hong Kong SAR. *Resour. Conserv. Recycl.* **2010**, *54*, 759–768. [CrossRef]
120. Smol, M.; Duda, J.; Czaplicka-Kotas, A.; Szołdrowska, D. Transformation towards circular economy (CE) in municipal waste management system: Model solutions for Poland. *Sustainability* **2020**, *12*, 4561. [CrossRef]
121. Vrabie, C. Converting municipal waste to energy through the biomass chain, a key technology for environmental issues in (Smart) cities. *Sustainability* **2021**, *13*, 4633. [CrossRef]
122. Vaida, D.; Lelea, D. Municipal Solid Waste Incineration: Recovery or Disposal. Case Study of City Timisoara, Romania. *Procedia Eng.* **2017**, *181*, 378–384. [CrossRef]

123. Quina, M.J.; Bontempi, E.; Bogush, A.; Schlumberger, S.; Weibel, G.; Braga, R.; Funari, V.; Hyks, J.; Rasmussen, E.; Lederer, J. Technologies for the management of MSW incineration ashes from gas cleaning: New perspectives on recovery of secondary raw materials and circular economy. *Sci. Total Environ.* **2018**, *635*, 526–542. [CrossRef]
124. Rogers, D.S.; Tibben-lembke, R. An examination of reverse logistics practices. *J. Bus. Logist.* **2001**, *22*, 129–148. [CrossRef]
125. Stucki, M.; Jattke, M.; Berr, M.; Desing, H.; Green, A.; Hellweg, S.; Laurenti, R.; Meglin, R.; Muir, K.; Pedolin, D.; et al. How life cycle–based science and practice support the transition towards a sustainable economy. *Int. J. Life Cycle Assess.* **2021**, *26*, 1062–1069. [CrossRef]
126. Yu, H.; Dai, H.; Tian, G.; Wu, B.; Xie, Y.; Zhu, Y.; Zhang, T.; Fathollahi-Fard, A.M.; Tang, H. Key technology and application analysis of quick coding for recovery of retired energy vehicle battery. *Renew. Sustain. Energy Rev.* **2021**, *135*, 110129. [CrossRef]
127. Bernon, M.; Cullen, J. An integrated approach to managing reverse logistics. *Int. J. Logist. Res. Appl.* **2007**, *10*, 41–56. [CrossRef]
128. Visser, H.M.; Van Goor, A.R. Recycling and reverse logistics. *Logist. Princ. Pract.* **2020**, *12*, 371–390.
129. Kalpana, D.; Cho, S.H.; Lee, S.B.; Lee, Y.S.; Misra, R.; Renganathan, N.G. Recycled waste paper-A new source of raw material for electric double-layer capacitors. *J. Power Sources.* **2009**, *190*, 587–591. [CrossRef]
130. Erdenebold, U.; Choi, D.H.; Ho, K.S.; Cheol, K.G.; Wang, J.P. A study on reduction of copper smelting slag by carbon for recycling into metal values and cement raw material. *Sustainability* **2020**, *12*, 1421.
131. Guo, S.; Shen, B.; Choi, T.M.; Jung, S. A review on supply chain contracts in reverse logistics: Supply chain structures and channel leaderships. *J. Clean. Prod.* **2017**, *144*, 387–402. [CrossRef]
132. Walker, T.R.; McGuinty, E.; Charlebois, S.; Music, J. Single-use plastic packaging in the Canadian food industry: Consumer behavior and perceptions. *Humanit. Soc. Sci. Commun.* **2021**, *8*, 80. [CrossRef]
133. Reverberi, M. The new packaged food products containing insects as an ingredient. *J. Insects as Food Feed.* **2021**, *7*, 901–908. [CrossRef]
134. Wang, C.; Hoang, Q.; Nguyen, T. Integrating the EBM Model and LTS(A,A,A) Model to Evaluate the Efficiency in the Supply Chain of Packaging Industry in Vietnam. *Axioms* **2021**, *10*, 33. [CrossRef]
135. Pareek, V.; Khunteta, A. Pharmaceutical packaging: Current trends and future. *Int. J. Pharm. Pharm. Sci.* **2014**, *6*, 480–485.
136. Sundqvist-Andberg, H.; Åkerman, M. Sustainability governance and contested plastic food packaging—An integrative review. *J. Clean. Prod.* **2021**, *306*, 127111. [CrossRef]
137. Marangoni Júnior, L.; Cristianini, M.; Padula, M.; Anjos, C.A.R. Effect of high-pressure processing on characteristics of flexible packaging for foods and beverages. *Food Res. Int.* **2019**, *119*, 920–930. [CrossRef]
138. Dixon-Hardy, D.W.; Curran, B.A. Types of packaging waste from secondary sources (supermarkets)—The situation in the UK. *Waste Manag.* **2009**, *29*, 1198–1207. [CrossRef] [PubMed]
139. Gopinathar, P.; Prabha, G.; Ravichandran, K. The Role of Packaging in Manufacturing-A Brief Understanding. *IOSR J. Bus. Manag.* **2016**, *18*, 1–7.
140. Albaar, N.; Budiastra, I.W.; Hariyadi, Y. Influence of Secondary Packaging on Quality of Carrots During Transportation. *Agric. Agric. Sci. Procedia.* **2016**, *9*, 348–352. [CrossRef]
141. Mahmoudi, M.; Parviziomran, I. Reusable packaging in supply chains: A review of environmental and economic impacts, logistics system designs, and operations management. *Int. J. Prod. Econ.* **2020**, *228*, 107730. [CrossRef]
142. Rana, S.; Siddiqui, S.; Goyal, A. Extension of the shelf life of guava by individual packaging with cling and shrink films. *J. Food Sci. Technol.* **2015**, *52*, 8148–8155. [CrossRef]
143. Van Esch, P.; Heller, J.; Northey, G. The effects of inner packaging color on the desirability of food. *J. Retail. Consum. Serv.* **2019**, *50*, 94–102. [CrossRef]
144. Sutrisna, A.; Vossenaar, M.; Poonawala, A.; Mallipu, A.; Izwardy, D.; Menon, R.; Tumilowicz, A. Improved information and educational messages on outer packaging of micronutrient powders distributed in Indonesia increase caregiver knowledge and adherence to recommended use. *Nutrients* **2018**, *10*, 747. [CrossRef] [PubMed]
145. Zhang, R.; Ma, X.; Shen, X.; Zhai, Y.; Zhang, T.; Ji, C.; Hong, J. PET bottles recycling in China: An LCA coupled with LCC case study of blanket production made of waste PET bottles. *J. Environ. Manage.* **2020**, *260*, 110062. [CrossRef] [PubMed]
146. Han, J.; Gu, L.Y.; Chen, D.R. Application of Innovative Design Thinking in Product Design∗ Intelligent Waste Paper Recycling Machine Design Case. *E3S Web Conf.* **2021**, *236*, 04062. [CrossRef]
147. Greenwood, S.C.; Walker, S.; Baird, H.M.; Parsons, R.; Mehl, S.; Webb, T.L.; Slark, A.T.; Ryan, A.J.; Rothman, R.H. Many Happy Returns: Combining insights from the environmental and behavioural sciences to understand what is required to make reusable packaging mainstream. *Sustain. Prod. Consum.* **2021**, *27*, 1688–1702. [CrossRef]
148. Joseph, B.; James, J.; Kalarikkal, N.; Thomas, S. Recycling of medical plastics. *Adv. Ind. Eng. Polym. Res.* **2021**, *4*, 199–208. [CrossRef]
149. Sasaki, Y.; Orikasa, T.; Nakamura, N.; Hayashi, K.; Yasaka, Y.; Makino, N.; Shobatake, K.; Koide, S.; Shiina, T. Optimal packaging for strawberry transportation: Evaluation and modeling of the relationship between food loss reduction and environmental impact. *J. Food Eng.* **2021**, *314*, 110767. [CrossRef]
150. Petris, G.; Petrone, S.; Campagnoli, P. Dynamic linear models. In *Dynamic Linear Models with R*; Springer: New York, NY, USA, 2009; pp. 31–84.

sustainability

Article

Assessment of Technological Developments in Data Analytics for Sensor-Based and Robot Sorting Plants Based on Maturity Levels to Improve Austrian Waste Sorting Plants

Karl Friedrich *, Theresa Fritz, Gerald Koinig, Roland Pomberger and Daniel Vollprecht

Chair of Waste Processing Technology and Waste Management, Department of Environmental and Energy Process Engineering, Montanuniversitaet Leoben, 8700 Leoben, Austria; theresa.fritz@unileoben.ac.at (T.F.); gerald.koinig@unileoben.ac.at (G.K.); roland.pomberger@unileoben.ac.at (R.P.); daniel.vollprecht@unileoben.ac.at (D.V.)
* Correspondence: karl.friedrich@unileoben.ac.at; Tel.: +43-3842-402-5139

Abstract: Sensor-based and robot sorting are key technologies in the extended value chain of many products such as packaging waste (glass, plastics) or building materials since these processes are significant contributors in reaching the EU recycling goals. Hence, technological developments and possibilities to improve these processes concerning data analytics are evaluated with an interview-based survey. The requirements to apply data analytics in sensor-based sorting are separated into different sections, i.e., data scope or consistency. The interviewed companies are divided into four categories: sorting machine manufacturers, sorting robot manufacturers, recycling plant operators, and sensor technology companies. This paper aims to give novel insights into the degree of implementation of data analytics in the Austrian waste management sector. As a result, maturity models are set up for these sections and evaluated for each of the interview partner categories. Interviewees expressed concerns regarding the implementation such as a perceived loss of control and, subsequently, a supposed inability to intervene. Nevertheless, further comments by the interviewees on the state of the waste management sector conveyed that data analytics in their processes would also be a significant step forward to achieve the European recycling goals.

Keywords: sensor-based sorting; robot sorting; data analytics; maturity model; recycling; waste treatment; waste management

Citation: Friedrich, K.; Fritz, T.; Koinig, G.; Pomberger, R.; Vollprecht, D. Assessment of Technological Developments in Data Analytics for Sensor-Based and Robot Sorting Plants Based on Maturity Levels to Improve Austrian Waste Sorting Plants. *Sustainability* **2021**, *13*, 9472. https://doi.org/10.3390/su13169472

Academic Editors: Sunil Kumar, Pooja Sharma and Deblina Dutta

Received: 17 June 2021
Accepted: 17 August 2021
Published: 23 August 2021

1. Introduction

Sensor-based sorting, being one of the newest technologies for the recycling industry, is hoped to improve waste sorting enough to lead the way into a digitalized future and subsequently help meet the goals presented by the EU in 2018. These are 70% for packaging by 2030 and municipal waste in 5-year steps to a minimum of 65% by 2035. In addition, landfilling of municipal waste must be ensured to decrease to a maximum of 10% [1].

In recent years sensor-based sorting became increasingly popular due to the numerous possible applications and advancements in sensor technology as micro technologies enabled mass production of low cost and high-reliability sensors [2,3]. Sensor-based sorting is a contactless automated separation of particles based on specific features. Applications of this method can vary depending on the complexity of the technical design and the number of sensors. The detected features include color, composition, density, and conductivity, and the detection units, while following similar principles, vary in their construction. Comparisons with the still widely applied manual sorting show that sensor-based systems can identify characteristics of waste components more accurately [4–6]. An additional advantage related to mechanical waste sorting is the reduced health risk for workers [7]. The option to combine sensors with different characterization principles [8] especially results in a better quality of the final product, higher product yield and improved valuable recovery [9], which are aspects that can correlate directly with a recycling plant's revenue [10,11]. Due to the dynamic development of sensor-based sorting in recent years, new areas for use were and still are found [12–14]. Some sensors detect the superficial properties of the material, others give information about internal characteristics [15]. Visible spectroscopy (VIS) and near-infrared-sensors (NIR) belong to the first group and the other group comprises X-ray-transmission, X-ray-fluorescence, and inductive sensors [16].

Industry 4.0, the IoT, and rapidly increasing digitalization will enable the individual stakeholders (companies, customers, products, among others) to share valuable information amongst themselves in real time [17]. At the same time, the use of IT and automation will ensure the processing, analysis, and collection of vast amounts of information [18].

Implementing Industry 4.0 into the existing value chain of producers and stakeholders, though necessary for remaining responsive and adaptive to increasingly dynamic markets [19], comes with its own set of challenges and demands. When implementing Industry 4.0 technologies, challenges such as implantability, embedment, flexibility, and, especially in the field of waste management, robustness need to be considered [20].

Among the four dimensions of Big Data, namely, the variety of data, the velocity of generation and analysis of new data, the value of data, and the volume of data [21], one of the most pressing issues when adapting an existing plant to an Industry 4.0 approach is the emergence of vast amounts of data that must be processed and transmitted.

Transmission under the current industrial wireless network protocol is infeasible due to the limited bandwidth, which is unsuited to the necessary transmission rates for large scale Industry 4.0 applications [19]. Industry 4.0 needs transmission protocols able to handle the expected increase in data transmission volume.

In addition to the transmission issue, data processing needs to be implemented in a manufacturing-specific manner to ensure high quality and fidelity in the processed data and cohesion among different data acquisition models has to be ensured to allow for big data analytics [22].

Lastly, Industry 4.0 calls for standardized communication protocols and interconnectivity. This increase in connectivity and ease of access through standardized connections leads to issues concerning cybersecurity. The need to protect critical infrastructure, sensitive manufacturing data, and classified information stored in local servers or cloud-based IT platforms [23] increase dramatically with the use of Industry 4.0 settings [24].

The central aspect of Industry 4.0, apart from gaining insight into current industry procedures, is determining the differences in handling data [25] along the sensor-based sorting value chain in waste management. This value chain starts with the sensor manufacturer, which produces the sensor. Next comes the sorting machines manufacturer or the sorting robot manufacturer who installs the sensor in his equipment. At the end of this value chain, the sorting plant operator shows up and installs the sorting machine or sorting robot in his plant.

The definition of data as well as the perceived important aspects and usages may vary between individual stakeholders, thus resulting in unrealized potential concerning the possibilities and advantages of a sound data analytics strategy. Therefore, this study aims to explore the different approaches to and goals of the data handling part of digitalization in each of the four stakeholder categories.

The scientific research questions that are answered in this paper are:

1. How mature is the sensor-based and robot sorting area in Austrian waste management in the use of data analytics?
2. Where are the current limitations in technologies or in the willingness to be able to use data analytics in sensor-based or robot sorting in Austria?
3. What are the risks and chances in the specific area of sensor-based and robot sorting in the Austrian waste management sector?

Scientific literature reviews were performed to find a suitable evaluation method, searching for approaches to similar overarching questions.

Graninger executed an interview-based survey in his master's thesis to monitor the current status of the interpretation, implementation and obstacles for Industry 4.0 in Austria's industrial sector. In Graninger´s study, 34 companies out of over 300 participated in an email-based survey, so the return rate was approximately 10%. Furthermore, only 26 of them filled out this survey completely. The expert interviews were evaluated with bar charts and key figures such as a score factor or the weighted average [26].

Another analogical study was brought up by the German federal ministry of economics and technology in 2013. An online survey to evaluate the innovation potentials of big data was created and sent out to companies over decision makers, providers, users and scientists. It is not stated how many surveys were sent or where the contacted companies are located, but it is mentioned that 185 assessments were returned. The evaluation was done with percentages in bar charts [27].

Schuhmacher et al. created a study for an Industry 4.0 maturity model with expert interviews, practitioner workshops, and literature research. It was evaluated with spider diagrams and weighing of influence factors [28]. A maturity level is a step with predefined characteristics, with each level having more advanced characteristics on the way toward a mature process. In the case of this study, a data analytics strategy embracing all later specified aspects was used.

The last reference study was published by Gonçalves et al. and evaluates the readiness for Industry 4.0 of manufacturing companies. An online self-check tool was created and sent to an unknown number of companies, of which a total of 602 companies responded [29].

All these studies only consider Industry 4.0 in general but do not consider sensor-based sorting as a special technology within Industry 4.0. Therefore, in this study, for the first time ever, a maturity level assessment for sensor-based sorting in waste treatment is carried out with a focus on the Austrian waste sorting sector.

2. Materials and Methods

The state-of-the-art in waste sorting plants compared with a literature review revealed that a lot of information on sensor-based sorting in waste treatment is not accessible in the literature and is only known and traded by industrial experts in this field. For this reason, instead of a literature review, expert interviews were selected as an appropriate methodology.

After analyzing the previous stated four studies [26–29], it was decided that an interview-based survey would fit best since more information may be gathered in a personal conversation than from evaluating answers to predefined survey questions alone.

The interview-based survey consisted of questions regarding data analytics in general and in sensor-based/robot sorting. Due to COVID-19, all the interviews were conducted via video calls from March 2020 until May 2020.

The interviewed stakeholder experts were separated into four categories along the sensor sorting value chain: sensor manufacturers, sorting machine manufacturers, sorting robot manufacturers, and sorting plant operators. These categories were selected because only they can provide original data, whereas other stakeholders such as public authorities or research institutions could only provide secondary data obtained from the same group of experts. According to the working hypothesis, the highest maturity level should occur at the sensor technology sector, and at every step of utilization it will decrease, i.e., the sorting robot manufacturer is technologically behind the sensor producer, and so on.

Twenty-eight stakeholder experts were contacted, but due to reduced working hours in many companies, 12 interviews were held. The interview length varied from 45 min to 2.5 h. These 12 interviewed stakeholder experts cover mainly the whole Austrian waste sorting sector, although the companies are located all in Europe, because their equipment is the most commonly installed in Austrian waste sorting plants. The interviewed stakeholder experts were two sensor manufacturers located in Europe, four sorting machine manufacturers located in Europe, two sorting robot manufacturers located in Europe, and four sorting plant operators located in Austria.

In this section, it has to be stated that two sensor manufacturers cover the Austrian waste sorting sector because some sorting machine manufacturers produce their own sensors for their sorting machines. The two sorting robot manufacturers also cover the Austrian waste sorting sector since there are only a few sorting robots installed currently.

At the beginning of the interviews, the interviewers introduced themselves (the Chair of Waste Processing Technology and Waste Management), the research area of sensor-based sorting in the industry, as well as the aim and the focus of the survey. Next, the interviewee introduced himself, described his job and responsibilities in his company, and had the opportunity to bring in some questions of interest for the study. An example for such a question would be how the acquired data in the assessment is processed, which was in most of the interviews as the first open question. After it was agreed that the acquired data is only allowed to be published in an anonymous way—which was the precondition for each of the companies to participate—the survey questioning itself started. In some cases, one answer flipped to another question, but it was decided to follow the survey strictly and discuss topics twice instead of assuming the risk of missing any information. Nevertheless, when additional questions came up for some answer, they were discussed and appended to the study's results.

The evaluation of the data acquired in the study is done with individual critical analysis for each of the expert interviews and graphically visualized with bar charts since the number of participants is straightforward and enables going into details with each of the interviewees.

The data analytics survey was primarily based on the doctoral thesis of Bernerstätter [30]. It consisted of general questions, a self-evaluation, and detailed questions, i.e., concerning the consistency and amount of data needed to calculate the degree of data analytics maturity [30]. Bernerstätter stated that a maturity for the use of data analytics cannot be determined with one overall maturity level that is detailed enough because the maturity for data analytics consists of many sectors which need to be determined individually to calculate an overall maturity level. These sectors are data collection, data provision and transfer, data formats, data encoding and presentation, data scope, data consistency, and data usage [30]. In his models, the maturity level 1 is the lowest level and the maturity level 4 the highest, which is also the basis for this study [30]. For this study there was also a new sector considered, which is the commitment to change, to bring in a perspective on whether applying data analytics is not a technical problem but a mental one when employees fear losing their jobs with increasing digitalization.

The first set of introductory questions covers data of sensor-based sorting systems, namely, which data are collected and where they are stored, and aims to determine a degree of occupation with the topic of the data in general Table 1.

The averaged data analytics maturity level is calculated via the summation of the answers divided by the number of questions, with a possible 0.5 gradation if the participants felt that the company was on the way to a higher level but not quite there yet. The self-assessment, which is an estimation of the overall maturity level based on the four possibilities given (Table 2) was done prior to the detailed questions which were used to calculate an average data analytics maturity level with all of the data analytic sectors (Table 3) to compare. Lastly, it was inquired if the industry experts trusted their recorded data.

Table 1. Questions concerning data collection by the sensor-based sorting system and the general approach to data.

What Data is Collected by the Sensor-Based Sorting System?	
Production data	83% (10/12)
Maintenance data	75% (9/12)
Quality data	58% (7/12)
Machine data	83% (10/12)
Other data	8% (1/12)
Where is the data recorded?	
Right at the plant	92% (11/12)
Measuring room	50% (6/12)
Not on site	8% (1/12)
Others	0% (0/12)
Has the company implemented a strategy for managing data?	83% yes (10/12)
Are data owners assigned for data governance?	50% yes (6/12)
Are efforts made as well to ensure high quality of transaction data?	75% yes (9/12)

Table 2. Self-assessed data analytics maturity level.

How Would You Assess the Degree of Maturity of Data Analysis for Sensor-Based Sorting or Robot Sorting in Your Company Using the Following Scale?	
Hardly any digitization in data analysis has been implemented. There is no actual concern about the subject.	1
An analysis of interrelationships has been implemented showing the reasons for an incident.	2
Partially automated recording and specific formatting standards have been implemented. However, there is no consistency across data sources.	3
Continuous data and information management have been implemented based on established standards. In addition, prescriptive analysis helps the system act autonomously and appropriately.	4

Table 3. Detailed data analytics maturity level to portray a more accurate state of the art.

Data Collection	
Data collection does not adhere to any standards and objectives and, in addition, is incomplete. Paper-recording predominates, the amount of data collected is generally relatively small.	1
Digital data collection is triggered manually or irregularly. Fault remedy measures and logic connecting the process generated and collected data are available.	2
Irregular predefined triggers constitute automated data collection. Manual records are regularly digitized.	3
No more manual data input, only confirmation of values is required. Automated data acquisition is made in regular intervals.	4
Data provision and transfer	
Data is not available in any format utilizable by analysis tools, so substantial data aggregation is not ensured.	1
Local server systems cause interface and compatibility problems. Manual transmission is sparse due to high effort and not in real-time.	2

Table 3. *Cont.*

Data provision and transfer	
A centralized database system prevents interface problems and enables real-time analysis. Unstructured data from measurement processes are immediately reduced to relevant characteristics.	3
Pre-processing steps are provided to immediately present data in a structured manner ready for analysis. Data is stored in a Data Warehouse.	4
Data formats	
It takes high effort to convert the data into a standard format.	1
Standard data formats are used (xls, PDF, …) but not consistently, so that compilation takes a lot of time and effort.	2
Data formats do not limit the common data stock. Large amounts of data can be stored.	3
Data formats are irrelevant because the file transfer passes through an interface straight to an analysis tool. Alternatively, file formats suitable for Big Data are available.	4
Data encoding and presentation	
Text-only or incomprehensible codes characterize this unstructured form of data collection.	1
Codes can be interpreted clearly and entries are comparable.	2
Unambiguous interpretability is standard; essential attributes are scaled metrically, enabling transformation into nominally scaled values.	3
Metadata facilitates the automatic interpretation of the standardized codes from all data sources.	4
Data scope	
Data collected is unstructured, partly irrelevant, and too little in number. Spreadsheet software is sufficient.	1
The amount of data collected is too large to be interpreted by staff. The recording period is at least nine months.	2
The recording period is at least one year.	3
For at least 1.5 years, data has been entirely recorded and its relevance checked by precise allocation to the relevant observation units.	4
Data consistency	
Manual recordings provide inadequate or no consistent time reference.	1
Consistent time reference cannot be ensured across data sources but can be achieved using time stamps.	2
Diagnostic purposes can be satisfied by a defined reliable interval between surveys (to provide forecasts). Consistent time reference is ensured even across data sources.	3
A consistent system ensures time stamp integrity and traceable quality by association with ID data (e.g., order numbers).	4
Data usage	
Data is not used, i.e., records are kept without interpretation, or no adjustments are performed after interpretation.	1
Individual records are converted into a format ready for interpretation. Problems with data quality/consistency are known but not remedied in a standardized way. The IT department is solely responsible.	2

Table 3. *Cont.*

Data usage	
Data is interpreted to remedy faults and to make decisions on a regular basis. Data management processes are documented and discussed with data protection and security. Data is considered a resource.	3
Both an archiving strategy and a disposal strategy are implemented. The use and expense of data can be financially evaluated. Data-based systems intervene in the process.	4
Commitment to change	
Staff and/or management resist real-time digital measurements, preferring paper-based recording or simultaneous digital and paper-based data recording. New technologies are faced with skepticism and no serious measures are taken to overcome resistance.	1
Individuals or mid-level management are voicing a desire for change. Change management is not systematic, but the relevance of data used as a resource is discussed.	2
Easy data access and fast interpretation, as well as automated process tracking, are key elements. Handling of data loss or insufficient data is improved. The entire management supports change projects and embraces new technologies.	3
New digital systems are embraced to support staff and to maintain and optimize the process. Change projects can be initiated top-down and bottom-up.	4

2.1. *Types of Data Recorded*

During the interviews, additional information about the nature of the data collected has been gathered. Despite varying amongst the different stakeholders, similar types of data are being collected across all participating companies. These types of recorded data and a detailed description to them is listed in Table 4. The data will be categorized into four groups, namely machine, production, maintenance, and quality data.

Table 4. Types of data collected across all participating companies.

Production Data
Occupation density Since the sorting efficiency is highly dependent on the occupation density (quote), many stakeholders opt to record the occupation density. The calculation is done by dividing the number of pixels detected by the area of the specific sorting aggregate. Throughput rate The throughput rate is defined as the amount of material in kg or m^3 passing through the sorting aggregate in a specified amount of time. Recording the throughput rate can help calibrate the sorting process to reach the ideal trade-off between yield and purity, highly dependent on the throughput rate [31].
Maintenance data
Operating hours Currently, the operating hours of the sorting equipment are being recorded. Nevertheless, so far, none of the interviewees intend to use this data set for advanced maintenance techniques such as prospective or predictive maintenance.
Quality data
Purity and Yield Purity is the quotient of valuables in the ejected material. This value, along with yield, is the defining factor for the evaluation of separation success. The yield is defined as the quotient material fraction mass (e.g., PET) in eject multiplied with the related eject concentration and divided by input mass, which is first multiplied with the concentration of the material fraction mass in the input (e.g., PET) [25].

Table 4. *Cont.*

Machine data
Object statistics The number of objects recognized by the sensor-based sorting setup defines the object statistics. Objects are defined as areas of coagulated pixels of a specified minimum area. Pixels statistics The number of pixels detected for each specified material. This statistic yields the basis for more advanced statistics such as area density or occupation density. Bad pixel replacement Sensors may exhibit defective pixels caused by production. Many sorting software packages come with the ability to exclude or filter those pixels to minimize their effect on the sorting efficiency. However, in most cases, the number of these faulty pixels is not recorded or not available to the software's user. Areal density By calculating the average mass of an object and correlating this with the average amount of pixels detected per object, e.g., a given PET bottle, the areal density of said material can be calculated. This measurement can be used to estimate the number of valuables in the input without the necessity of costly hand sorting or input analysis. Detection rate The detection rate defines the number of correctly identified pixels and objects with a custom sorting model relative to the standard settings of the given sorting aggregate. Valve activity According to the questioned stakeholders, the activation statistics of the pressurized air valves are being saved in most machine statistics. These may be used to recognize one-sided loading of the sorting aggregate in addition to the pixel statistics.
Other data
When the customer wants to record individual data in his sensor-based sorting machines, this option can be additionally enabled. This data could be, e.g., the used spare parts or how many remote maintenance accesses have been performed since the commissioning of the sensor-based sorting machine.

2.2. Validation of Results

In order to validate the answers of the companies, site visits were conducted at every second company. During these site visits, selected sorting machines and sensors were conducted to confirm the given maturity level.

3. Results

Since 28 stakeholder experts were contacted and 12 interviews were held, a return rate of 43% could be achieved. As not all the 12 participants own a sensor-based sorting system directly, some answers refer to industry partners or customers. Most of the data collected regards the production and machine data, and nearly all of it is collected right at the plant. The introductory questions showed that most of the surveyed companies have a data managing strategy implemented, but only half have a designated person responsible for it. Transaction data, meaning the continual evaluation of data quality, was important to 75% of the participants. Questions and answers are listed in Table 1.

The self-assessed data analytics maturity level compared to the calculated average for each sector is shown in Figure 1. Sensor manufacturers have estimated their overall maturity level approximately one degree lower than the assessment resulted, and the same was true of the sorting plant operators. The sorting machine manufacturers' self-assessed maturity level is lower than calculated for two stakeholders and higher for another two stakeholders. In contrast, the sorting robot manufacturers self-assessed themselves higher than the calculation result. The average for all of the stakeholders would be an overall data maturity level between 2.0 and 3.0, which would also be the similar to the self-assessed average.

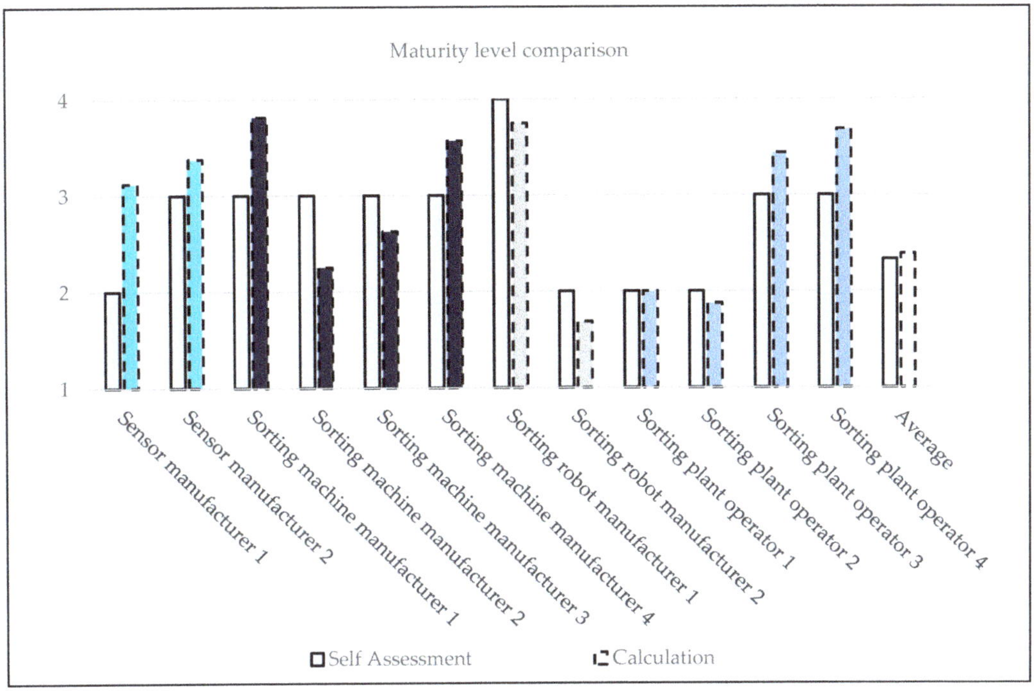

Figure 1. Data analytics maturity level split up into the individual answers of all participants and comparison to the self-assessment.

It was mentioned by the stakeholder experts that smaller and younger companies often do not have the means to build a data management strategy yet and, in addition, presently do not need it. However, all the questioned companies found data analytics to be an important topic that cannot be overlooked in the future and pledged to improve their approach to data handling. The following text will go into detail about which maturity level was calculated for each sector from the overall calculated data analytics maturity level. Figure 2 shows the results for each data analytics sector split up for each stakeholder expert.

Primarily following the data analytics maturity assessment of Bernerstätter [24], the first question for the calculated data analytics maturity level concerns data collection. Most of the participants were identified as being on the third step or higher. The area of data collection was generally considered to be the most digitized, especially for three sorting machine manufacturers and one sorting robot manufacturer, which reached level 4. Table 3 contains detailed descriptions of all maturity levels for each sector. The third level for data collection is described as 'Irregular predefined triggers constitute automated data collection. Manual records are regularly digitized'. Data use is evidently not automated in plants to optimize efficiency. It is only used manually to optimize the sorting machine efficiency in periodic maintenance or troubleshooting when sorting machines face problems.

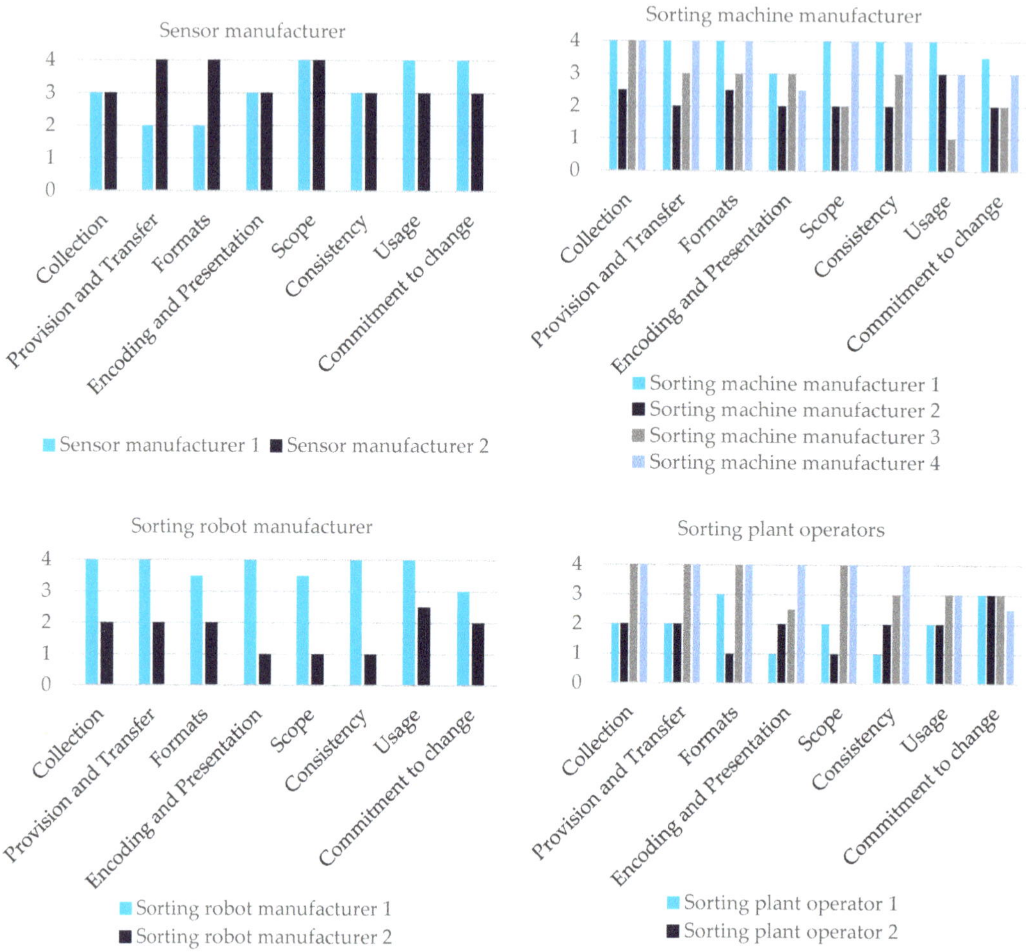

Figure 2. Comparison of the stakeholder answers in maturity levels from one to four in each data analytics category.

Question 2 concerns data provision and transfer and was estimated to be a level 4 for half of the interviewed stakeholder experts through the sensor-based sorting value chain. The description of the fourth level is 'Pre-processing steps are provided to immediately present data in a structured manner ready for analysis. Data is stored in a Data Warehouse'.

Data formats, as a sector of data analytics, show similar results to data provision and transfer instead of sorting machine manufacturer 2, which faces a maturity level of 2.5. The fourth maturity level, which is the dominant one, is defined as 'Data formats are irrelevant because the file transfer passes through an interface straight to an analysis tool. Alternatively, file formats suitable for Big Data are available'.

Generally, there is no visible trend seen for all four stakeholder categories in data encoding and presentation. The most established maturity level in this sector was level 3 (four times), which is defined as 'Unambiguous interpretability is standard; essential attributes are scaled metrically, enabling transformation into nominally scaled values'.

On the data scope, half of the participated stakeholders are on the fourth step with the description 'For at least 1.5 years, data has been entirely recorded and its relevance checked by precise allocation to the relevant observation units'. The maturity level 3 could not be achieved by any interviewed stakeholder, and all of them had either a lower or a higher level.

Regarding data consistency, four participants are on the fourth level. The sensor manufacturers are both on second step, defined by the following statement: 'Consistent time reference cannot be ensured across data sources, but can be achieved using time stamps'. An interesting result is that the maturity level of all four interviewed sorting plant operators varies from the lowest level to the highest level.

Consistent with the other maturity levels, the data usage also strongly depends on each interviewed stakeholder individually. The most determined maturity level was level 3 at five stakeholders. The definition for this level is 'Data is interpreted to remedy faults and to make decisions on a regular basis. Data management processes are documented and discussed with data protection and security. Data is considered a resource'.

The last maturity level deviates from the work of Bernerstätter [24] but is also considered to be of interest and concerns the commitment to change. This maturity level does not deal with data but is essential to be considered on the way to a digitized future that does not start and stop at the IT department. Most of the participating stakeholder experts took a second to think about this part and changed their answers at least once. Only sensor manufacturer 1 found himself on the fourth maturity level and sorting machine manufacturer 1 was on the way to the fourth. The other stakeholder experts consider themselves to be on the way to the third, at the second level, or in between. Maturity level 3 is described as follows: 'Easy data access and fast interpretation, as well as automated process tracking, are key elements. Handling of data loss or insufficient data is improved. The entire management supports change projects and embraces new technologies.'

The last question, "Do you trust the recorded data?", received positive answers for 11 out of 12 interviewees (92%), emphasizing the need to verify data permanently. Considering the different maturity levels for each sector and each stakeholder, it cannot be claimed to determine a trend for each sector. However, advancement can be attempted for the whole European waste management industry as the interviewed stakeholders in the categories of sorting machine manufacturers and sorting robot manufacturers hold a considerable share of the market in Europe. Nonetheless, of interest, Figure 2 shows the individual sectors of data analytics in sensor-based sorting and the commitment to change for each stakeholder.

Finally, as a supplementary question, it was asked where there are currently still barriers to the use of data analytics in sensor-based sorting. Ten out of 12 participants stated that, currently, no mathematical relationships or models between the recorded data had been investigated. Whether there can be mathematical models, e.g., describing the influence of the processed data on one another, would first have to be examined. Furthermore, the area of validity for newly found relationships in the recorded data is still not exactly known. Since these mathematical relationships in recorded data are still not investigated on an industrial level, these 10 participants see the use of data analytics as a risk, which can either be a chance or a hazard to a machine and, in the end, may weaken its performance instead of optimizing it. It would be a significant step to investigate the mathematical relationships in the recorded sensor-based sorting machine data to handle this industry's risk correctly.

Furthermore, although the influences of different machine settings are known, they have not yet been investigated on a level that a sensor-based sorting machine can automatically adapt its sorting settings to the material flow to achieve the best sorting results. These settings would be, e.g., the illuminance of the used emitter(s), the used pressure for ejecting, the minimum object area and object height that is discharged, or the delay time for the activation of the compressed air nozzles.

At last, the stakeholders are interested in making sensor-based and robot sorting processes more efficient, either by improving the identification to characterize more particles correctly or by improving sorting efficiency with, e.g., mathematical models.

4. Discussion

The introductory questions generally show high interest in keeping the quality of data high with a minimal tendency to monitor machine and production data in contrast to data concerning the maintenance and quality of the product. In this chapter, the research questions of the study are discussed and interpreted.

4.1. How Mature Is the Sensor-Based and Robot Sorting Area in Austrian Waste Management in the Use of Data Analytics?

In Figure 1, the comparison between the self-assessed and averaged data analytics maturity level, sensor manufacturers and sorting plant operators have estimated their overall maturity level lower than it was calculated in the assessment. Two sorting machine manufacturers self-assessed lower than the results of the assessment and two self-assed higher than the results. Sorting robot manufacturers tend to self-assess themselves a bit higher than the calculated maturity level. The overall average data maturity level for all stakeholders would be between 2.0 and 3.0, which would also be the similar to the self-assessed average.

The maturity levels of each stakeholder in each data analytics sector differ from each other with slight to no correlations, and there is no derivable trend, as can be seen in Figure 2. The maturity level of each data analytic sector strongly depends on the company itself, so the stakeholder categories need to be analyzed individually.

4.1.1. Sensor Manufacturers

Sensor manufacturer 1 has been in the market for waste sorting sensors for years and has a broader product portfolio than the sensor manufacturer 2. Sensor manufacturer 2 has a slighter product portfolio, which might be the reason that the data analytic sectors are in the scope and the usage higher for 1. In provision and transfer, as well as for the formats, the maturity level might be higher for sensor manufacturer 2 since all their sensor portfolio is new and they have already thought about the relevance of these sectors in their product development. Meanwhile, sensor manufacturer 2 still has also "older" sensors in their equipment, which are not supplied with functions of the higher maturity levels. It can be said that new developed sensors are mostly supplied with the opportunity to provide data so that they can be used in sorting plants to develop a smart waste sorting plant.

4.1.2. Sorting Machine Manufacturers

For the sorting machine manufacturers, it can be seen that number 1 and number 4 are the leaders for all of the technical categories. The reason for this might be that these two companies are far older than the other two, so the global size of the companies as well as the amount of sold sorting machines result directly in a high maturity level for using data analytics in sensor-based sorting.

Sorting machine manufacturers 2 and 3 are in the lower maturity levels for the sectors, especially sorting machine manufacturer 3, which has the maturity level of 1 in data usage: 'Data is not used, i.e., records are kept without interpretation, or no adjustments are performed after interpretation'. Taking a closer look on the company itself, it can be determined that this company supplies mostly smaller plants with their equipment, which might be the reason for their lower level: that customers do not favor this option was one of the answers that was given during the interviews. If the customer would have a demand for these options, they would of course integrate such opportunities in their new sorting machine generations. This leads to the next statement, which is that larger sorting machine manufacturers are on a higher data analytics maturity level in nearly all of the sectors than the smaller ones.

4.1.3. Sorting Robot Manufacturers

For the sorting robot manufacturers, the same statement as for the sensor manufacturers is valid, but in the other way around. Sorting robots are quite new technologies in the waste management branch, so they are developed in a way that data analytics can be used in smart waste sorting plants. The main difference between sorting robot manufacturer 1 and sorting robot manufacturer 2 is that manufacturer 1 developed his robots so that it can be easily integrated in a plant and all of the data can be elected and used by other plant equipment. That is not the intention of sorting robot manufacturer 2: he does not want to share all the data from the robot with other machines, he only provides predefined selected data, which are mostly only finished calculations of objects and pixel statistics. It can be stated that sorting robots are able to provide data so that it can be used in sorting plants to develop a smart waste sorting plant, but this depends—as is also valid for the sensor manufacturers and the sorting machine manufacturers—on which data and how far the supplier is willing to hand over the access to his customer/sorting plant operator.

4.1.4. Sorting Plant Operators

The maturity level results of the category of the sorting plant operators shows that sorting plant operators 3 and 4 are further developed than the others. Sorting plant operator 4 is one of the largest waste sorting plants in Austria, which leads to this high maturity level in each category. Sorting plant operator 3 is has new sorting lines and old sorting lines installed and is also much bigger compared to the other sorting plants in Austria. The two smaller sorting plant operators 1 and 2 are not sorting fractions. They only sort out contaminants for waste, which is thermally treated after the sorting. A high maturity level of the data analytic sectors is not required for them since the sorting task is not to obtain a maximized pure sorted output product. They focus is on the legal threshold values for contaminants, which requires, in the worst case, a second sorter to reach the threshold values, but no intelligent plant, which works with cascade connections, uses intelligent circuits or scavenger concepts. In the case of the smaller sorting plants, the investment in a high digitalization level is not required since the tasks are different. For the sorting plant operators, it can be said that there are two main factors: one is the goal of the sorting tasks (high purity of output product or depose contaminants) and the size of the plant, measured in the yearly throughput rate.

In summary, it can be said that new developed sensors are able to provide all requirements to use data analytics in sensor-based and robot sorting. In any case, whether all of these options can be used depends strongly on the knowledge and willingness to share data of the sorting machine or sorting robot manufacturer. Here, as it can be seen in Figure 2, the data analytics sector's commitment to change will be most important for the future. When these two criteria are fulfilled, the last criterion is whether the sorting plant operator wants or needs new innovations to achieve better sorting results as well as the plant size.

4.2. Where Are the Current Limitations in Technologies or in the Willingness to Be Able to Use Data Analytics in Sensor-Based or Robot Sorting in Austria? What Are the Risks and Chances in the Specific Area of Sensor-Based and Robot Sorting in the Austrian Waste Management Sector?

Supplementary questioning discovered the unused potential for further use of data analytics by developing mathematical models and the use of machine learning algorithms. However, the realization of this potential is inhibited by concerns about the reliability of these machine learning technologies. In addition, interviewees voiced their concerns about diminishing control over their machinery, which could lead to adverse effects on the sorting success without them being able to intervene promptly to alleviate the problem. These are viable concerns and must be dealt with in further evaluation of the applicability of machine learning based on mathematical models in the waste processing industry. Simultaneously, further studies have to be conducted to assess the essential machine parameters, e.g., the intensity of the emitters, the used pressure for ejecting, the minimum object area and object height that is discharged or the delay time for the activation of the compressed air nozzles to be controlled. Integrating data analysis systems and intelligent machinery control algorithms backed by mathematical models successfully into the processes would be a significant step into the future for the waste processing industry. The main objective of all stakeholders is to make sensor-based and robot sorting processes more efficient by improving either the identification or the mechanical operation with mathematical models.

Author Contributions: Conceptualization, K.F. and D.V.; methodology, K.F.; validation, K.F., T.F. and G.K.; formal analysis, K.F. and T.F.; investigation, K.F. and T.F.; data curation, T.F.; writing—original draft preparation, K.F. and T.F.; writing—review and editing, K.F., T.F. and G.K.; visualization, T.F.; supervision, D.V.; project administration, R.P. All authors have read and agreed to the published version of the manuscript.

Funding: This research received no external funding.

Data Availability Statement: All determined and acquired data is listed in the text of this manuscript.

Conflicts of Interest: The authors declare no conflict of interest.

References

1. European Union. Implementation of the Circular Economy Action Plan. Report: COM (2019) 190 Final. 2019. Available online: https://eur-lex.europa.eu/legal-content/EN/TXT/?uri=CELEX%3A52019DC0190 (accessed on 23 August 2021).
2. Kanoun, O.; Trankler, H.-R. Sensor Technology Advances and Future Trends. *IEEE Trans. Instrum. Meas.* **2004**, *53*, 1497–1501. [CrossRef]
3. Sparks, D. Application of MEMS Technology in Automotive Sensors and Actuators. *Proc. IEEE* **1998**, *86*, 1747–1755. [CrossRef]
4. Wotruba, H. Stand der Technik der Sensorgestützten Sortierung. *Berg-Und Hüttenmänn. Mon.* **2008**, *153*, 221–224. [CrossRef]
5. Pretz, T.; Julius, J. Stand der Technik und Entwicklung bei der Berührungslosen Sortierung von Abfällen. (State-of-the-Art and Developments in Contactless Waste Sorting). *Österr. Wasser-Und Abfallwirtsch.* **2008**, *60*, 105–112. [CrossRef]
6. Bonello, D.; Saliba, M.A.; Camilleri, K.P. An Exploratory Study on the Automated Sorting of Commingled Recyclable Domestic Waste. *Procedia Manuf.* **2017**, *11*, 686–694. [CrossRef]
7. Robert, G.; Marcin, P.; Marek, M. Analysis of Picked up Fraction Changes on the Process of Manual Waste Sorting. *Procedia Eng.* **2017**, *178*, 349–358. [CrossRef]
8. Chahine, K.; Ghazal, B. Automatic Sorting of Solid Wastes Using Sensor Fusion. *Int. J. Eng. Technol.* **2017**, *9*, 4408–4414. [CrossRef]
9. Küppers, B.; Chen, X.; Seidler, I.; Friedrich, K.; Raulf, K.; Pretz, T.; Feil, A.; Pomberger, R.; Vollprecht, D. Influences and Consequences of Mechanical Delabelling on Pet Recycling. *Detritus* **2019**, *6*, 39–46. [CrossRef]
10. Friedrich, K.; Möllnitz, S.; Holzschuster, S.; Pomberger, R.; Vollprecht, D.; Sarc, R. Benchmark Analysis for Plastic Recyclates in Austrian Waste Management. *Detritus* **2019**, 105–112. [CrossRef]
11. Friedrich, K.; Holzschuster, S.; Fritz, T.; Pomberger, R.; Aldrian, A. Benchmark Analysis for Recycled Glass in Austrian Waste Management. *Detritus* **2020**, 87–98. [CrossRef]
12. Curtis, A.; Sarc, R. Real-Time Monitoring of Volume Flow, Mass Flow and Shredder Power Consumption in Mixed Solid Waste Processing. *Waste Manag.* **2021**, *131*, 41–49. [CrossRef]
13. Rahman, W.; Islam, R.; Hasan, A.; Bithi, N.I.; Hasan, M.; Rahman, M.M. Intelligent Waste Management System Using Deep Learning with IoT. *J. King Saud Univ.-Comput. Inf. Sci.* **2020**. [CrossRef]
14. Sarc, R.; Curtis, A.; Kandlbauer, L.; Khodier, K.; Lorber, K.; Pomberger, R. Digitalisation and Intelligent Robotics in Value Chain of Circular Economy Oriented Waste Management—A Review. *Waste Manag.* **2019**, *95*, 476–492. [CrossRef]
15. Flamme, S.; Hams, S.; Zorn, M. *Sensortechnologien in der Kreislaufwirtschaft. Converence Transprict of the 14th Recy-& DepoTech-Konferenz*; AVAW Eigenverlag: Leoben, Austria, 2018; pp. 787–792, ISBN 978-3-200-05874-3.
16. Beel, H. Sortierung von schwarzen kunststoffen nach ihrer polymerklasse mit hyperspectral-imaging-technologie (sorting of black plastics to their polymer types with hyper-spectral-imaging-technology). In *Recycling und Rohstoffe Band 10, Proceedings of Recycling und Rohstoffe*; Thomé-Kozmiensky, K.J., Goldmann, D., Eds.; TK-Verlag: Neuruppin, Germany, 2017; pp. 175–191, ISBN 978-3-944310-34-3.
17. Brozzi, R.; Forti, D.; Rauch, E.; Matt, D.T. The Advantages of Industry 4.0 Applications for Sustainability: Results from a Sample of Manufacturing Companies. *Sustainability* **2020**, *12*, 3647. [CrossRef]
18. Erhart, W. Digitale Geschäftsmodelle und Schnelle Innovationszyklen in der Traditionellen Industrie. Am Beispiel: Konzeptionierung und Implementierung Eines Digitalen, Mehrseitigen Geschäftsmodells in der Verwertungs- und Entsorgungsbranche (Digital Business Models and Fast Innovation Cycles in the Traditional Industry. Concrete Example: Conceptual Design and Implementation of a Digital, Multilateral Business Model in the Recycling and Disposal Industry). Master's Thesis, FH Campus 02, Graz, Austria, 2017.
19. Antikainen, M.; Uusitalo, T.; Kivikytö-Reponen, P. Digitalisation as an Enabler of Circular Economy. *Procedia CIRP* **2018**, *73*, 45–49. [CrossRef]
20. Vaidya, S.; Ambad, P.; Bhosle, S. Industry 4.0—A Glimpse. *Procedia Manuf.* **2018**, *20*, 233–238. [CrossRef]
21. Wang, S.; Wan, J.; Li, D.; Zhang, C. Implementing Smart Factory of Industrie 4.0: An Outlook. *Int. J. Distrib. Sens. Netw.* **2016**, *12*, 3159805. [CrossRef]

22. Witkowski, K. Internet of Things, Big Data, Industry 4.0—Innovative Solutions in Logistics and Supply Chains Management. *Procedia Eng.* **2017**, *182*, 763–769. [CrossRef]
23. Thoben, K.-D.; Wiesner, S.; Wuest, T. "Industrie 4.0" and Smart Manufacturing—A Review of Research Issues and Application Examples. *Int. J. Autom. Technol.* **2017**, *11*, 4–16. [CrossRef]
24. Ivanov, D.; Sokolov, B.; Ivanova, M. Schedule Coordination in Cyber-Physical Supply Networks Industry 4.0. *IFAC-PapersOnLine* **2016**, *49*, 839–844. [CrossRef]
25. Rüßmann, M.; Lorenz, M.; Gerbert, P.; Waldner, M.; Engel, P.; Harnisch, M.; Justus, J. *Industry 4.0: The Future of Productivity and Growth in Manufacturing Industries*; Boston Consulting Group: Boston, MA, USA, 2015.
26. Graninger, G. Industrie 4.0 in der Österreichischen Industrie—Interpretation, Umsetzung, Hindernisse (Industry 4.0 in the Austrian Industry-Interpretation, Implementation, Obstacles). Master's Thesis, Montanuniversität Leoben, Leoben, Austria, 2017.
27. Markl, V.; Löser, A.; Hoeren, T.; Krcmar, H.; Hemsen, H.; Schermann, M.; Gottlieb, M.; Buchmüller, C.; Uecker, P.; Bitter, T. *Innovationspotentialanalyse für die Neuen Technologien für das Verwalten und Analysieren von Großen Datenmengen (Big Data Management)*; Bundesministerium für Wirtschaft und Technologie: Berlin, Germany, 2013; p. 44.
28. Schumacher, A.; Nemeth, T.; Sihn, W. Roadmapping towards Industrial Digitalization Based on an Industry 4.0 Maturity Model for Manufacturing Enterprises. Procedia CIRP (2019) 79. In Proceedings of the 12th CIRP Conference on Intelligent Computation in Manufacturing Engineering, Gulf of Naples, Italy, 18–20 July 2018; pp. 409–414. [CrossRef]
29. Machado, C.G.; Winroth, M.; Carlsson, D.; Almström, P.; Centerholt, V.; Hallin, M. Industry 4.0 Readiness in Manufacturing Companies: Challenges and Enablers towards Increased Digitalization. Procedia CIRP 81 (2019). In Proceedings of the 52nd CIRP Conference on Manufacturing Systems (CMS), Ljubljana, Slovenia, 12–14 June 2019; pp. 1113–1118. [CrossRef]
30. Bernerstätter, R. Reifegradmodell zur Bewertung der Inputfaktoren für Datenanalytische Anwendungen-Konzeptionierung am Beispiel der Schwachstellenanalyse (Maturity Model to Evaluate the Input Factors for Data Analytics Application-Conceptional-Design Exampled on the Weak Point Analysis). Ph.D. Thesis, Montanuniversität Leoben, Leoben, Austria, 2019.
31. Küppers, B.; Schlögl, S.; Friedrich, K.; Lederle, L.; Pichler, C.; Freil, J.; Pomberger, R.; Vollprecht, D. Influence of Material Alterations and Machine Impairment on throughput Related Sensor-Based Sorting Performance. *Waste Manag. Res.* **2020**, *39*, 122–129. [CrossRef] [PubMed]

in its distribution in the soil is due to evaporation processes, which in turn lead to changes in soil engineering properties. Therefore, many engineering and environmental problems are directly or indirectly related to the evaporation. In arid and semiarid regions, evaporation is one of the key driving forces in soil salinization and desertification, which not only bringing serious harm to agricultural production and the ecological environment, but also weakening the engineering properties of the soils [15–17]. Research on bioretention in the past focused primarily on water quality purification and hydrological flow effect and not much attention has been given to evapotranspiration, which is an important aspect in bioretention performance. The moisture content in the packing layers directly affects the detention and storage efficiency of bioretention [18]. Previous studies noted that the evapotranspiration process is affected by external climatic factors such as temperature, relative humidity, and wind speed, together with internal factors such as surface water content, permeability coefficient, and packing structure [14,19–21].

Soil water evaporation involves many disciplines, such as geology, geotechnical, water conservancy, roads, environment, atmosphere, etc. Especially in recent years, under the influence of global climate change, extreme arid climate occurs frequently, related disasters are becoming increasingly prominent, and economic losses are becoming much larger. Therefore, a large number of soil moisture evaporation studies on soil and water conservation in arid and semiarid areas have been carried out worldwide [22,23]. In the process of operation, the moisture content in the bioretention packing layer is constantly changing, and the water content of the packing layer directly affects the stagnation-stage evaporation rate [24,25]. Therefore, it is very important to study the water evaporation mechanism in bioretention for the objective evaluation of its hydrologic effect. Current work focused on the study of the common packing profile in Kunshan city, a model sponge city in China. An artificial climate chamber was used to investigate the effect of environmental factors and bioretention structure on the evaporation efficiency of bioretention. A water evaporation resistance model in bioretention was established in this current study. The evaporation model of bioretention was used to consider the hydrological effect of the bioretention calculated after the effect of water evaporation, which was more consistent with the actual measured runoff reduction. Current research findings can be used to evaluate the hydrological effect of the bioretention and to prepare guidelines for the design and implementation in local environments. Therefore, based on abovementioned information, key objectives of the current work follow: (i) design and establish a water evaporation test system for bioretention; (ii) prepare an experimental system consisting of an artificial climate chamber, a bioretention cell unit, a weighing unit, a water supply unit, and a signal acquisition unit; (iii) simulate evaporation rate from the bioretention under different temperature and humidity conditions, with or without a submergence zone; and (iv) measure, collect, process, and analyze the evaporation data through real-time monitoring.

2. Materials and Methods

2.1. Experimental Equipment

The system consists of the following parts: artificial climate chamber, bioretention unit, weighing unit, water supply unit, and signal sampling unit, as shown in Figure 1. The weighing unit is made up of a loading cell and weight base. The water supply unit consists of a plastic water tank and a pump. The signal sampling unit consists of a data acquisition device and computer.

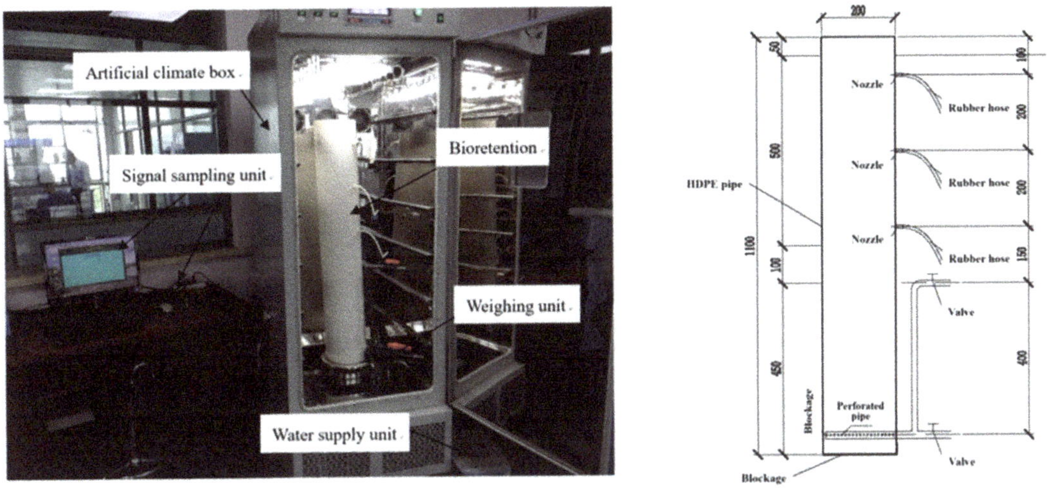

Figure 1. The experimental bioretention unit (laboratory and schematic diagram).

2.1.1. Artificial Climate Chamber

The trial was undertaken in a PQX-500 artificial climate chamber. It was controlled by the computer, which has system functions, e.g., light, constant temperature, and automatic wetness control. It can be used for seed germination, plant cultivation, cultivation of microorganisms and insects, and small animal husbandry.

Device specifications were as follows:

- Temperature: 0–50 °C;
- Temperature fluctuation: $\pm 0.3 \sim \pm 1.0$ °C
- Relative humidity: 30–98%
- Relative humidity fluctuation: $\pm 5 \sim \pm 7$%

2.1.2. Bioretention Simulation Test

The bioretention model can simulate the operating conditions with or without a submerged zone, as shown in Figure 1. While the upper valve remained open at running time and if the inferolateral valve was closed, the bioretention had a 450 mm submerged zone. If the inferolateral valve was open, there was no submerged zone.

The bioretention experimental unit was made of white high-density polyethylene (HDPE) pipes. The bioretention unit was constructed as follows: total height 1100 mm, diameter 200 mm, intermediate layer 100 mm, and height of the submerged zone 450 mm. The permeability coefficient of packing could be measured by three piezometer tubes with different heights, located at the sidewall of the bioretention unit. The internal surface of the equipment was sanded with a fine grit wheel, which increased the roughness of the inner wall to prevent rain runoff from flowing along the inner wall.

2.2. Sustrate Packing Structure

The packing structure of the bioretention unit has a filter layer, intermediate layer, and submerged zone, which was designed according to the local catchment, climatic conditions, and soil characteristics of the experimental area in Kunshan city, China. The permeability coefficient of the filter layer was 140 mm/h. The compounding ratio of different materials forming the filter layer follows: coarse sand—48%, medium size sand—20%, fine sand—30%, and soil—2%. The intermediate layer was packed with coarse sand, and its height was 100 mm. The submerged zone was 450 mm high and made of gravel. All the materials are shown in Figure 2.

Figure 2. Different packing materials used in the bioretention unit: (**a**) coarse sand; (**b**) medium size sand; (**c**) fine sand; (**d**) soil; (**e**) intermediate layer packing; and (**f**) gravel.

2.3. Methods

The bioretention simulated runoff in the trial experiment was computed by the storm strength formula for Kunshan. Designed storm strength (i) could be calculated by Equation (1), which was given in a report published by the Kunshan city government:

$$i = \frac{9.5336(1 + 0.5917\,\mathrm{l\,g}\,T_M)}{(t + 5.9828)^{0.6383}} \quad (1)$$

where; T_M—storm recurrence period (yr), t—rainfall duration (min), and i—designed storm strength.

According to five years of recent data from the Kunshan government, the following parameters were used: bioretention area was 10% of total catchment area, synthetic runoff coefficient was 0.8, storm recurrence period (T_M) was 2 yr, and rainfall duration (t) was 60 min. According to the Kunshan city storm strength analysis formula, the total runoff was calculated to be 0.012 m^3. The three main parts of the trial experiments are discussed in following sections.

2.3.1. Evaluation of Bioretention Evaporation Capacity

The evaporation capacity in bioretention was the possible maximum evaporation in certain weather conditions when water is supplied without limitation. Petri dishes with pure water were first placed in the artificial climate box with set temperature and relative humidity, and then their weight was measured after a period of time. The evaporation capacity was calculated based on the evaporation, evaporation time, and the surface area of the Petri dish, as given in Equation (2):

$$E = \frac{4000(M - m)}{\rho \pi d^2 t} \quad (2)$$

where E—evaporation capacity, mm/d; M—initial mass of Petri dish with water, g; m—subsequent mass of the Petri dish after evaporation, g; ρ—water density, approximately 1 g/cm^3; D—diameter of the Petri dish, mm; and t—period of evaporation, d.

To improve the accuracy of the determination, Petri dishes should be set in the artificial climate box for more than 10 h and then weighed several times to obtain an average. The temperature was set at 20, 25, 30, 35, and 40 °C, and the relative humidity was set at 30%, 40%, 50%, 60%, and 70%, as shown in Table 1.

Table 1. Parameters used to determine the water surface evaporation rate.

No.	Temperature (°C)	Relative Humidity (%)
1	20	30
2	20	40
3	20	50
4	20	60
5	20	70
6	25	30
7	25	50
8	25	70
9	30	30
10	30	40
11	30	50
12	30	60
13	30	70
14	35	30
15	35	50
16	35	70
17	40	30
18	40	40
19	40	50
20	40	60
21	40	70

2.3.2. Determination of Evaporation Efficiency in Bioretention

Evaporation efficiency of bioretention can be affected by external weather factors such as temperature, relative humidity, and wind speed, and internal factors such as permeability coefficient, with or without a submerged zone and packing structure [20,23]. This trial attempted to change the external and internal factors of bioretention to compare the evaporation efficiency, which includes temperature, relative humidity, and with or without a submerged zone. Temperature was set at 20, 30, and 40 °C, and relative humidity was set at 30%, 50%, and 70%, as shown in Table 2.

Table 2. Parameter settings used for the evaporation trial.

No.	Submerged Zone	Temperature (°C)	Relative Humidity (%)
1	With a 450 mm-high submerged zone	20	30
2		20	50
3		20	70
4		30	30
5		30	50
6		30	70
7		40	30
8		40	50
9		40	70
10	Without submerged zone	20	30
11		30	30
12		40	30

Each set of evaporation test stayed in the artificial climate box for 50 h. After each test, the bottom valve was opened to empty the water in the submerged zone. The bioretention unit was left unused for some time, after which simulated runoff was added for further analysis.

2.3.3. Determination of the Surface Water Content in Bioretention

In the process of evaporation in bioretention, it would gradually form a layer of dry soil in the surface of the packing layer that prevents evaporation, where the water content would apparently change [26,27]. Therefore, surface water content was an important parameter of evaporation capacity in the bioretention unit. In the process of evaporation, the water content from the 0~1 cm surface sand samples was measured during the bioretention. The samples were collected every 2 h in the first 6 h of evaporation, and then collected every 4~6 h. The water content can be calculated by comparing the weights of samples before and after heating in an oven at 105 °C for more than 24 h. The final water content was the average of two samples taken at each time point.

3. Results and Discussion

3.1. Effect of Temperature on Bioretention Evaporation

3.1.1. Effect of Temperature on Evaporation Accumulation

At the same relative humidity, the cumulative evaporation under different temperatures was analyzed and compared, as presented in Figure 3.

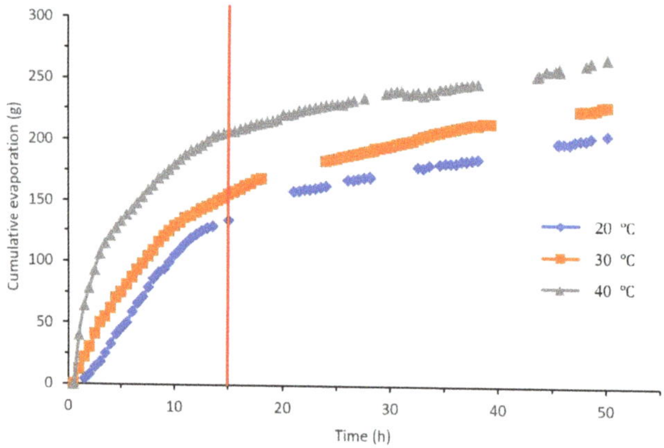

Figure 3. Cumulative evaporation at different temperatures (30% relative humidity, with submerged zone).

Results reveal that at 30% relative humidity, cumulative evaporation rates from bioretention with a submerged zone were 205, 229, and 268 g at ambient temperatures of 20, 30, and 40 °C, respectively, after evaporating for 50 h. Compared to the cumulative evaporation in 20 °C, cumulative evaporation increased 11.71% and 30.73%, respectively, for 30 and 40 °C. The gap in cumulative evaporation among different temperatures was apparent in the first 15 h evaporation period, and cumulative evaporation of bioretention was 134, 156, and 207 g in 20, 30, and 40 °C. Compared to the cumulative evaporation at 20 °C, the cumulative evaporation increased 16.42% and 54.48%, respectively, for 30, 20, 30, and 40 °C. It can be noted from the findings that temperature had substantial influence on the accumulation of evaporation during the initial stage of evaporation [28], but the difference of cumulative evaporation was relatively constant for different temperatures.

Findings presented in Figure 4 reveal the cumulative trend in evaporation from bioretention without a submerged zone for 30% relative humidity. The cumulative evaporation from bioretention without a submerged zone was 169, 200, and 216 g for ambient temperatures of 20, 30, and 40 °C, respectively, after 50 h evaporation. Compared to the cumulative evaporation at 20 °C, cumulative evaporation increased 18.34% and 27.8%, respectively, at 30 and 40 °C. The cumulative evaporation of bioretention was 106, 148, and 164 g for 20, 30, and 40 °C, respectively, after 15 h evaporation. Compared to the cumulative evaporation at 20 °C, cumulative evaporation increased 28.38% and 54.73%, respectively, for 30 and 40 °C.

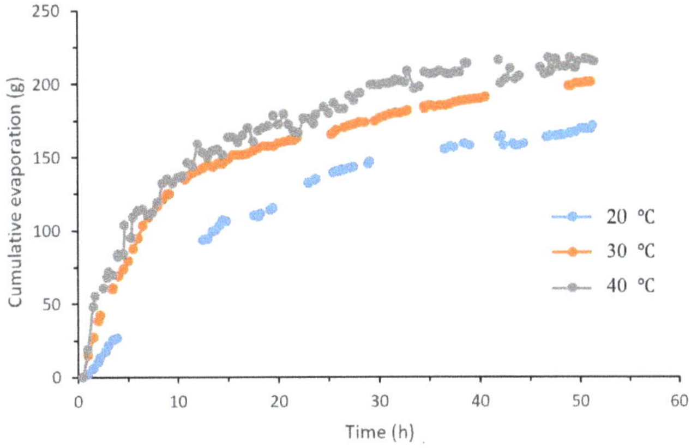

Figure 4. Cumulative evaporation for different temperatures (30% relative humidity, without submerged zone).

Therefore, the cumulative evaporation from bioretention increased significantly with the temperature increment for the same relative humidity. Most of the increase in cumulative evaporation occurred in the first and second stages, which means cumulative evaporation was mainly influenced by temperature, one of the external environmental factors. However, in the third stage, the difference in cumulative evaporation among different temperatures was steady, which reflects the fact that temperature had a minor impact on the cumulative evaporation. Because evaporation was weak in the third stage, which was mainly determined by vapor supply from the lower layer to the upper layer, the water content was therefore low in the surface packing of the bioretention unit [16].

3.1.2. Effect of Temperature on Evaporation Rate

According to the cumulative evaporation data collected by the sensor, the evaporation rate at different time scale was calculated to compare the relationship between temperature and evaporation rate at the same relative humidity, as presented in Figure S1. Results show the relationship between evaporation rate and temperature in bioretention could be concluded as follows: (a) the evaporation capacity at 20, 30, and 40 °C were 10.52, 15.21, and 19.91 mm/d, respectively, as the red line shown. The evaporation rates at 20, 30, and 40 °C after 2 h were 9.75, 14.52, and 19 mm/d, respectively; (b) the evaporation capacity in 20, 30, and 40 °C were 7.84, 11.19, and 14.54 mm/d, respectively. The evaporation rate at 20, 30, and 40 °C after 1.5 h were 5.92, 8.41, and 10.1 mm/d respectively; (c) the evaporation capacity at 20, 30, and 40 °C were 10.52, 15.21, and 19.91 mm/d, respectively. The evaporation rate at 20, 30, and 40 °C after 1.5 h were 7.54, 14.1, and 17.62 mm/d, respectively. The evaporation rate was close to the evaporation capacity, as the water content was almost at saturation in the early evaporation [16,27]. At this stage, the evaporation rate was determined mainly by external weather conditions, which have less to do with packing structure and permeability coefficient of bioretention.

As the evaporation proceeded, the evaporation rate decreased rapidly while the surface water content decreased constantly. Results were as follows: (a) the evaporation rate at 20, 30, and 40 °C after 15 h were 3.3, 3.4, and 2.5 mm/d, respectively, decreased 66.15%, 76.58%, and 86.84% compared to the initial evaporation rate; (b) the evaporation rate at 20, 30, and 40 °C after 15 h were 3.7, 2.8, and 2.68 mm/d, respectively, decreased

37.5%, 66.71%, and 73.47% compared to the initial evaporation rate; (c) the evaporation rate at 20, 30, and 40 °C after 15 h were 2.3, 1.9, and 2.4 mm/d, respectively, decreased 76.41%, 86.91%, and 87.37% compared to the initial evaporation rate. The reduction of evaporation rate at higher temperature was larger than lower temperature, which means the higher the temperature, the faster the descending speed of the evaporation rate, as the water cannot be supplied in time from the lower layer to the upper layer.

At later stage of evaporation, the curves of evaporation rate stay steady that were low values. The evaporation could be reduced due to the formation of a dry layer on the bioretention surface, in which the water content was very low [26,27]. Results reveal that the evaporation rates at 20, 30, and 40 °C in 50 hour's duration were 1.06, 0.99, and 1.1 mm/d, respectively, whereas the evaporation rate at 20, 30, and 40 °C after 50 h were 0.83, 0.92, and 0.82 mm/d, respectively, equaling 14.02%, 10.94%, and 8.12% of the initial evaporation rate. In this stage, the evaporation rate was mainly determined by water content of packing, packing structure, and permeability coefficient of bioretention, which had minimal influence by external weather conditions.

Variation in the evaporation rate at different temperatures and trend in the same relative humidity are presented in Figure S1. In the early evaporation stage, the evaporation rate was higher and close to the evaporation capacity. This trend occurred due to the saturation in the water content of packing in the early evaporation. This trend reflects that there was no limitation in water supply for evaporation. At this stage, the evaporation rate was mainly determined by external weather conditions, such as temperature, relative humidity, and wind speed. As the evaporation proceeded, the evaporation rate fell quickly. After 15 h, the evaporation rate decreased by more than 60% of the initial evaporation rate. In addition, the higher the temperature, the faster the descending speed of evaporation rate. This results from high temperature accelerating the descending speed of surface water content. In later evaporation, the variation of evaporation rate stayed low and steady. After 50 h, the evaporation rate of bioretention remained at approximately 1 mm/d (0.82~1.1 mm/d) at different relative humidity. In this stage, the evaporation becomes difficult as a dry layer formed in the surface of the bioretention unit. The evaporation rate was mainly determined by water content of packing, packing structure, and permeability coefficient of bioretention, which has less to do with external weather conditions.

3.2. *Effect of Relative Humidity on the Evaporation*

The relative humidity can affect the diffusion and exchange of water vapor in the packing surface because of the difference in relative humidity between the inside and outside of the packing. When the relative humidity was low, the evaporation rate was high as the fast speed of diffusion and exchange of vapor; when the relative humidity was high, the evaporation rate was low as the slow speed of diffusion and exchange of vapor [28,29]. While the relative humidity rises to a certain extent, the surface evaporation nearly stops.

3.2.1. Analysis of Relative Humidity Effect on the Accumulation of Evaporation

At the same temperature, this section compares cumulative evaporation under different relative humidity. Results revealed that when the temperature was 20 °C, cumulative evaporation of bioretention with a submerged zone were 205, 159, and 146 g at relative humidity 30%, 50%, and 70%, respectively, after evaporating for 50 h. Compared to the cumulative evaporation at relative humidity 30%, cumulative evaporation decreased 22.44% and 28.78%, respectively, at relative humidity 50% and 70%. The gap of cumulative evaporation among different temperatures was obvious in the first 15 h. The cumulative evaporation of bioretention was 136, 90, and 54 g at relative humidity 30%, 50%, and 70%, respectively, after evaporating for 15 h. Compared to the cumulative evaporation at relative humidity 30%, cumulative evaporation decreased 33.82% and 60.29%, respectively, at relative humidity 50% and 70%. It can be seen that relative humidity has substantial influence on the accumulation of evaporation at the initial stage of evaporation, but the difference of cumulative evaporation was relatively constant for different relative humidity.

When the temperature was 30 °C, cumulative evaporation of bioretention with a submerged zone was 228 and 137 g for relative humidity 30% and 50%, respectively, after evaporating for 50 h. Compared to the cumulative evaporation for relative humidity 30%, cumulative evaporation decreased 39.91% for relative humidity 50%. The gap of cumulative evaporation among different temperatures was obvious in the first 15 h. The cumulative evaporation of bioretention was 155 and 80 g for relative humidity 30% and 50%, respectively, after evaporating for 15 h. Compared to the cumulative evaporation for relative humidity 30%, cumulative evaporation decreased 48.39% for relative humidity 50%. It can be seen that relative humidity has substantial influence on the accumulation of evaporation at the initial stage of evaporation, but the difference of cumulative evaporation was relatively constant for different relative humidity. In case where the temperature was 40 °C, cumulative evaporation of bioretention with a submerged zone was 271 and 249 g for relative humidity 30% and 50%, respectively, after evaporating for 50 h. Compared to the cumulative evaporation for relative humidity 30%, cumulative evaporation decreased 8.12% for relative humidity 50%. The gap of cumulative evaporation among different temperatures was obvious in the first 15 h. The cumulative evaporation of bioretention was 207 and 167 g for relative humidity 30% and 50%, respectively, after evaporating for 15 h. Compared to the cumulative evaporation for relative humidity 30%, cumulative evaporation decreased 19.32% for relative humidity 50%. It can be seen that relative humidity has substantial influence on the accumulation of evaporation at the initial stage of evaporation, but the difference of cumulative evaporation was relatively constant for different relative humidity.

The cumulative evaporation of bioretention decreases significantly with the relative humidity increment at the same temperature. Because the gap of vapor between the bioretention packing and air decrease, it leads to the decline of evaporation rate, while the relative humidity increases [28,29]. Most of the increase in cumulative evaporation occurs in the early evaporation, which means cumulative evaporation was mainly influenced by relative humidity. However, in the later evaporation, the cumulative evaporation becomes steady, which means relative humidity had minor impact on it. Earlier study also pointed out that the occurrence and maintenance of soil evaporation must meet three conditions: (1) continuous heat supply to meet the consumption of evaporation latent heat; (2) vapor pressure in the atmosphere must be lower than the vapor pressure on the soil surface, and there is a relative humidity difference; and (3) continuous water supply to the evaporating surface in the soil [30].

3.2.2. Effect of Relative Humidity on Evaporation Rate

According to the cumulative evaporation data collected by the sensor, the evaporation rate of different timings was calculated to compare the relationship between relative humidity and evaporation rate at 20 °C, as presented in Figure S2. When the temperature was 20 °C, the evaporation capacity for relative humidity 30%, 50%, and 70% were 10.52, 7.84, and 5.16 mm/d, respectively, as shown by the red line. The evaporation rate for relative humidity 30%, 50%, and 70% after 1.5 h was 7.13, 5.64, and 4.31 mm/d, respectively. In the early evaporation, the evaporation rate was close to the potential evaporation rate.

As the evaporation proceeded, the evaporation rate was falling fast while the surface water content decreased constantly. The evaporation rate for relative humidity 30%, 50%, and 70% after 15 h was 4.41, 4.12, and 2.65 mm/d, respectively, and decreasing 38.15%, 26.95%, and 38.52%, respectively, compared to the initial evaporation rate. Evaporation rate decreased quickly in a short time. In later evaporation, the evaporation rate remained steady with low values. The evaporation rate for relative humidity 30%, 50%, and 70% after 50 h were 1.13, 0.86, and 0.82 mm/d, respectively, equaling 15.85%, 15.25%, and 19.03% of the initial evaporation rate. In this stage, the evaporation rate was mainly determined by the water content of packing, packing structure, and permeability coefficient of bioretention, which had less to do with external weather conditions.

3.3. *Effect of Submerged Condition on the Evaporation*

3.3.1. Analysis of Submerged Zone Effect on the Accumulation of Evaporation

Findings for comparison of the cumulative evaporation trend in bioretention with or without a submerged zone at similar temperature and relative humidity (20 °C and 30%) are presented in Figure S3a. The cumulative evaporation in bioretention with a submerged zone was higher than without a submerged zone. After 50 h, the cumulative evaporation rate in bioretention with a submerged zone was 251 g, which was 15.67% more than the cumulative evaporation in bioretention without a submerged zone, which was 217 g. In addition, the cumulative evaporation in bioretention with a submerged zone became higher than without a submerged zone between the 10th to 20th hours, as revealed by the red line shown in the graph. At the 10th hour, the cumulative evaporation in bioretention with and without a submerged zone was 107 and 75 g, which was 32 g in difference. At the 20th hour, the cumulative evaporation in bioretention with and without a submerged zone was 161 and 116 g, which shows a difference of 45 g. In the later evaporation stage, the difference between cumulative evaporation under the two conditions was relatively stable.

In the case of similar temperature and relative humidity (30 °C and 30%), results comparing cumulative evaporation in bioretention with and without a submerged zone are presented in Figure S3b. The cumulative evaporation in bioretention with a submerged zone was higher than without the submerged condition. After 50 h, the cumulative evaporation in bioretention with a submerged zone was 237 g, which was 18.5% more than the cumulative evaporation in bioretention without a submerged zone, which was 200 g. In addition, the cumulative evaporation in bioretention with a submerged zone became larger than without a submerged zone between the 10th and 20th hours, as presented by the red line in the figure. At the 10th hour, the cumulative evaporation in bioretention with and without a submerged zone was 137 and 133 g, which was 4 g in difference; at the 20th hour, the cumulative evaporation in bioretention with and without a submerged zone was 181 and 159 g, which shows a difference of 22 g. In the later evaporation stage, the difference between cumulative evaporation under abovementioned two conditions was relatively stable.

In the case of similar temperature and relative humidity (40 °C and 30%), comparison results of cumulative evaporation in bioretention with or without a submerged zone are presented in Figure S3c. The cumulative evaporation in bioretention with a submerged zone was higher than without a submerged zone. After 50 h, cumulative evaporation in bioretention with a submerged zone was 249 g, which was 15.28% more than bioretention without a submerged zone (216 g). In addition, the cumulative evaporation in bioretention with a submerged zone became higher than without a submerged zone between the 5th and 15th hour, as presented by the red line in the figure. At the 5th hour, the cumulative evaporation in bioretention with and without a submerged zone was 115 and 106 g, which was 9 g in difference; at the 15th hour, the cumulative evaporation in bioretention with and without a submerged zone were 187 and 160 g, a difference of 27 g. In the later evaporation stage, the difference between cumulative evaporation trends under two different conditions was relatively stable.

The cumulative evaporation in bioretention with a submerged zone was higher than without a submerged zone in the similar temperature and relative humidity conditions. In the early evaporation stage, it had minor impact on the cumulative evaporation in the case of presence of a submerged zone in the bioretention unit. As the evaporation advanced, the cumulative evaporation in bioretention with a submerged zone became larger than without a submerged zone. This trend occurred due the sharp decline in surface water content. The surface water content and the water supply ability of packing were the main contributory factors that determined the cumulative evaporation, while the influence of temperature and relative humidity was the secondary factor [16]. The water from the submerged zone in bioretention can be supplied to the upper layer through diffusion and capillary, reflecting that the water supply ability of bioretention with a submerged zone was better than bioretention without a submerged zone [30–32].

3.3.2. Effect of Submerged Zone on Evaporation Rate

Results presented in Figure S4a reveal that under the 20 °C and 30% temperature and relative humidity condition, the evaporation capacity in bioretention was 10.52 mm/d. The evaporation rates in bioretention with and without a submerged zone after 1 h were 10.2 and 9.1 mm/d. In the early evaporation stage, the evaporation rate was close to its potential. As the evaporation proceeded, the evaporation rate revealed a sharp decline while the surface water content showed constant decrease. The downward trend in evaporation rate with a submerged zone was slower than without a submerged zone. The evaporation rates in bioretention with and without a submerged zone after 10 h were 6.5 and 4.5 mm/d, decreasing 36.27% and 50.52% compared to the initial evaporation rate. Evaporation rate decreases quickly in a short interval. In the later evaporation stage, the evaporation rates stay steady and were close between bioretention with and without a submerged zone. The evaporation rates in bioretention with and without a submerged zone after 50 h were 1.08 and 1.06 mm/d, which were 10.59% and 11.65% of the initial evaporation rate. In this stage, presence or absence of a submerged zone in bioretention had negligible impact.

The evaporation capacity in bioretention was 15.21 mm/d when the temperature and relative humidity were 30 °C and 30%, as presented Figure S4b. The evaporation rate in bioretention with and without a submerged zone after 1 h was 15.28 and 15 mm/d, respectively. In the early evaporation stage, the evaporation rate was close to potential evaporation. As the evaporation proceeded, the evaporation rate showed a sharp decrease while there was a constant decrease in the surface water content. The downward trend in evaporation rate with a submerged zone was slower than without a submerged zone. The evaporation rates in bioretention with and without a submerged zone after 15 h were 3.44 and 1.53 mm/d, decreasing 77.49% and 89.8%, respectively, compared to the initial evaporation rates. It reveals a sharp decrease in evaporation rates in a shorter duration, which is consistent with the trend revealed in other studies [29,32–34].

Results showed that in the later evaporation stage, the evaporation rate revealed a steady trend, which is a very similar trend between bioretention with and without a submerged zone. Evaporation rate in bioretention with and without a submerged zone after 50 h was 0.91 and 0.87 mm/d, which was 5.96% and 5.8% of the initial evaporation rate. In this stage, it was not important whether there was a submerged zone in bioretention. Findings show that when the temperature and relative humidity were 40 °C and 30%, respectively, the evaporation capacity was 19.91 mm/d, as presented in Figure S4c. The evaporation rate in bioretention with and without a submerged zone after 1 h was 19.82 and 20.06 mm/d, respectively. In the early evaporation stage, the evaporation rate was close to the potential evaporation rate. As the evaporation process started in bioretention, the evaporation rate showed a sharp decline while the surface water content decreased constantly. The downward trend in evaporation rate with a submerged zone was slower than without a submerged zone. The evaporation rates in bioretention with and without a submerged zone after 10 h were 5.61 and 3.5 mm/d, decreasing 71.73% and 82.55% compared to the initial evaporation rates and reflects a sharp decrease in a short time.

The evaporation rates reflected a steady trend in later evaporation, which was close between bioretention with and without a submerged zone [32–34]. The evaporation rate in bioretention with and without a submerged zone after 50 h was 1.11 and 1.02 mm/d, which were 5.6% and 5.1% of the initial evaporation rate. In this stage, it was not important whether there was a submerged zone in the bioretention. The evaporation rate was close to the potential evaporation rate in the early evaporation, whether in presence or absence of a submerged zone in bioretention. As the evaporation proceeded, the evaporation rate fell fast, but the downward trend of the evaporation rate with a submerged zone was slower than without a submerged zone [35]. The water supply in bioretention with a submerged zone was better than bioretention without a submerged zone, as the water from the submerged zone in bioretention can supply the upper layer through diffusion and capillary process.

4. Conclusions

Results revealed that the bioretention evaporation rate was in accordance with the soil evaporation, which can be divided into three stages, namely, the constant rate stage, the falling rate stage, and the residual stage. In the first two stages, the environmental factors were the main factors affecting water evaporation of the bioretention. The evaporation rate at the initial stage of evaporation was close to the maximum evaporation capacity under the environment, and then the evaporation rate decreased rapidly. After 15 h, the evaporation rate decreased more than 60%, and the higher the temperature, the faster the evaporation rate decreased. The evaporation rate of the third stage was low and stable. The evaporation rate was about 1 mm/d (0.82~1.1 mm/d), and the surface layer formed a dry soil layer. The cumulative evaporation of the bioretention with the submerged zone was higher than that without a submerged zone, and the cumulative evaporation after 50 h was 16.48% higher. In the second stage of evaporation, the decreasing amplitude of the evaporation capacity of the bioretention with a submerged zone was also relatively slow, whereas, the water in the bottom of the submerged zone recharged the upper layer by capillary action and water vapor diffusion.

Supplementary Materials: The following are available online at https://www.mdpi.com/article/10.3390/su14031286/s1, Figure S1: The relationship between temperature and evaporation rate, Figure S2: The relationship between relative humidity and evaporation rate (20 °C, with submerged zone), Figure S3: The influence of submerged zone on accumulation of evaporation, Figure S4: The influence of submerged zone on evaporation rate.

Author Contributions: Conceptualization, D.F. and T.Z.; methodology, J.Q. and S.M.; formal analysis, J.Q.; investigation, J.Q. and T.Z.; resources, S.M. and R.P.S.; data curation, J.Q. and T.Z.; writing—original draft preparation, J.Q. and T.Z.; writing—review and editing, J.Q. and R.P.S.; visualization, J.Q. and R.P.S.; supervision, D.F. and R.P.S.; project administration, J.Q. and D.F.; funding acquisition, S.M. All authors have read and agreed to the published version of the manuscript.

Funding: This research was funded by National Natural Science Foundation of China, grant number 41807514; Natural Science Foundation of Jiangsu Province, grant number BK20170682; and Priority Academic Program Development of the Jiangsu Higher Education Institution, Jiangsu Province, China.

Institutional Review Board Statement: Not applicable.

Informed Consent Statement: Not applicable.

Data Availability Statement: Not applicable.

Conflicts of Interest: The authors declare no conflict of interest.

References

1. Ali, W.; Takaijudin, H.; Yusof, K.W.; Osman, M.; Abdurrasheed, A.S.I. The common approaches of nitrogen removal in bioretention system. *Sustainability* **2021**, *13*, 2575. [CrossRef]
2. Osman, M.; Yusof, K.W.; Takaijudin, H.; Goh, H.W.; Malek, M.A.; Azizan, N.A.; Ghani, A.A.; Abdurrasheed, A.S. A review of nitrogen removal for urban stormwater runoff in bioretention system. *Sustainability* **2019**, *11*, 5415. [CrossRef]
3. Gülbaz, S.; Kazezyılmaz-Alhan, C.M. Hydrological model of LID with rainfall-watershed-bioretention system. *Water Resour. Manag.* **2017**, *31*, 1931–1946. [CrossRef]
4. Gulbaz, S.; Kazezyılmaz-Alhan, C.M.; Temür, R. Development of an empirical formula for estimation of bioretention outflow rate. *Water SA* **2019**, *45*, 209–215. [CrossRef]
5. Goor, J.; Cantelon, J.; Smart, C.C.; Robinson, C.E. Seasonal performance of field bioretention systems in retaining phosphorus in a cold climate: Influence of prolonged road salt application. *Sci. Total Environ.* **2021**, *778*, 146069. [CrossRef]
6. Qian, J.; Miao, S.; Tapper, N.; Xie, J.; Ingleton, G. Investigation on airport landscape cooling associated with irrigation: A case study of Adelaide airport, Australia. *Sustainability* **2020**, *12*, 8123. [CrossRef]
7. Maliva, R.G. Low impact development and rainwater harvesting. In *Anthropogenic Aquifer Recharge*; Springer: Cham, Switzerland, 2020; pp. 765–825. [CrossRef]
8. Qin, Y. Urban flooding mitigation techniques: A systematic review and future studies. *Water* **2020**, *12*, 3579. [CrossRef]
9. Spraakman, S.; Rodgers, T.; Monri-Fung, H.; Nowicki, A.; Diamond, M.; Passeport, E.; Thuna, M.; Drake, J. A need for standardized reporting: A scoping review of bioretention research 2000–2019. *Water* **2020**, *12*, 3122. [CrossRef]

10. Yang, F.; Fu, D.; Liu, S.; Zevenbergen, C.; Singh, R.P. Hydrologic and pollutant removal performance of media layers in bioretention. *Water* **2020**, *12*, 921. [CrossRef]
11. Shrestha, P.; Salzl, M.T.; Jimenez, I.J.; Pradhan, N.; Hay, M.; Wallace, H.R.; Abrahamson, J.N.; Small, G.E. Efficacy of spent lime as a soil amendment for nutrient retention in bioretention green stormwater infrastructure. *Water* **2019**, *11*, 1575. [CrossRef]
12. Shahrokh Hamedani, A.; Bazilio, A.; Soleimanifar, H.; Shipley, H.; Giacomoni, M. Improving the treatment performance of low impact development practices—Comparison of sand and bioretention soil mixtures using column experiments. *Water* **2021**, *13*, 1210. [CrossRef]
13. Sang, M.; Huang, M.; Zhang, W.; Che, W.; Sun, H. A pilot bioretention system with commercial activated carbon and river sediment-derived biochar for enhanced nutrient removal from stormwater. *Water Sci. Technol.* **2019**, *80*, 707–716. [CrossRef] [PubMed]
14. Jiang, C.; Li, J.; Li, H.; Li, Y. Experiment and simulation of layered bioretention system for hydrological performance. *J. Water Reuse Desalin.* **2019**, *9*, 319–329. [CrossRef]
15. Yuan, C.; Feng, S.; Wang, J.; Huo, Z.; Ji, Q. Effects of irrigation water salinity on soil salt content distribution, soil physical properties and water use efficiency of maize for seed production in arid Northwest China. *Int. J. Agric. Biolog. Eng.* **2018**, *11*, 137–145. [CrossRef]
16. Liu, Q.; Zhao, R.; Miao, J.; Wang, J.; Jia, D. Effect of diatomite on soil evaporation characteristics. *Environ. Earth Sci.* **2021**, *80*, 219. [CrossRef]
17. Hou, L.; Zhou, X.; Wang, S. Numerical analysis of heat and mass transfer in kiwifruit slices during combined radio frequency and vacuum drying. *Int. J. Heat Mass Transf.* **2020**, *154*, 119704. [CrossRef]
18. Sandhu, J.S.; Takhar, P.S. Verification of hybrid mixture theory based two-scale unsaturated transport processes using controlled frying experiments. *Food Bioprod. Process.* **2018**, *110*, 26–39. [CrossRef]
19. Kaya, Y.Z.; Zelenakova, M.; Üneş, F.; Demirci, M.; Hlavata, H.; Mesaros, P. Estimation of daily evapotranspiration in Košice City (Slovakia) using several soft computing techniques. *Theor. Appl. Climatol.* **2021**, *144*, 287–298. [CrossRef]
20. Valipour, M. Importance of solar radiation, temperature, relative humidity, and wind speed for calculation of reference evapotranspiration. *Arch. Agron. Soil Sci.* **2015**, *61*, 239–255. [CrossRef]
21. Shokri, N.; Lehmann, P.; Or, D. Characteristics of evaporation from partially wettable porous media. *Water Resour. Res.* **2009**, *45*, W02415. [CrossRef]
22. Fox, M.J. A technique to determine evaporation from dry stream beds. *J. Appl. Meteorol.* **2010**, *7*, 697–701. [CrossRef]
23. Denisov, Y.M.; Sergeev, A.; Bezborodov, G.; Bezborodov, Y.G. Moisture evaporation from bare soils. *Irrig. Drain. Syst.* **2002**, *16*, 175–182. [CrossRef]
24. Idso, S.B.; Reginato, R.J.; Jackson, R.D.; Kimball, B.A.; Nakayama, F.S. The three stages of drying of a field soil. *Soil Sci. Soc. Am. J.* **1974**, *38*, 831–837. [CrossRef]
25. Chapman, C.; Horner, R.R. Performance assessment of a street-drainage bioretention system. *Water Environ. Res.* **2010**, *82*, 109–119. [CrossRef]
26. An, N.; Tang, C.-S.; Xu, S.-K.; Gong, X.-P.; Shi, B.; Inyang, H.I. Effects of soil characteristics on moisture evaporation. *Eng. Geol.* **2018**, *239*, 126–135. [CrossRef]
27. Verdú González, A.M.C.; Mas Serra, M.; Josa March, R.; Ginovart Gisbert, M. The effect of a prototype hydromulch on soil water evaporation under controlled laboratory conditions. *J. Hydrol. Hydromech.* **2020**, *68*, 404–410. [CrossRef]
28. Sterlyagov, A.N.; I Nizovtsev, M.; Borodulin, V.Y.; Letushko, V.N. The effect of air relative humidity on the evaporation temperature of water-ethanol droplets. In *Journal of Physics: Conference Series, Proceedings of the 5th All-Russian Scientific Conference Thermophysics and Physical Hydrodynamics with the School for Young Scientists (TPH-2020), 13–20 September 2020, Yalta, Crimea*; IOP Publishing Ltd.: Bristol, UK, 2020; Volume 1675, p. 012058. [CrossRef]
29. Roger, K.; Sparr, E.; Wennerström, H. Evaporation, diffusion and self-assembly at drying interfaces. *Phys. Chem. Chem. Phys.* **2018**, *20*, 10430–10438. [CrossRef]
30. Hillel, D. *Introduction to Environmental Soil Physics*, 1st ed.; Elsevier: Amsterdam, The Netherlands, 2003. [CrossRef]
31. Tamm, T.; Nõges, T.; Järvet, A.; Bouraoui, F. Contributions of DOC from surface and ground flow into Lake Võrtsjärv (Estonia). *Hydrobiologia* **2008**, *599*, 213–220. [CrossRef]
32. He, K.; Qin, H.; Wang, F.; Ding, W.; Yin, Y. Importance of the submerged zone during dry periods to nitrogen removal in a bioretention system. *Water* **2020**, *12*, 876. [CrossRef]
33. Zhang, J.; Singh, R.P.; Liu, Y.; Fu, D. Design and operation of submerged layer in bioretention for enhanced nitrate removal. *J. Water Supply Res. Technol.—AQUA* **2019**, *68*, 744–756. [CrossRef]
34. Wei, D.; Singh, R.P.; Liu, J.; Fu, D. Effect of alternate dry-wet patterns on the performance of bioretention units for nitrogen removal. *Desalination Water Treat.* **2017**, *59*, 295–303. [CrossRef]
35. Singh, R.P.; Zhao, F.; Ji, Q.; Saravanan, J.; Fu, D. Design and performance characterization of roadside bioretention systems. *Sustainability* **2019**, *11*, 2040. [CrossRef]

sustainability

MDPI

Article

Research on Recycling Strategies for New Energy Vehicle Waste Power Batteries Based on Consumer Responsibility Awareness

Jiajing Fan [1,*], Hao Teng [2] and Yibo Wang [3]

[1] Chinese-German Institute for Applied Engineering, Zhejiang University of Science and Technology, Hangzhou 310023, China
[2] School of Economics and Management, Zhejiang University of Science and Technology, Hangzhou 310023, China
[3] School of Information and Electronic Engineering, Zhejiang University of Science and Technology, Hangzhou 310023, China
* Correspondence: fanjiajing@zust.edu.cn

Abstract: Due to the limited service life of new energy vehicle power batteries, a large number of waste power batteries are facing "retirement", so it will soon be important to effectively improve the recycling and reprocessing of waste power batteries. Consumer environmental protection responsibility awareness affects the recycling of waste power batteries directly. Therefore, under the two recycling modes of new energy vehicle manufacturers and third-party recycling enterprises, this study analyzes the impact of consumer environmental protection responsibility awareness on the recycling price of waste power batteries and profit in the supply chain. The influence of factors such as recycling income, recycling input cost, and black-market recycling prices on consumer awareness of responsibility is also analyzed. Through theoretical research, it was found that: Under the model that third-party recycling enterprises are responsible for recycling, it can obtain better overall supply chain benefits; consumer environmental protection responsibility awareness and recycling benefits are positively correlated with supply chain benefits overall; and recycling benefits have a certain role in promoting consumer awareness of responsibility, while the increase in informal recycling prices inhibits consumer awareness of responsibility.

Keywords: NEV; waste power batteries; consumer environmental protection responsibility awareness; recycling strategy; reverse supply chain

Citation: Fan, J.; Teng, H.; Wang, Y. Research on Recycling Strategies for New Energy Vehicle Waste Power Batteries Based on Consumer Responsibility Awareness. *Sustainability* **2022**, *14*, 10016. https://doi.org/10.3390/su141610016

Academic Editors: Sunil Kumar, Pooja Sharma and Deblina Dutta

Received: 27 June 2022
Accepted: 10 August 2022
Published: 12 August 2022

1. Introduction

With growing environmental problems, all countries have focused on the new energy vehicle industry and expect to alleviate the global energy crisis and environmental problems through the promotion of new energy vehicles (NEV). In China, the NEV industry is developing rapidly and it has been the largest producer and seller of NEVs in the world for more than 6 years, since 2015. While the production and sales of NEVs are booming, the new energy vehicle power batteries of the first batch are facing "retirement". It is estimated that, by 2025, the "retired" power batteries in China will be up to 780,000 tons [1]. Owing to the high technical requirements and high cost of power battery recycling, the Ministry of Industry and Information Technology of China has announced a "whitelist" of 26 enterprises that have agreed to the *Industry Specifications for Comprehensive Utilization of Waste Power Batteries for New Energy Vehicles* to improve recycling efficiency. Meanwhile, by the end of September 2021, 171 new energy vehicle manufacturers and comprehensive utilization enterprises have set up 9985 recycling service networks across the country to ensure the effective recycling of power batteries. However, under such circumstances, only about 20% of the power batteries on the market are recycled by "whitelisted" enterprises in formal channels, which means that most of the waste power batteries are recycled in informal channels. This results in a greatly reduced recycling rate of waste power batteries and is likely to cause a series of environmental problems and safety hazards. Therefore, it is necessary to effectively improve consumer environmental protection responsibility awareness and improve the recycling rate of waste power batteries in a formal channel.

According to the *"Resource Continuation: Research Report on the Circular Economy Potential of New Energy Vehicle Batteries in 2030"*, released by the international environmental protection organization Greenpeace and the China Environmental Protection Federation on 29 October 2020, the total amount of decommissioned power batteries for passenger electric vehicles worldwide will reach 12.85 million tons in 2021–2030, and the market scale of recycling and reuse will exceed 100 billion yuan [2]. The decommissioning of new energy vehicle batteries is a global phenomenon. The European Union, the United States, Japan, and other countries started earlier in the recycling of lead–acid batteries and lithium batteries, and the established recycling system has achieved good results [3]. For example, US power battery recycling laws involve federal, state, and local governments at all levels, and laws at all levels complement and regulate each other. It guides retailers and consumers (referring to new energy vehicle owners, the same below) through the power battery recycling price mechanism. In 2016, the European Union mandated that member states must recycle at least 45% of waste batteries, and the processing and utilization rate of lead, nickel, and isolators should not be less than 50% [4]; German law stipulates that producers, consumers, and recyclers should take corresponding responsibilities and obligations in the power battery recycling industry chain and emphasizes the extended producer responsibility system; Japan has promulgated laws and regulations in basic law, comprehensive law, and special law [5], and it has established a battery recycling system of "battery production-sales-recycling-renewable processing" [6]. Judging from the fact that most developed countries have formulated different levels of laws and regulations for power battery recycling and particularly emphasize the obligations and roles of consumers, we can see that consumers are extremely important in the process of power battery recycling. In China, the State Council promulgated *the Implementation Plan for the Extended Producer Responsibility System (EPR system)* (General Office of the State Council. Implementation plan of extended producer responsibility system (EB/OL). (25 December 2016). http://www.gov.cn/zhengce/content/2017-01/03/content_5156043.htm e.g., (accessed on 6 April 2022)) in 2017 [7] to ensure the effective recycling of power batteries, which stipulates that new energy vehicle manufacturers and battery production enterprises are responsible for the recycling of power batteries. Since 2017, the Ministry of Industry and Information Technology has issued eight policies for power battery recycling management to promote the construction of a new energy vehicle power battery recycling system [8]. In July 2021, the National Development and Reform Commission and other ministries of China jointly issued *the Circular on Issuing the 14th Five-Year Plan for the Development of Circular Economy* (National Development and Reform Commission. Development plan of circular economy in the 14th five year plan (EB/OL). (1 July 2021). https://www.ndrc.gov.cn/xxgk/zcfb/ghwb/202107/t20210707_1285527.html?code=&state=123 e.g., (accessed on 6 April 2022)) [9], which proposed action: strengthening the construction of traceability management platform and improving the traceability management system of new energy vehicle power batteries. Although the framework of the power battery recycling system in China is becoming more and more mature [10], the law still mainly focuses on new energy vehicle manufacturers and battery manufacturers, and it ignores the other stakeholders in the power battery recycling process, especially consumers, who are the source of power battery recycling.

In research on power battery recycling strategies, the recycling rate is one of the most important parameters. Many researchers regard the recycling rate as a fixed parameter [11–14] or take the recycling rate (or recycling amount) as the linear/nonlinear function of the recycling price. For example, Li and Mu [15], based on the research results of Heydari [16], took the recovery rate as a linear function of the recovery price to analyze the pricing strategy and its impact on the recycling rate with or without government regulation; Lu [17], considering the dual risks of market demand and quality in power battery recycling, took the recycling price as a function of the recycling rate to analyze the optimal recycling price under decentralized and centralized decisions. In addition, many researchers have focused on analyzing other factors affecting the recycling rate of power batteries, such as government supervision and subsidy policy [18,19], socio-ecological environment, economic conditions and recycling technology [20], supply chain cooperation degree and recycling technology [14], government subsidies and channel selection [21], etc. Obviously, the current research ignores the influence of consumer environmental protection responsibility awareness on the recycling rate. Consumer environmental protection responsibility awareness means that consumers should not affect the living environment of others with their own consumption, which has an important impact on the selection of recycling channels, recycling rates, and recycling prices [22], but only increasing the recycling price is not effective in increasing consumers' environmental protection responsibility awareness of recycling [15,23]. Therefore, in research on power battery recycling strategies, it is necessary to take the consumer environmental protection responsibility awareness as one of the main factors affecting the recycling rate of power batteries in formal channels.

The recycling entities of the waste power batteries of new energy vehicles generally include new energy vehicle manufacturers, retailers, third-party recycling enterprises, battery manufacturers, etc. Savaskan et al. first studied the selection of the optimal recycling channel in a closed-loop supply chain and concluded that, under this assumption, retailer recycling is better than manufacturer recycling or third-party recycling [24]. Then, Savaskan et al. concluded that the degree of competition between retailers will have an impact on the manufacturer's recycling channel selection decision to a certain extent [25]. Sun et al. analyzed the impact of recycling price and sales volume on channel selection [26]. Hong et al. considered the significant impact of advertising on consumers and studied the impact of advertising on the selection of recycling channels and recycling pricing decisions [27]. Li et al. (2016) investigated the impact of different recycling channels on the profits of each supply chain member under a decentralized structure and identified the conditions and equilibrium characteristics of each recycling channel selection [28]. Chen and Tian found that the recycling price and sales price have an important impact on the choice of recycling mode for manufacturers and retailers [29]. Zhou considered the impact of collection efforts and the quality of recycled products, and the result showed that recycling by manufacturers or retailers depends on the cost-saving level of remanufacturing [30]. Zhang et al. analyzed the impact of environmental benefits, economic benefits, and social welfare on recycling channels [31]. Chen et al. studied the optimal selection of recycling channels based on mutual win–win situations in the supply chain [32]. Gong studied government funding policies with respect to ecological design levels, the recycling of power batteries as a government funding policy of reward factors, and the choice of power battery closed-loop supply chain recycling channels [33]. Although researchers have considered the many influencing factors of recycling channels, there is little research on the impact of consumer environmental protection responsibility awareness on channel selection.

In summary, consumers are the source of waste power battery recycling and play a crucial role in the recycling rate in formal channels. However, previous researchers did not take consumer factors into account. Therefore, this paper aims at the reverse supply chain formed by consumers, new energy vehicle manufacturers, and third-party recycling enterprises. It determines the pricing strategy and recycling channel strategy of the reverse supply chain based on consumer environmental protection responsibility awareness and explores the impact of different factors on consumers' sense of responsibility.

2. The Pricing Model of the Reverse Supply Chain of Waste Power Batteries

2.1. Problem Description

According to the *EPR*, new energy vehicle manufacturers and battery manufacturers are mainly responsible for power battery recycling. Ding [34] comprehensively considered factors, such as economic profits, recycling costs, and resource utilization, and believed that it is more appropriate to adopt the manufacturer alliance mode to recycle power batteries. Yao and Jiang [35] proposed a battery recycling mode based on new energy vehicle enterprises, which is conducive to recycling power batteries from consumers and solving the problem of the irregular battery recycling market. This paper establishes a three-level reverse supply chain composed of consumers, new energy vehicle manufacturers, and third-party recycling enterprises, as shown in Figure 1. New energy vehicle manufacturers and third-party recycling enterprises can participate in power battery recycling, and the latter is responsible for the disposal of waste power batteries. Consumers, as the source of power battery recycling, can recycle waste power batteries in formal or informal channels, but both channels will be regulated by the government.

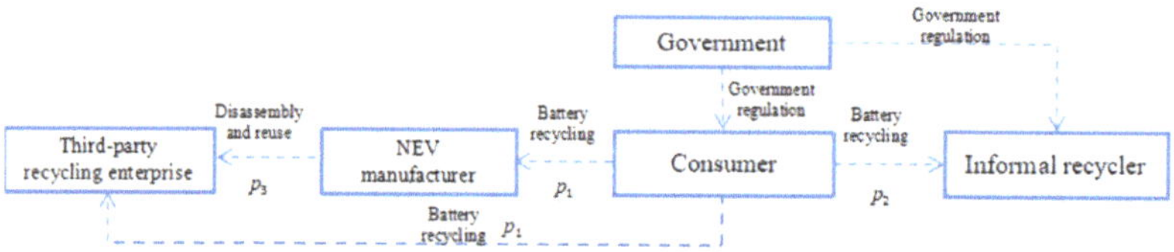

Figure 1. Reverse supply chain of power battery recycling.

2.2. Model Assumptions

(1) There are two recycling paths in the reverse supply chain. One is that the manufacturer is responsible for recycling waste power batteries from consumers at the price of p_1 and then sells them to a third-party recycling enterprise at the price of p_3, which is responsible for processing; second, the third-party recycling enterprises directly obtain waste power batteries from consumers at the price of p_1.

(2) Whether consumers will recycle waste power batteries through formal channels is mainly affected by two factors, namely, consumer environmental protection responsibility awareness and recycling price. Assuming that consumer environmental protection responsibility awareness is β, $\beta \in (0,1]$. If $\beta = 1$, consumers are willing to recycle waste power batteries in formal channels no matter what the recycling price is. The waste power batteries can be sold to NEV manufacturers or third-party recycling enterprises at the price of p_1, and they can also be sold to an informal recycler at the price of p_2. The closer p_1 is to p_2, the more the formal channels are preferred by consumers. When $p_1 \geq p_2$, regardless of consumer environmental protection responsibility awareness, they will always choose the formal channels. Therefore, this paper believes that the recycling rate (r) at which consumers choose in formal channels to recycle waste power batteries is $1-(1-\beta)\left(1-\frac{p_1}{p_2}\right)$. When the sales volume of NEV is Q, assuming that its formal recycling volume is $Q\left[1-(1-\beta)\left(1-\frac{p_1}{p_2}\right)\right]$, then the informal recycling volume will be $Q(1-\beta)\left(1-\frac{p_1}{p_2}\right)$.

(3) The government has a certain supervision role over consumers. If consumers are found to recycle waste power batteries illegally, they will be punished and fined. The probability of punishment as set as t, and the fine is set as δ.

(4) Recycling channels could be offered by NEV manufacturers or third-party recycling enterprises, both of which are not limited in terms of recycling capacity, while their recycling investment is related to the recycling rate, r. Drawing on the research results of [36], we set recycling investment as $\frac{1}{2}d_m\left(1-(1-\beta)\left(1-\frac{p_1}{p_2}\right)\right)^2$ or $\frac{1}{2}d_r\left(1-(1-\beta)\left(1-\frac{p_1}{p_2}\right)\right)^2$, where d_m represents the manufacturer's cost coefficient for recycling waste power batteries and d_r represents the third-party recycling enterprises' cost coefficient for recycling waste power batteries.

(5) The ability of third-party recycling enterprises is not limited. The treatment and reuse of waste batteries can obtain recycling income, ω. Meanwhile, because power batteries are recycled and reused, the cost of new energy vehicle manufacturers can be reduced, a factor which is set as $\Delta\varepsilon$.

2.3. Single-Channel Recycling Decision Model for New Energy Manufacturers

2.3.1. Decentralized Decision Model

Under decentralized decision-making, the profits of manufacturers, third-party recycling enterprises, and consumers are as follows:

$$\pi_m{}^{md} = Q\left(1-(1-\beta)\left(1-\frac{p_1}{p_2}\right)\right)(p_3-p_1+\Delta\varepsilon)-\frac{1}{2}d_m\left(1-(1-\beta)\left(1-\frac{p_1}{p_2}\right)\right)^2 \quad (1)$$

$$\pi_r{}^{md} = Q\left(1-(1-\beta)\left(1-\frac{p_1}{p_2}\right)\right)(\omega-p_3) \quad (2)$$

$$\pi_k{}^{md} = \left(1-(1-\beta)\left(1-\frac{p_1}{p_2}\right)\right)p_1+\left((1-\beta)\left(1-\frac{p_1}{p_2}\right)\right)(p_2-t\delta) \quad (3)$$

The first item in Formula (1) is the income from the recycling of waste power batteries by NEV manufacturers, and the second item is the investment required to carry out the recycling of waste power batteries; Formula (2) represents the income from the recycling of waste power batteries by third-party recycling enterprises; and Formula (3) represents the consumer's income.

NEV manufacturers determine the recycling prices with the goal of maximizing their own profits, it can be obtained:

$$p_1{}^{md^*} = \frac{p_2[Qk_1+\beta Qp_2-2\beta k_2]}{2(1-\beta)k_2},\ p_3{}^{md^*} = \frac{(1-\beta)(\omega-\Delta\varepsilon)-p_2\beta}{2(1-\beta)}.$$

where: $k_1=(1-\beta)(\omega+\Delta\varepsilon)$; $k_2=2Qp_2+(1-\beta)d_m$.

According to $\left(p_1^{md^*}, p_3^{md^*}\right)$, the optimal profits of NEV manufacturers, third-party recycling enterprises, and consumers can be obtained, respectively, as:

$$\pi_m{}^{md^*} = \frac{Q^2(k_1+\beta p_2)^2}{8k_2(1-\beta)},\ \pi_r{}^{md^*} = \frac{Q^2(k_1+\beta p_2)^2}{4k_2(1-\beta)},\text{ and}$$

$$\pi_k{}^{md^*} = \frac{Q^2p_2(k_1+\beta p_2)(Qk_1-2\beta k_2+Q\beta p_2)}{4(1-\beta)k_2^2}+\frac{Q(2k_2-Qk_1-4Q\beta p_2)}{2k_2}(p_2-t\delta).$$

2.3.2. Centralized Decision-Making Model

Under centralized decision-making, the reverse supply chain profits are as follows:

$$\pi_{sc}{}^{mc} = Q\left(1-(1-\beta)\left(1-\frac{p_1}{p_2}\right)\right)(\omega-p_1+\Delta\varepsilon)-\frac{1}{2}d_m\left(1-(1-\beta)\left(1-\frac{p_1}{p_2}\right)\right)^2 \quad (4)$$

According to the maximization of the reverse supply chain profits, the following can be obtained: $p_1{}^{mc^*} = \frac{p_2[Qk_1+\beta Qp_2-\beta k_2]}{(1-\beta)k_2}$, $\pi_{sc}{}^{mc^*} = \frac{Q^2(k_1+\beta p_2)^2}{2k_2(1-\beta)}$.

Proposition 1. *Under the NEV manufacturer recycling mode, the optimal value of the recycling price will be* $\left(p_1^{md^*}, p_3^{md^*}\right)$. *Under decentralized decision-making, the profits of manufacturers and third-party recycling enterprises will be* $\left(\pi_m^{md^*}, \pi_r^{md^*}\right)$, *and the supply chain profits will be* $\frac{3Q^2(k_1+\beta p_2)^2}{8k_2(1-\beta)}$. *In the case of a centralized decision, the optimal value of the recycling price is* $p_1{}^{mc^*}$, *and the reverse supply chain profits will be* $\frac{Q^2(k_1+\beta p_2)^2}{2k_2(1-\beta)}$. *It is obvious that centralized decisions can achieve higher profits than decentralized decisions, and the consumer profits under centralized decisions are also higher than those under decentralized decisions.*

2.4. *Single-Channel Recycling Model for Third-Party Recycling Enterprises*

2.4.1. Decentralized Decision-Making Model

When the third-party recycling enterprises are responsible for recycling waste power batteries, the profits of NEV new energy vehicle manufacturing will be:

$$\pi_m{}^{rd} = Q\left(1-(1-\beta)\left(1-\frac{p_1}{p_2}\right)\right)\Delta\varepsilon \tag{5}$$

The profits of third-party recycling enterprises are as follows:

$$\pi_r{}^{rd} = Q\left(1-(1-\beta)\left(1-\frac{p_1}{p_2}\right)\right)(\omega-p_1)-\frac{1}{2}d_r\left(1-(1-\beta)\left(1-\frac{p_1}{p_2}\right)\right)^2 \tag{6}$$

Under decentralized decision-making, $p_1{}^{rd^*} = \frac{p_2[Qk_3-\beta(k_4-Qp_2)]}{(1-\beta)k_4}$; $\pi_m{}^{rd^*} = Q^2\Delta\varepsilon\frac{(k_3+p_2\beta)}{k_4}$; $\pi_r{}^{rd^*} = \frac{Q^2(k_3+\beta p_2)^2}{2k_4(1-\beta)}$, where $k_3 = (1-\beta)\omega$ and $k_4 = 2Qp_2+(1-\beta)d_r$.

2.4.2. Centralized Decision Model

Under centralized decision-making, the reverse supply chain profits are as follows:

$$\pi_{sc}{}^{rc} = Q\left(1-(1-\beta)\left(1-\frac{p_1}{p_2}\right)\right)(\omega-p_1+\Delta\varepsilon)-\frac{1}{2}d_r\left(1-(1-\beta)\left(1-\frac{p_1}{p_2}\right)\right)^2 \tag{7}$$

We can thus obtain the following: $p_1{}^{rc^*} = \frac{p_2[Qk_1-\beta k_4+\beta Qp_2]}{(1-\beta)k_4}$ and $\pi_s{}^{rc^*} = \frac{Q^2(k_1+\beta p_2)^2}{2k_4(1-\beta)}$.

Proposition 2. *In the single-channel recycling of third-party recycling enterprises, the optimal value of the recycling price obtained under decentralized decision-making is* $p_1^{rd^*}$*, the profits of manufacturers and third-party enterprises will be* $\left(\pi_m^{rd^*}, \pi_r^{rd^*}\right)$*, and the supply chain profits will be* $\frac{Q^2(k_3+\beta p_2)(k_1+\beta p_2+(1-\beta)\Delta\varepsilon)}{2k_4(1-\beta)}$*. The optimal value of the recycling price obtained under centralized decision-making is* $p_1{}^{rc^*}$*, and the total supply chain profits will be* $\pi_{sc}{}^{c^*} = \frac{Q^2(k_1+\beta p_2)^2}{2k_4(1-\beta)}$*. Higher profits will be obtained under centralized decisions than under decentralized decisions, and as will the profits for consumers.*

3. Model Analysis

Based on the assumed parameters used to analyze the impact of different parameters on the recycling price, p_1, recycling rate, r, and the profits of consumers and supply chain, the hypothesis is as follows:

$$p_2 = 500,\ \omega = 1200,\ \Delta\varepsilon = 200,\ \delta = 100,\ d_r = 5000,\ d_m = 5000,\ Q = 1000,\ t = 0.5.$$

3.1. *Analysis of Recycling Pricing,* p_1

According to Figure 2, the recycling price, p_1, decreases when consumer environmental protection responsibility awareness, β, gradually increases. At the same time, when β reaches a certain value, the value of p_1 will be 0, which means consumers are still willing to recycle waste power batteries through formal channels even if they are not paid. According to Figure 2a, the value of p_1 will also gradually increase when the value of the recycling income, ω, increases. The main reason is that when the recycling income, ω, increases, third-party recycling enterprises are willing to feed more profits back to consumers, that is, by raising the recycling price, p_1, so as to improve the recycling rate and form a virtuous circle. According to Figure 2b,c, the recycling cost coefficient, d_m, and the sales volume, Q, have little effect on the recycling price, p_1. According to Figure 2d, it can be seen that the value of p_1 will decrease as the value of p_2 increases. This is mainly due to the fact that, when informal recycling prices, p_2, continue to rise, fewer and fewer consumers are willing to recycle waste power batteries through formal channels, resulting in a decline in the profits of recyclers, which, in turn, leads to the value of p_1 decreasing.

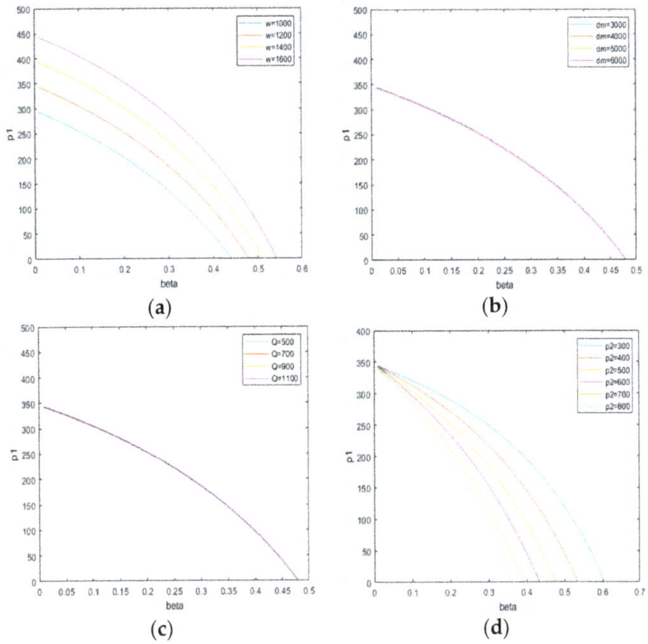

Figure 2. Influence of parameter β, ω, d_m, Q, p_2 on the recycling price, p_1. (**a**) Influence of parameter β and ω on the recycling price, p_1. (**b**) Influence of parameter β and d_m on the recycling price, p_1. (**c**) Influence of parameter β and Q on the recycling price, p_1. (**d**) Influence of parameter β and p_2 on the recycling price, p_1.

It can be seen from Figure 3 that, when β and ω are determined and $d_m = d_r$, then ${p_1}^{md^*} < {p_1}^{rd^*} < {p_1}^{mc^*} = {p_1}^{rc^*}$, which means that the optimal recycling price under the decentralized decision-making of manufacturer recycling is the lowest, the optimal recycling price of centralized decision-making is significantly higher than that of decentralized decision-making(consistent with the conclusion of Proposition 1 and Proposition 2), and the difference of ${p_1}^{mc^*}$ and ${p_1}^{rc^*}$ is determined by d_m and d_r.

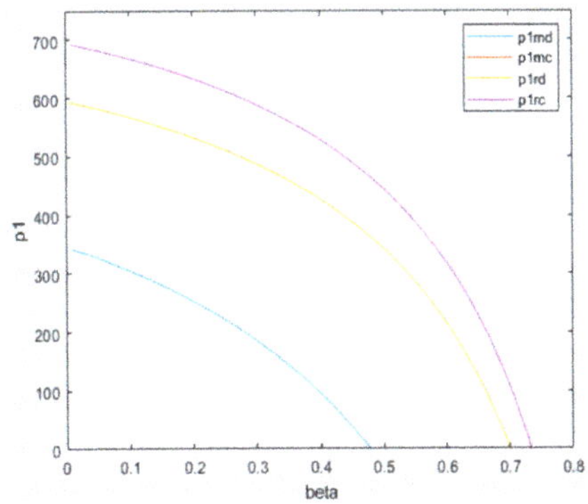

Figure 3. Comparison of recycling price changes under different recycling modes.

3.2. Analysis of the Recycling Rate, r

It can be seen from Figure 4 that, when the consumer environmental protection responsibility awareness, β, gradually increases, the recycling rate, r, initially decreases, and after β reaches the lowest point, the recycling rate, r, gradually rises. This is mainly because the recycling rate is $r = 1 - (1-\beta)\left(1 - \frac{p_1}{p_2}\right)$; when the value of β is lower, the increase in β leads to a decrease in the recycling price, p_1, and a decrease in $(1-\beta)$ is greater than the increase in $\left(1 - \frac{p_1}{p_2}\right)$, which leads to a decrease in the recycling rate, r. However, when the value of β is larger, consumers are willing to recycle waste power batteries through formal channels even though the recycling price is very low, even if it tends toward 0. When $\left(1 - \frac{p_1}{p_2}\right) = 1$, then $r = \beta$, which means the recycling rate, r, increases linearly with the value of β. According to Figure 2a, when the value of the recycling income, ω, increases, the recycling rate, r, increases gradually. This is mainly because when the value of recycling income, ω, increases, the recycling price, p_1, can be appropriately increased, thereby increasing the recycling rate, r. However, when the recycling rate reaches the lowest point, no matter how ω changes, the recycling rate is $r = \beta$; this is because after β increases to a certain value, the recycling price becomes $p_1 = 0$. From Figure 2b,c, it can be seen that the recycling cost coefficient, d_m, and the sales volume, Q, have little effect on the recycling rate, r. According to Figure 2d, the recycling rate, r, is negatively correlated with p_2; thus, it is easy to understand why, when the value of p_2 increases, more consumers will naturally tend to sell waste power batteries through informal channels for a greater profit.

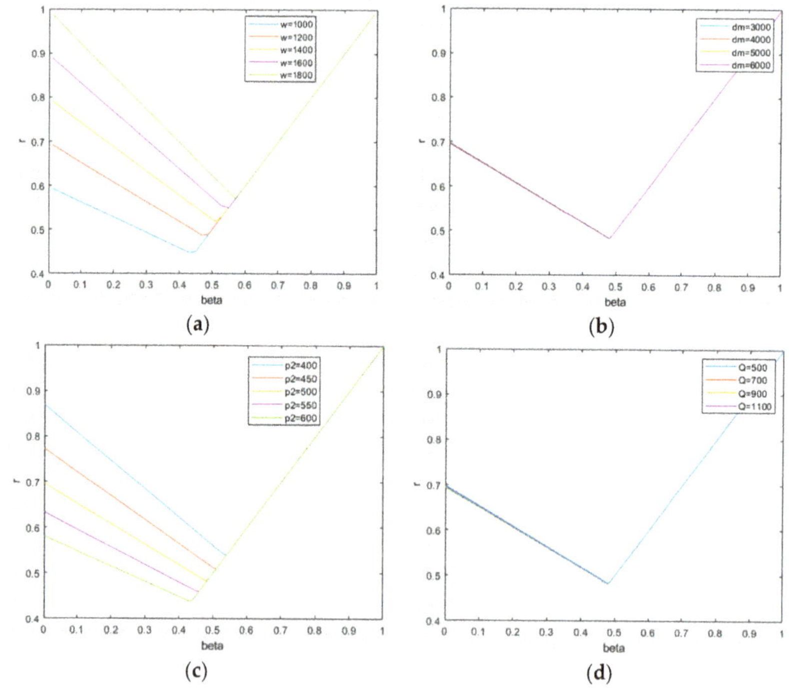

Figure 4. Influence of parameter β, ω, d_m, Q, p_2 on the recycling rate, r. (**a**) Influence of parameter β and ω on the recycling rate, r. (**b**) Influence of parameter β and d_m on the recycling rate, r. (**c**) Influence of parameter β and p_2 on the recycling rate, r. (**d**) Influence of parameter β and Q on the recycling rate, r.

3.3. Analysis of Consumer Profits

Comparing Figures 2 and 5, it can be seen that the changing trend in consumer expected profit, π_k, is basically the same as that of recycling price, p_1. As the consumer environmental protection responsibility awareness, β, gradually increases, the recycling price, p_1, gradually decreases, and π_k also gradually decreases. It can be seen from Figure 5a that recycling income, ω, is positively correlated with π_k. When the value of recycling income, ω, continues to increase, the value of p_1 will also gradually increase, and so does the value of π_k. It can be seen from Figure 2b,c that the cost coefficient, d_m, and the sales volume, Q, have little effect on π_k. According to Figure 5d,e, at first, π_k decreases with the increase in p_2 and then increases with the increase in p_2, which is mainly due to $\pi_k = rp_1 + (1-r)(p_2 - t\delta)$, and p_2 with p_1 and r are both negatively correlated. When p_2 increases, rp_1 decreases and $(1-r)(p_2 - t\delta)$ increases. When the value of p_2 is small, rp_1 decreases more than $(1-r)(p_2 - t\delta)$, so π_k decreases with the increase in p_2. When p_2 increases to a certain value, rp_1 decreases less than $(1-r)(p_2 - t\delta)$, so π_k increases as p_2 increases.

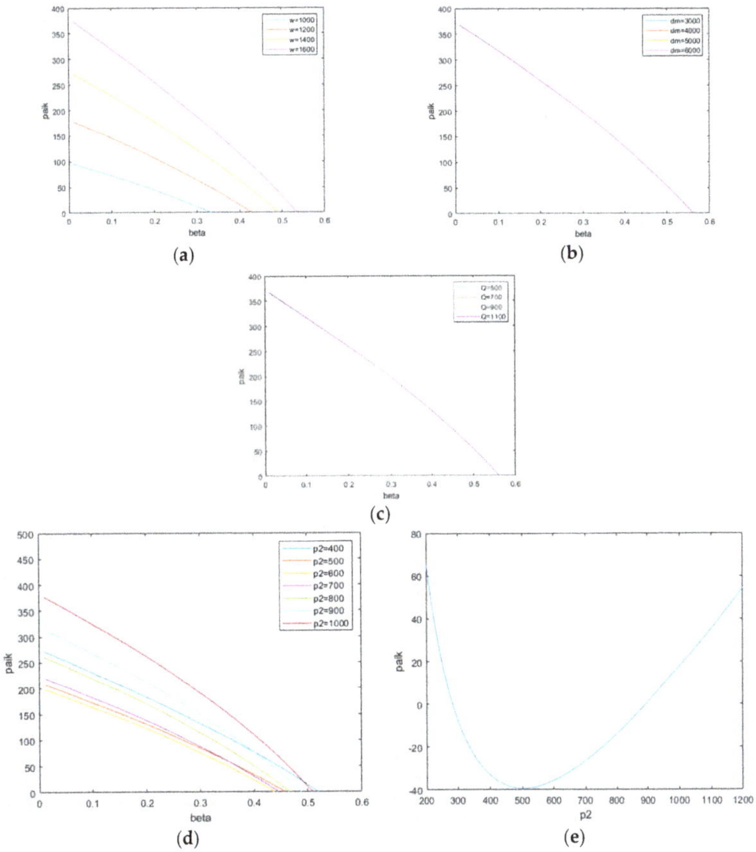

Figure 5. The effect of parameter β, d_m, Q, p_2 on the consumer expected profit, π_k. (**a**) The effect of parameter β and ω on the consumer expected profit, π_k. (**b**) The effect of parameter β and d_m on the consumer expected profit, π_k. (c) The effect of parameter β and Q on the consumer expected profit, π_k. (**d**) The effect of parameter β and p_2 on the consumer expected profit, π_k. (e) The effect of parameter p_2 on the consumer expected profit, π_k.

Meanwhile, it can be seen from Figure 6 that, when β and ω are determined and $d_m = d_r$, then ${\pi_k}^{md^*} < {\pi_k}^{rd^*} < {\pi_k}^{mc^*} = {\pi_k}^{rc^*}$. That means the expected profits of consumers, π_k, in centralized decision-making are higher than those of decentralized decision-making under the two recycling modes. In the recycling mode of third-party recycling enterprises, the expected profit of consumers, π_k, is higher than that of the manufacturer recycling mode (consistent with the conclusion of Proposition 1 and Proposition 2). In centralized decision-making, the expected profit of consumers π_k under the two modes is basically the same, the difference is mainly caused by the recycling cost coefficients d_m and d_r.

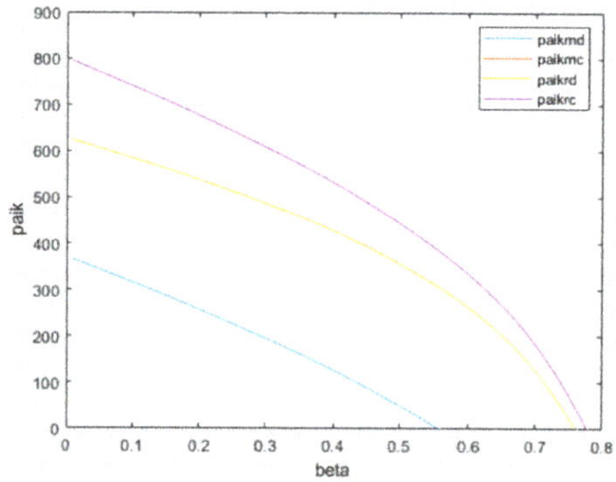

Figure 6. Comparison of consumer profits under different recycling modes.

3.4. Analysis of Profit of the Supply Chain

According to Figure 7a, the profits of the supply chain, π_{sc}, increase with an increase in recycling income ω, which is mainly due to the fact that $\pi_{sc} = G\omega^2$ (according to Table 1, it can be seen that, when other parameters are determined, G is a constant). When the value of β is less than a certain value, the change in the profits of the supply chain, π_{sc}, is not very large, but when β is greater than this value, the change in the profits of the supply chain π_{sc} increases significantly. The main reason is $\pi_{sc} = A\frac{Q^2((1-\beta)(\omega+\Delta\varepsilon)+\beta p_2)^2}{(2Qp_2+(1-\beta)d_m)(1-\beta)}$ (according to Table 1, A is a constant value under different recycling modes); when the value of β is too large, the value of $(2Qp_2 + (1-\beta)d_m)(1-\beta)$ tends toward 0, which leads to a significant increase in the value of π_{sc}. At the same time, when ω is small, the profits of the supply chain, π_{sc}, increase with the increase in β, but the increase is gentle at first and then rises sharply. When the value of ω increases to a certain value, the profit of the reverse supply chain π_{sc} decreases with the increase in β, and after β reaches a minimum value, π_{sc} increases as an increase in β. It can be seen from Figure 7b that the cost coefficient d_m has little effect on π_{sc}. From Figure 7c, it can be seen that π_{sc} increases with the increase in sales volume, Q, and its curve shape is similar to ω. According to Figure 7d,e, at first, π_{sc} decreases with the increase in p_2; when $p_2 = \frac{(1-\beta)[Q(\omega+\Delta\varepsilon)-\beta d_m]}{\beta Q}$, π_{sc} reaches the lowest value, and then it increases with the increase in p_2.

It can be seen from Figure 8 that, when β and ω are determined and $d_m = d_r$, then ${\pi_{sc}}^{md^*} < {\pi_{sc}}^{rd^*} \approx {\pi_{sc}}^{mc^*} = {\pi_{sc}}^{rc^*}$, and the profit of the supply chain, π_{sc}, obtained though decentralized decision-making under the manufacturer recycling mode is the smallest (consistent with the conclusion of Proposition 1 and Proposition 2), while the profits of the other three modes are not significantly different, which shows that this system can obtain greater profits under the third-party recycling enterprise recycling mode.

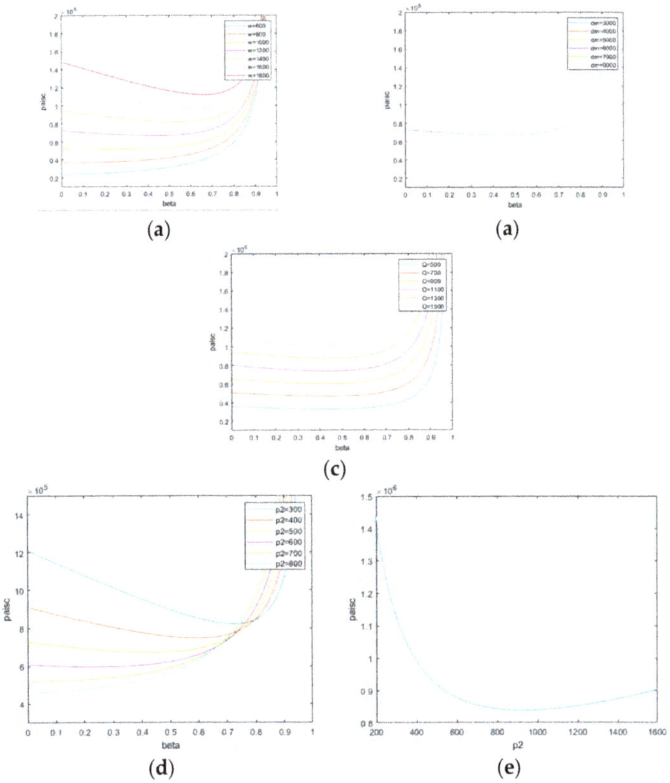

Figure 7. Impact of parameter β, ω, d_m, Q, p_2 on the π_{sc} of the reverse supply chain. (**a**) Impact of parameter β and ω on the π_{sc}. of the reverse supply chain. (**b**) Impact of parameter β and d_m on the π_{sc} of the reverse supply chain. (**c**) Impact of parameter β and Q on the π_{sc} of the reverse supply chain. (**d**) Impact of parameter β and p_2 on the π_{sc} of the reverse supply chain. (**e**) Impact of parameter p_2 on the π_{sc} of the reverse supply chain.

Table 1. Recycling prices and profits of the supply chain under different recycling modes.

Modes	p_1	π_m	π_r	π_{sc}	π_k
Manufacturer recycling (decentralized decision-making)	$\frac{p_2[Qk_1+\beta Qp_2-2\beta k_2]}{2(1-\beta)k_2}$	$\frac{Q^2(k_1+\beta p_2)^2}{8k_2(1-\beta)}$	$\frac{Q^2(k_1+\beta p_2)^2}{4k_2(1-\beta)}$	$\frac{3Q^2(k_1+\beta p_2)^2}{8k_2(1-\beta)}$	$\frac{Qp_2(k_1+\beta p_2)(Qk_1-2\beta k_2+Q\beta p_2)}{4(1-\beta)k_2^2}+\frac{(2k_2-Qk_1-4Q\beta p_2)}{2k_2}(p_2-t\delta)$
Manufacturer recycling (centralized decision-making)	$\frac{p_2[Qk_1+\beta Qp_2-\beta k_2]}{(1-\beta)k_2}$			$\frac{Q^2(k_1+\beta p_2)^2}{2k_2(1-\beta)}$	$\frac{Qp_2(k_1+\beta p_2)(Qk_1+Q\beta p_2-\beta k_2)}{(1-\beta)k_2^2}+\frac{(k_2-Qk_1-Q\beta p_2)}{k_2}(p_2-t\delta)$
Third-party recycling (decentralized decision-making)	$\frac{p_2[Qk_3-\beta(k_4-Qp_2)]}{(1-\beta)k_4}$	$Q^2\Delta\varepsilon\frac{(k_3+p_2\beta)}{k_4}$	$\frac{Q^2(k_3+\beta p_2)^2}{2k_4(1-\beta)}$	$\frac{Q^2(k_3+\beta p_2)(k_1+(1-\beta)\Delta\varepsilon+\beta p_2)}{2k_4(1-\beta)}$	$\frac{Qp_2(k_3+\beta p_2)(Qk_3+Q\beta p_2-\beta k_4)}{(1-\beta)k_4^2}+\frac{(k_4-Qk_3-Q\beta p_2)}{k_4}(p_2-t\delta)$
Third-party recycling (centralized decision-making)	$\frac{p_2[Qk_1-\beta k_4+\beta Qp_2]}{(1-\beta)k_4}$			$\frac{Q^2(k_1+\beta p_2)^2}{2k_4(1-\beta)}$	

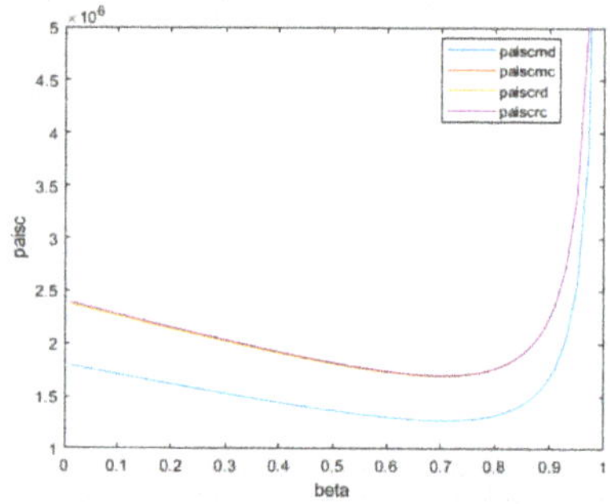

Figure 8. Comparison of profits under two recycling modes.

3.5. *Analysis of the Influencing Factors of Consumer Environmental Protection Responsibility Awareness, β*

Sections 3.1–3.4 mainly analyze the influence of different parameters on the recycling price (p_1) recycling rate (r), and the profits of supply chain (π_{sc}), but at the same time, consumer environmental protection responsibility awareness, β, is also related to ω, p_2, and cost coefficients d_m or d_r. Under the two recycling modes, the value of consumer environmental protection responsibility awareness, β, can be seen in Table 2 based on the maximization of profit in the reverse supply chain.

Table 2. Values of consumer environmental protection responsibility awareness.

Recycling Mode	β
Manufacturer recycling (decentralized/centralized decision-making)	$\frac{\left[-2Q(\omega+\Delta\varepsilon)^2+4Qp_2(\omega+\Delta\varepsilon)-2Qp_2{}^2+2(\omega+\Delta\varepsilon)d_m-d_mp_2\right]}{2\left[-Q(\omega+\Delta\varepsilon)^2+2Qp_2(\omega+\Delta\varepsilon)-Qp_2{}^2+(\omega+\Delta\varepsilon)d_m-d_mp_2\right]}+\frac{\sqrt{4Q^2p_2{}^4+d_m{}^2p_2{}^2+8Q^2p_2{}^3+4Q^2p_2{}^2(\omega+\Delta\varepsilon)^2-4Qp_2{}^2d_m(\omega+\Delta\varepsilon)+4Qd_mp_2{}^3}}{2\left[-Q(\omega+\Delta\varepsilon)^2+2Qp_2(\omega+\Delta\varepsilon)-Qp_2{}^2+(\omega+\Delta\varepsilon)d_m-d_mp_2\right]}$
Third-party recycling (decentralized decision-making)	$\frac{\left[-4Q(\omega+\Delta\varepsilon)\omega-4Q\omega\Delta\varepsilon+4Qp_2\omega+4Q(\omega+\Delta\varepsilon)-4Qp_2{}^2+2d_rp_2+3\omega d_r+4\Delta\varepsilon d_r\right]}{2\left[-2Q(\omega+\Delta\varepsilon)\omega-2Q\omega\Delta\varepsilon+4Qp_2\omega+2Q(\omega+\Delta\varepsilon)+6Qp_2{}^2+2d_rp_2+3\omega d_r+2\Delta\varepsilon d_r\right]}+\frac{\sqrt{16Q^2p_2{}^4+4d_r{}^2p_2{}^2+9\omega^2d_r{}^2+32Q^2p_2{}^3\omega+40Qp_2{}^2\omega d_r+24\omega^2Qp_2d_r+16Qp_2{}^3d_r+12p_2\omega d_r{}^2+32Q^2\omega p_2{}^2(\omega+\Delta\varepsilon)-16Q^2p_2{}^2(\omega+\Delta\varepsilon)-16Qp_2{}^2\Delta\varepsilon d_r}}{2\left[-2Q(\omega+\Delta\varepsilon)\omega-2Q\omega\Delta\varepsilon+4Qp_2\omega+2Q(\omega+\Delta\varepsilon)+6Qp_2{}^2+2d_rp_2+3\omega d_r+2\Delta\varepsilon d_r\right]}$
Third-party recycling (centralized decision-making)	$\frac{\left[-2Q(\omega+\Delta\varepsilon)^2+4Qp_2(\omega+\Delta\varepsilon)-2Qp_2{}^2+2(\omega+\Delta\varepsilon)d_r-d_rp_2\right]}{2\left[-Q(\omega+\Delta\varepsilon)^2+2Qp_2(\omega+\Delta\varepsilon)-Qp_2{}^2+(\omega+\Delta\varepsilon)d_r-d_rp_2\right]}+\frac{\sqrt{4Q^2p_2{}^4+d_r{}^2p_2{}^2+8Q^2p_2{}^3+4Q^2p_2{}^2(\omega+\Delta\varepsilon)^2-4Qp_2{}^2d_r(\omega+\Delta\varepsilon)+4Qd_rp_2{}^3}}{2\left[-Q(\omega+\Delta\varepsilon)^2+2Qp_2(\omega+\Delta\varepsilon)-Qp_2{}^2+(\omega+\Delta\varepsilon)d_r-d_rp_2\right]}$

According to Figure 9a, consumer environmental protection responsibility awareness, β, is positively correlated with recycling income, ω; that is, when the value of ω increases, the β will also increase. When third-party recycling enterprises obtain more profits from recycling waste power batteries, this will inevitably increase the publicity of recycling or feedback on some of the recycling profits to consumers, thereby increasing consumer environmental protection responsibility awareness. It is obvious that when the value of ω is small, the marginal effect brought by ω is better, and the increase in β is large; however, as the value of ω increases, the marginal effect gradually decreases, and the increase in β gradually decreases. It can be seen from Figure 9b that, when the sales volume, Q, of NEV increases, the value of β also increases. When the Q increases from 500 to 2000, β increases from 0.4382 to 0.4427, only by 0.0045, which means that Q has little effect on β. It can also be seen from Figure 9c that when d_m increases, β will decrease, but the overall effect is not large. Under normal circumstances, when d_m is larger, it indicates that the recycler has invested more cost into recycling channels and recycling efforts, which should increase the value of β, but as the cost of recycling investment is too large, the recycling profits fed back to consumers will decline, which will affect the decrease in β. Therefore, under the action of this positive–negative mechanism, the change in d_m has little effect on β. From Figure 9d, it can be seen that β and p_2 are negatively correlated. When the value of p_2 increases, the value of β is declining, which is consistent with the way of thinking of rational individuals;

that is, when higher prices can be obtained under informal channels, there will be more shakeups in environmental protection awareness.

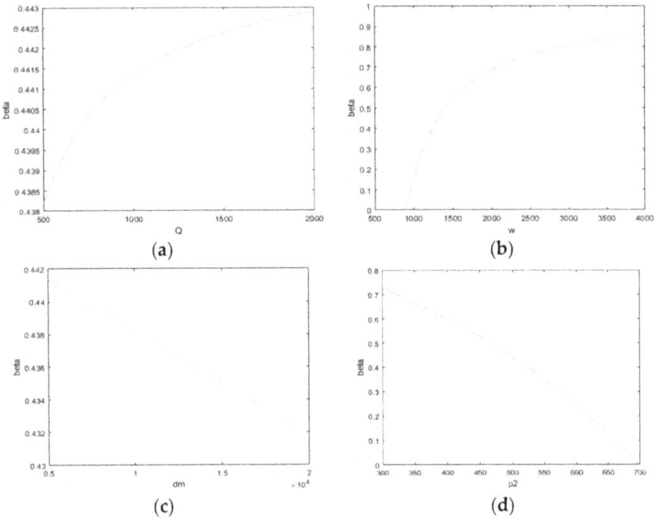

Figure 9. The relationship between β and some parameters. (**a**) The relationship between β and Q. (**b**) The relationship between β and ω. (**c**) The relationship between β and d_m. (**d**) The relationship between β and p_2.

4. Conclusions

The recycling of waste power batteries for NEV has always attracted the attention of the business community and academia. This paper studies the reverse supply chain of waste power battery recycling, which is composed of consumers, NEV manufacturers, and third-party recycling enterprises, and analyzes how different parameters affect the recycling price, recycling rate, consumer expected profit, and supply chain profit under the two recycling modes of NEV manufacturers and third-party recycling enterprises. It comes to the following conclusions: (1) According to the Proposition 1 and Proposition 2, under the two recycling modes, the supply chain profits obtained under decentralized decision-making are lesser than those under centralized decision-making. (2) Under the recycling mode of third-party recycling enterprises, the profits of the supply chain are higher than those under the NEV manufacturer mode. (3) Consumer environmental protection responsibility awareness, β, is negatively correlated with recycling price, p_1, and consumer expected profit, π_k. At the same time, when $p_1 > 0$, β is negatively correlated with the recycling rate, r, and when $p_1 = 0$, β is linearly and positively correlated with recycling rate, r. β is basically positively correlated with supply chain profit, π_{sc}. (4) Recycling income, ω, is positively correlated with recycling price, p_1; recycling rate, r; consumer expected profit, π_k; and supply chain profit, π_{sc}. (5) The recycling price, p_2, through informal recycling channels is negatively correlated with the recycling price, p_1, and the recycling rate, r, while the impact on the expected profit of consumers, π_k, and the supply chain profit, π_{sc}, decreases first and then rises. (6) NEV sale volume, Q, and recycling cost coefficient, d_m, have little effect on recycling price, p_1; recycling rate, r; consumer expected profit, π_k; and supply chain profit, π_{sc}. (7) Recycling income, ω, has a positive effect on β, while an increase in recycling price, p_2, through informal recycling channels inhibits β.

According to the above conclusions, although different parameters have certain influences on recycling prices, recycling rates, consumer profits, and supply chain profits, the main influences are consumer environmental protection responsibility awareness, β; recycling income, ω; and recycling price, p_2, through informal recycling channels. Therefore, the following suggestions may improve the recycling rate of waste power batteries through formal channels:

(1) Continuously improving consumer environmental protection responsibility awareness through positive publicity and regulatory constraints. The government, new energy manufacturers, and 4S stores can publicize the method and role of power battery recycling and disposal, as well as the harm caused by improper disposal to the environment through advertising and other means, and they can encourage consumers to recycle the waste power batteries of NEVs in a formal way. Government departments can strengthen the supervision and management of consumers according to the gradually improved national monitoring system of NEVs and the comprehensive management platform for the recycling and traceability of power batteries. If consumers are found to dispose of waste power batteries in violation of regulations, they can be fined or dismissed from public office, etc.

(2) By strengthening investment in scientific research, the recycling and processing income of waste power batteries can be continuously improved. The government can provide subsidies originally given to NEV sale enterprises to the waste power battery recycling industry, which can help recycling enterprises increase investment in scientific research. Meanwhile, third-party recycling enterprises can cooperate with universities and research institutes to carry out technical research on effectively designing a closed-loop supply chain for the gradient utilization of NEV power battery recycling, strengthening the selective separation and purification technology of various metals, etc., thus improving the economic value of waste power battery recycling and enhancing consumer environmental protection responsibility awareness and the recycling rate.

(3) The government should strictly prohibit waste lithium-ion batteries from entering the informal market and crack down on informal workshops so as to encourage the healthy development of the waste power battery recycling industry to gradually get on the right track and enter into a new era.

In addition, there are some shortcomings in this paper. This paper concludes that the unified recycling and processing of waste power batteries by third-party recycling enterprises can form economies of scale and improve recycling income. However, third-party recycling companies can recycle waste power batteries from different brands of NEV, so it is necessary to adopt a responsibility-sharing contract to better divide the responsibilities between different entities and establish a closed-loop supply chain coordination mechanism for NEV waste power batteries to promote the effective recycling of waste power batteries, which is also one of the directions of follow-up research.

Author Contributions: Model Construction and Solution, J.F.; Model Analysis, J.F. and H.T.; Literature Review, H.T.; Software Programming, Y.W. All authors have read and agreed to the published version of the manuscript.

Funding: This research received no external funding.

Conflicts of Interest: The authors declare no conflict of interest.

References

1. Li, Z. Why is 90% of the battery missing? How to break the cascade utilization of battery recycling? *Resour. Recycl.* **2020**, *8*, 33–36.
2. China Merchants Securities. In-Depth Report on Power Battery Recycling and Gradient Utilization Industry. Available online: http://chuneng.bjx.com.cn/news/20190801/996967.shtml (accessed on 12 March 2020).
3. Lyu, X.; Xu, Y.; Sun, D. An Evolutionary Game Research on Cooperation Mode of the NEV Power Battery Recycling and Gradient Utilization Alliance in the Context of China's NEV Power Battery Retired Tide. *Sustainability* **2021**, *13*, 4165. [CrossRef]
4. Ma, X.M.; Ma, Y.; Zhou, J.P. The Recycling of Spent Power Battery: Economic Benefits and Policy Suggestions. *IOP Conf. Ser. Earth Environ. Sci.* **2018**, *159*, 012017. [CrossRef]
5. Zhao, S.; Xu, N.; Qiao, Y.; Yang, B. Suggestions on speeding up recycling of power battery for new energy vehicles in China. *Chin. J. Eng. Sci.* **2018**, *20*, 144. [CrossRef]
6. Power Battery Recycling in Europe and America Has Become a System. When Will China's Recycling System Take Shape? Available online: https://www.chuandong.com/news/news209836.html (accessed on 29 July 2022).
7. General Office of the State Council. Implementation Plan of Extended Producer Responsibility System (EB/OL). 25 December 2016. Available online: http://www.gov.cn/zhengce/content/2017-01/03/content_5156043.htm (accessed on 6 April 2022).

8. Yang, Y.K.; Zhang, B.; Liang, Y.L.; Lu, Q. Analysis on the current situation of waste power battery recycling policy and technology. *Automob. Accessories* **2021**, *1*, 50–51.
9. National Development and Reform Commission. Development Plan of Circular Economy in the 14th Five Year Plan (EB/OL). 1 July 2021. Available online: https://www.ndrc.gov.cn/xxgk/zcfb/ghwb/202107/t20210707_1285527.html?code=&state=123 (accessed on 6 April 2022).
10. Wang, F. "Internet + second-hand", power battery recycling welcome good policy. *21st Century Economic Report*, 23 July 2021.
11. Hao, S.S.; Dong, Q.Y.; Li, J.H. Analysis and tendency on the recycling mode of used EV batteries based on cost accounting. *Environ. Sci. China* **2021**, *41*, 4745–4755.
12. Shen, H.; Liu, J.; Zhao, X.G.; Liu, Q. Research on new energy vehicle battery recycling strategy in the post-subsidy era. *J. Xi'an Technol. Univ.* **2020**, *40*, 455–463.
13. Xie, J.Y.; Le, W.; Guo, B.H. Pareto equilibrium of new energy vehicle power battery recycling based on extended producer responsibility. *Manag. Sci. China* **2022**. [CrossRef]
14. Mu, D.; Yang, J.; Li, X. The influence of enterprise cooperation within closed-loop supply chain on power battery recovery and recycling of new energy vehicles. *Supply Chain Manag.* **2021**, *1*, 54–67.
15. Li, X.; Mu, D. Mu Recycling price decision and coordinated mechanism of electrical vehicle battery closed-loop supply chain. *Soft Sci.* **2018**, *32*, 124–129.
16. Heydari, J.; Govindan, K.; Jafari, A. Reverse and Closed Loop Supply Chain Coordination by Considering Government Role. *Transp. Res. Part D Transp. Environ.* **2017**, *52*, 379–398. [CrossRef]
17. Lu, C.; Zhao, M.Y.; Tao, J.; Liu, C.; Yu, J. Pricing strategy and coordination mechanism of power battery recycling under the dual risks from demand and quality. *Oper. Res. Manag. Sci.* **2020**, *29*, 195–203.
18. Choi, Y.; Rhee, S.W. Current status and perspectives on recycling of end-of-life battery of electric vehicle in Korea (Republic of). *Waste Manag.* **2020**, *106*, 261–270. [CrossRef] [PubMed]
19. Qiu, Z.G.; Zheng, Y.; Xu, Y.Q. Closed-loop supply chain recycling subsidy strategy of new energy vehicle power battery: An analysis based on evolutionary games. *Bus. Res.* **2020**, *8*, 28–36.
20. Natkunarajah, N.; Scharf, M.; Scharf, P. Scenarios for the return of lithium-ion batteries out of electric cars for recycling. *Procedia Cirp* **2015**, *29*, 740–745. [CrossRef]
21. Liu, K.; Wang, C. The impacts of subsidy policies and channel encroachment on the power battery recycling of new energy vehicles. *Int. J. Low-Carbon Technol.* **2021**, *16*, 770–789. [CrossRef]
22. Xiong, Z.K.; Liang, X.P. Study on recycling mode of closed-loop supply chain considering consumer environmental awareness. *Soft Sci.* **2014**, *28*, 61–66.
23. Jia, Y.L.; Li, C.B. *Recycling Modes of Power Batteries of New Energy Vehicles Based on Principal Behavior*; IEEE: Toronto, ON, Canada, 2018.
24. Savaskan, R.C.; Bhattacharya, S.; Van Wassenhove, L.N. Closed-loop supply chain models with product manufacturing. *Manag. Sci.* **2004**, *50*, 239–252. [CrossRef]
25. Savaskan, R.C.; Van Wassenhove, L.N. Reverse channel design: The case of competing retailers. *Manag. Sci.* **2006**, *52*, 1–14. [CrossRef]
26. Sun, J.Y.; Teng, C.X.; Chen, Z.B. Channel selection model of remanufacturing closed-loop supply chain based on buy-back price and sale quantity. *Syst. Eng. Theory Pract.* **2013**, *33*, 3079–3086.
27. Hong, X.; Xu, L.; Du, P.; Wang, W. Joint advertising, pricing and collection decisions in a closed-loop supply chain. *Int. J. Prod. Econ.* **2015**, *167*, 12–22. [CrossRef]
28. Li, X.J.; Ai, X.Z.; Tang, X.W. Research on recycling channels of remanufactured products under competitive supply chain. *J. Manag. Eng.* **2016**, *30*, 90–98.
29. Chen, J.; Tian, D.G. Selection of the Recycling Mode Based on Closed-loop Supply Chain Model. *Chin. J. Manag. Sci.* **2017**, *25*, 88–97.
30. Zhou, X.; Xiong, H.; Chen, X. Reverse channel selection in closed-loop supply chain based on quality of recycled products. *Control Decis.* **2017**, *32*, 193–202.
31. Zhang, C.; Lyu, R.; Li, Z.; MacMillen, S.J. Who should lead raw materials collection considering regulatory pressure and technological innovation? *J. Clean. Prod.* **2021**, *298*, 126762. [CrossRef]
32. Chen, J.; Mei, J.; Cao, J. Decision making of hybrid recycling channels selection for closed-loop supply chain with dominant retailer. *Comput. Integr. Manuf. Syst.* **2021**, *27*, 954–964.
33. Gong, B.; Gao, Y.; Liu, Z.; Cheng, Y.; Zheng, X. Selection of recycling channels in thepower battery closed-loop supply chain under government fund policy. *Comput. Integr. Manuf. Syst.* **2021**. Available online: http://kns.cnki.net.ez.zust.edu.cn/kcms/detail/11.5946.tp.20210911.1259.004.html (accessed on 29 July 2022).
34. Ding, X.; Ma, Y. Multiple attribute decision making problem in power battery recycling mode selection. *Tongji Daxue Xuebao/J. Tongji Univ.* **2018**, *46*, 1312–1318.
35. Yao, Y.; Jiang, Q. Recovery mode analysis of vehicle spent power battery. *Trends Overv.* **2019**, *12*, 91–94.
36. Ma, L.; Liu, Y.J.; Zhu, H. Design of dual-channel recycling contract for new energy vehicle batteries from the perspective of closed-loop supply chain. *Sci. Technol. Manag. Res.* **2021**, *40*, 184–192.

sustainability

MDPI

Article

Treatment of Sugarcane Vinasse Using Heterogeneous Photocatalysis with Zinc Oxide Nanoparticles

Jacqueline Roberta Tamashiro [1], Iara Souza Lima [2], Fábio Friol Guedes de Paiva [1], Lucas Henrique Pereira Silva [1], Daniela Vanessa Moris de Oliveira [1], Oswaldo Baffa [2] and Angela Kinoshita [1,*]

1 Pró-Rectory of Research and Graduate Studies, Graduate Program in Environment and Regional Development, University of Western São Paulo—Unoeste, Raposo Tavares km 572, Presidente Prudente 19067-175, Brazil

2 Physics Department, Universidade de São Paulo—FFCLRP-USP, Ribeirão Preto 13900-000, Brazil

* Correspondence: angela@unoeste.br

Abstract: Vinasse is the main by-product of the ethanol industry; for each liter of ethanol, 13 to 18 L of vinasse is generated. Vinasse is composed of 93% water and 7% organic and inorganic solids and has an acidic pH and a high concentration of macro- and micronutrients used by plants, which is the reason for its widespread application in soil fertigation. However, over time, excessive direct discharge of vinasse into the soil causes damage, such as salinization and groundwater contamination. In this study, we used heterogeneous photocatalysis with zinc oxide nanoparticles (ZnO-NPs) to reduce chemical oxygen demand (COD) and biochemical oxygen demand (BOD) and as an antimicrobial treatment. ZnO-NPs were synthesized by the precipitation of zinc sulfate heptahydrate and sodium hydroxide, resulting in nanoparticles with a size of 21.6 ± 0.3 nm and an energy bandgap of 2.6 eV. Microscopic examinations revealed that *Saccharomyces cerevisiae* microorganisms are present in vinasse and that the minimum inhibitory concentration for the ZnO-NPs is 1.56 g/L. Photocatalysis with 40 mg/L of ZnO-NPs for 4 h of exposure to sunlight resulted in COD and BOD reduction efficacies of 17.1% and 71.7%, respectively. This study demonstrates the viability of using ZnO-NPs in vinasse treatment, contributing to sustainable applications and reducing the environmental impacts of fertigation.

Keywords: advanced oxidation process; vinasse management; ZnO; ZnO-NPs; effluent quality

Citation: Tamashiro, J.R.; Lima, I.S.; Paiva, F.F.G.d.; Silva, L.H.P.; Oliveira, D.V.M.d.; Baffa, O.; Kinoshita, A. Treatment of Sugarcane Vinasse Using Heterogeneous Photocatalysis with Zinc Oxide Nanoparticles. *Sustainability* **2022**, *14*, 16052. https://doi.org/10.3390/su142316052

Academic Editors: Sunil Kumar, Pooja Sharma and Deblina Dutta

Received: 31 October 2022
Accepted: 29 November 2022
Published: 1 December 2022

1. Introduction

Vinasse is an effluent produced in agro-industrial activities and is generated in the sugarcane ethanol production process. Sugarcane vinasse is composed of micronutrients used by plants, namely, potassium (2.53–2100 mg·L^{-1}), calcium (354.8 mg·L^{-1}), and magnesium (169.9 mg·L^{-1}) [1], as well as dissolved organic carbon (4000–14,000 mg·L^{-1}) [2,3]. Vinasse has a low pH (3–4.5) [4], a high chemical oxygen demand (COD) (>150,000 mg O_2/L) [5], and a 30–70% biochemical oxygen demand (BOD) [6]. In ethanol production, for each liter of ethanol produced, 13–18 L of vinasse is generated, amounting to a volume of $2.38 \times 10^7/m^3$ per year [7–9]. Vinasse is used in fertigation; however, indiscriminate disposal into the soil and continued exposure to vinasse have environmental impacts, such as saturation with potassium, sulfates, and metal ions; salinization; nutrient leaching; and the permanent acidification of soils and water resources [8,10–12]. Studies show that residues and by-products from sugar and ethanol production potentially impact water resources, soils, and the atmosphere [4,13,14]. The degree of impact depends on the concentrations released into the environment, the duration of vinasse management, and the resilience of natural systems. Therefore, technologies for its treatment are needed to prevent side-effects of vinasse use. Conventional treatment methods have already been applied in vinasse management, such as filtration, adsorption, biological oxidation, coagulation–flocculation, and pH adjustment [15,16]. Effluent treatment is complex, in general, and, except for anaerobic digestion, which reaches COD reduction rates higher than 80% [17], greater efficacy can be achieved by combining processes [6], such as advanced oxidative processes (AOPs) mediated by semiconductors, ozonation, Fenton and photo-Fenton reactions, electrochemical remediation, and other processes based on ultraviolet (UV) light irradiation [16,18–20]. However, combined processes are more complex and have higher costs.

In photocatalytic reactions, the electron in the semiconductor absorbs a photon with greater energy than the bandgap. The electron is excited from the valence band to the conduction band, producing an electron pair, creating a hole (h+) in the valence band, where oxidation and/or reduction of the adsorption substrates begins. In aqueous solutions, the hydroxyl ion (OH·) reacts with the catalyst surface and generates an oxidizing agent for effluent treatment [21].

The main semiconductors used in heterogeneous photocatalysis are catalysts in the form of metal oxides, such as zinc oxide (ZnO), titanium dioxide (TiO_2), tin dioxide (SnO_2), sulfates, selenium, and telluride minerals, such as cadmium sulfide (CdS), zinc sulfide (ZnS), and cadmium selenide (CdSe) [22–24]. The photocatalytic properties of zinc oxide nanoparticles (ZnO-NPs) have been extensively studied due to their chemical stability, high thermal conductivity, energy bandgap compatible with UV-Vis, and low production costs compared to other metallic oxides [25,26]. In recent years, heterogeneous photocatalysis using ZnO-NPs as semiconductors have been used in the decontamination of food waste, as well as pharmaceutical, textile industry, and domestic effluents. Kee et al. [16] studied the efficacy of COD reduction and discoloration of anaerobically digested vinasse with photocatalytic degradation under adjusted conditions. Samples were processed using a 0.45 μm membrane filter and pH levels modified by sodium hydroxide (NaOH) and sulfuric acid (H_2SO_4) to achieve photocatalytic degradation in an alkaline condition (pH 10). When the catalyst dosage of ZnO-NPs increased from 0.5 to 2.0 g/L, the efficacy of decolorization and COD reduction also increased in the solution by 99.29% and 83.40%, respectively.

This work provides an approach for sugarcane vinasse treatment without the need for complex additional infrastructures and energy expenditures. We used heterogeneous photocatalysis mediated by zinc oxide nanoparticles with sunlight irradiation. The work was divided into three parts: (i) the synthesis and characterization of zinc oxide nanoparticles (ZnO-NPs); (ii) sugarcane vinasse treatment by heterogeneous photocatalysis, comparing different energy sources; and (iii) microbiological studies of vinasse and antimicrobial effects of ZnO-NPs.

2. Materials and Methods

2.1. ZnO-NP Synthesis and Characterization

The synthesis of zinc oxide nanoparticles (ZnO-NPs) was performed by precipitation. Solutions of 0.1 M zinc sulfate heptahydrate ($ZnSO_4.7H_2O$) and 0.4 M sodium hydroxide (NaOH) were used in the proportions of 1:4 M. The $ZnSO_4.7H_2O$ solution was added (at a rate of 1 mL/min) to the NaOH solution, which was kept under constant stirring (800 rpm) for 30 min at 21 °C. The white product obtained was washed with deionized water, filtered with 3 μm filter paper under negative pressure, and heated to 60 °C for 1 h until a constant mass was reached. The chemical reaction is described by Equation (1).

$$ZnSO_4 \cdot 7H_2O + 2NaOH \overset{\Delta t}{\rightarrow} ZnO + Na_2SO_4 + 8H_2O \quad (1)$$

ZnO-NPs were characterized using X-ray diffraction (XRD) with a Shimadzu–XRD 6000, λ = 1.54984 Å. The ZnO-NPs samples were spread on a carbon tape for the morphological analyses performed via field emission scanning electron microscopy (FEG-SEM), using JEOL-JSM-7500F, PC-SEM software v.2.1.0.3., and transmission electron microscopy (TEM) with a JEOL-JEM-100 CXII microscope. The samples were dropped on the surface of a copper TEM grid and dried before the analysis. Size distributions from TEM images were acquired with ImageJ software [27]. The UV-Vis absorption spectra for colloidal dispersions were recorded with an Ultrospec 2100 Pro spectrophotometer (Amersham Pharmacia).

2.2. *Analytical Methods and Performance Assessment*

Sugarcane vinasse from *Saccharum* spp. was collected at a sugarcane mill in the municipality of Osvaldo Cruz, São Paulo State, Brazil. Chemical oxygen demand (COD) was determined by the closed reflux colorimetric 5220D method [28] and biochemical oxygen demand (BOD) was determined by the 5-day 5210B and 5210D respirometric methods [29] before and after vinasse treatment with photocatalysis. Moreover, potential of hydrogen (pH) measurement was performed using a Quimis pHmeter-Q400MT at 21 °C.

The percentage of reduction in COD values after treatment was calculated according to Equation (2), where *CODi* stands for the average initial COD and *CODf* represents the average of final values. All experiments were performed in triplicate.

$$\% \text{ COD reduction } \frac{(CODi - CODf)}{CODi} \times 100 \tag{2}$$

The percentage of BOD reduction was calculated using Equation (3), where *BODi* is the initial biochemical oxygen demand and *BODf* is the final biochemical oxygen demand measured at the end of 5 days [11,29].

$$\% \text{ BOD reduction } \frac{(BODi - BODf)}{BODi} \times 100 \tag{3}$$

One-way ANOVA (analysis of variance) and the Tukey test were used to determine the statistical significance of differences between COD values after 4 h of treatment. Differences were considered statistically significant when $p < 0.05$.

2.3. *Photocatalysis of Vinasse Using ZnO-NPs*

The heterogeneous photocatalysis experiments on vinasse were performed with two concentrations of ZnO-NPs (40 mg/L and 1560 mg/L) and using three radiation sources: UV-C (Osram Puritec germicidal, 18 W), white halogen light (Osram Haloline, 80 W), and sunlight. Solar exposure data at the moment of the experiment were collected in situ from the Unoeste Meteorological Station [30]. The UV-C lamp power was similar to that in other reports in the literature [31].

Table 1 presents a description of each treatment. All processes were monitored every 1 h until the completion of a 4 h period.

Table 1. Description of vinasse treatments through photocatalysis.

Sample	ZnO-NPs (mg/L)	Radiation Source
Control	0	n/a
T1	0	Sunlight
T2	40	Sunlight
T3	1560	Sunlight
T4	0	UV-C
T5	40	UV-C
T6	1560	UV-C
T7	0	White halogen light
T8	40	White halogen light
T9	1560	White halogen light

Note: n/a: not applied.

The photocatalysis experiment under UV-C and white halogen light was performed with a volume of 250 mL of vinasse and ZnO-NPs, with an exposure surface area of 113.04 cm^2, according to Table 1, using a photocatalytic reactor (Figure 1a). The solution was kept under constant agitation (300 rpm) and a constant temperature during the experiment.

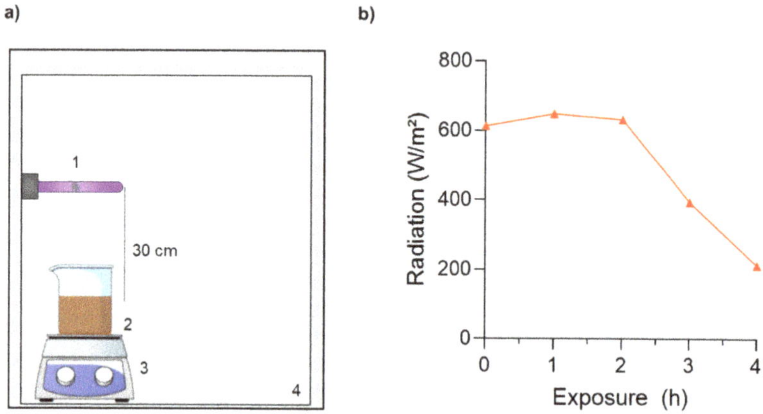

Figure 1. (**a**) Schematic representation of the photocatalytic reactor (54 cm wide × 80 cm high and 47 cm deep). 1: UV-C or white halogen lamp; 2: Sample; 3: Magnetic stirrer; 4: Wall isolation. (**b**) Sunlight exposure data (W/m^2) at the time and place of the experiment.

The photocatalysis under sunlight exposition was made using 250 mL vinasse and ZnO-NPs, with the same exposure surface area (113.04 cm^2) (Table 1). The samples were exposed directly to sunlight for 4 h, between 11h00 and 15h00 [30], at the following coordinates: latitude: −22.116009, S22°6′57.63168″; longitude: −51.450153, W51°27′0.55116″ (Figure 1b). In this experiment, the samples (vinasse and ZnO-NPs) remained at rest.

For COD monitoring, aliquots of 1 mL were extracted from the total solution every hour until the 4 h period was completed. COD values were measured without sedimented material. For BOD measurement, the entire volume (vinasse and nanoparticles) was homogenized and used in the experiment.

2.4. Susceptibility of Microorganisms to ZnO-NPs

The Gram method for staining was used to investigate which microorganisms were mostly present in the vinasse [32]. The antimicrobial activity of ZnO-NPs against microorganisms in vinasse was quantitatively measured by the minimum inhibitory concentration (MIC) assay using Eucast/BrCast 7.3.2 [33].

For the microdilution of ZnO-NPs, 12 tubes were used containing a negative control (C−) with 50 mg of ZnO-NPs and 1 mL of dimethyl sulfoxide (DMSO) and a positive control (C+) with 2 mL of in natura vinasse. The remaining 10 tubes contained serial microdilutions of ZnO-NPs (Figure 2a). For the serial microdilutions, firstly, a volume of 1 mL of sterile distilled water was placed in each tube. Subsequently, 1 mL of ZnO-NP solution from C− was transferred to the first tube, resulting in a solution with a half concentration of ZnO-NPs. The same procedure was repeated for the other tubes, allowing a serial dilution in the tubes, ranging from 25 mg/mL to 0.049 mg/mL. After this process, 1 mL of vinasse was added to each tube, totaling 2 mL of solution (Figure 2a). Figure 2b shows the aliquots of 100 μL transferred to microplates with 96 wells. The plates with the microdilutions were incubated in an oven for 24 ± 2 h at 35 ± 2 °C [33].

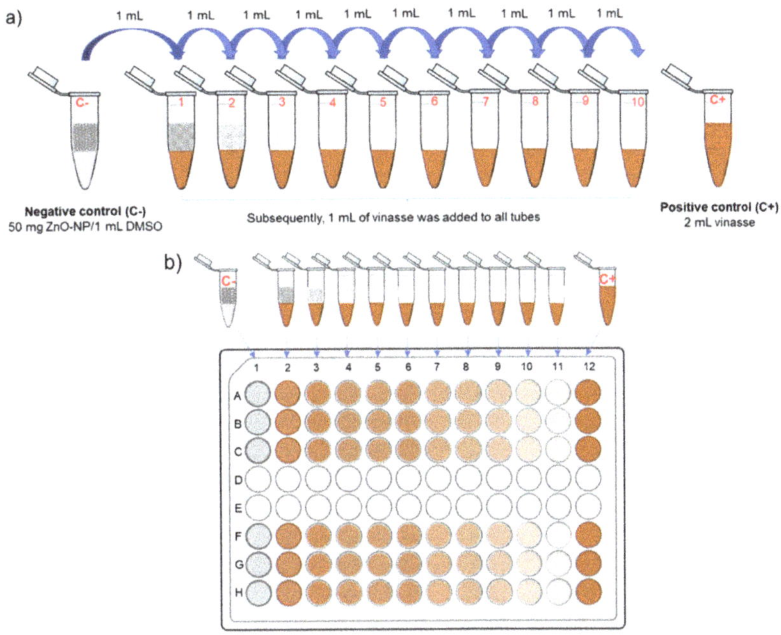

Figure 2. Serial dilutions of ZnO-NPs: (**a**) microdilution from 50 mg ZnO-NPs/mL DMSO with further addition of vinasse; (**b**) 96-well plate to determine minimum inhibitory concentrations (MICs) of ZnO-NPs. C+: positive control containing vinasse in natura, corresponding to growth of viable cells; C−: negative control with ZnO-NPs and DMSO corresponding to absence of viable cells.

For the microbial growth analysis, 20 µL of 0.01% resazurin solution and 20 µL of sterile distilled water were added to each well. The plates were then incubated again in an oven for 1 h at 35 °C. With this technique, the blue color represents the absence of viable cells, while the pink/red color represents their presence. To aid in the determination, since the vinasse is colored, a plugin of ImageJ software was used [27] which indicates the RGB coordinates of each color. The reddish coloration was found with this procedure, and then the breakpoint was determined as the previous concentration and confirmed with CFU counting. The tests were performed in sextuplicate.

The viability of microorganisms in vinasse after the treatments was tested by the drop plate method and CFU counting [33]. The microdilution samples were removed from the oven after incubation (24 ± 2 h at 35 ± 2 °C) and shaken in a Vortex mixer, variable speed, at 2000 rpm for 15 s for seeding in agar Sabouraud plates. Each plate was divided into five sections, and three aliquots of 30 µL of the corresponding microtube were placed in each section. After preservation for 48 h in an oven at 35 °C, the CFUs were quantified with ImageJ software [27], and the non-parametric Kruskal–Wallis and Student–Newman–Keuls tests were applied. The inhibition efficacy in % (η) is defined in Equation (4) [34]:

$$\eta = \frac{N1 - N2}{N1} \times 100 \tag{4}$$

where η corresponds to the inhibition efficacy (%), $N1$ is the CFU number of C+, and $N2$ is the number of colonies formed after ZnO-NP treatment.

3. Results and Discussion

3.1. Synthesis and Characterization of ZnO-NPs

Figure 3a,b show FEG-SEM images of ZnO-NPs synthesized by precipitation. Figure 3c shows a TEM image of the nanoparticles, while Figure 3d presents a histogram of their size distribution.

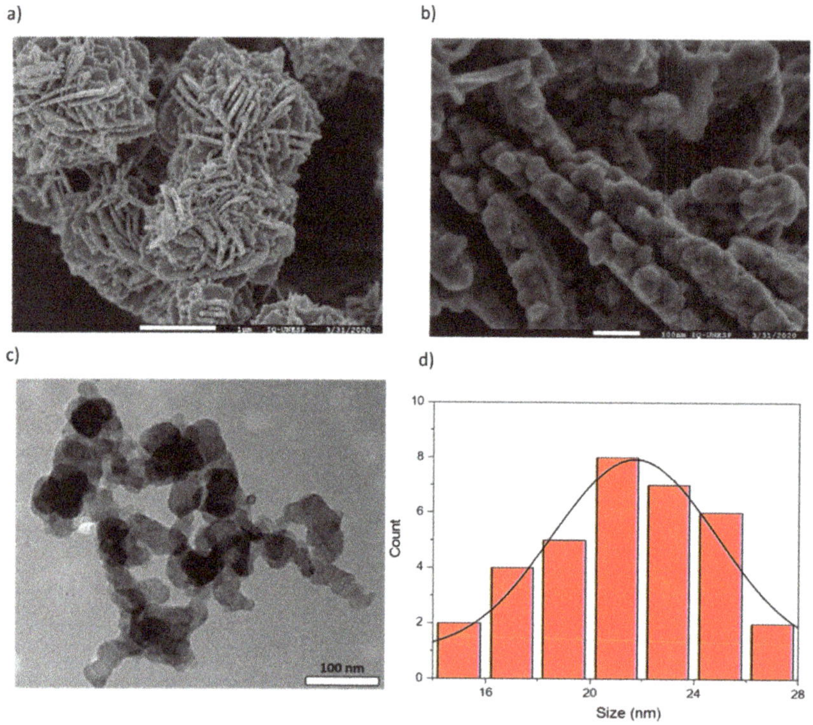

Figure 3. FEG-SEM images of the synthesized ZnO-NPs at magnifications of (**a**) 25 and (**b**) 150×. (**c**) TEM image of ZnO-NPs. (**d**) Histogram of size distribution of nanoparticles (from TEM) obtained by Gaussian fitting. Average size: 21.6 ± 0.3 nm.

Figure 4 shows a diffractogram (XRD) of the ZnO-NPs, with the XRD pattern structure with a sharp and well-defined diffraction peak according with the reference catalog (JCPDS n. 36-1451), suggesting the formation of nanoparticles with highly crystalline structures. The average crystalline diameter of the ZnO particles was obtained by the Scherrer equation, Equation (5), for the diffraction peak at (101).

$$d_{XRD} = \frac{0.9\lambda}{\beta Cos\theta} \quad (5)$$

where λ is the wavelength (nm) of the X-ray, θ is the angle of the Bragg diffraction at the (101) plane, and β is the full width at half-maximum (FWHM) of the diffraction peak at the (101) plane [35].

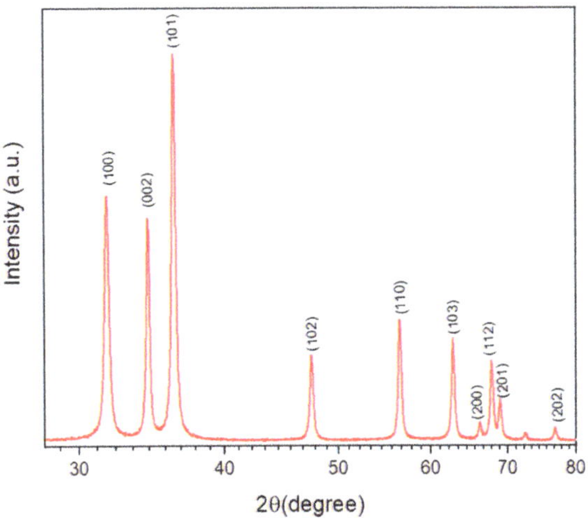

Figure 4. ZnO-NP diffractogram. The peaks are coincident with the crystal structure attributed to wurtzite (JCPDS n° 36-1451) with plane Miller indices: (100) 31.7°, (002) 34.5°, (101) 36.3°, (102) 47.5°, (110) 56.6°, (103) 63.0°, (200) 68.9°, (112) 67.7°, (201) 69.2°, and (202) 76.7°.

The size of the ZnO-NPs was 21.5 ± 0.3 nm, which agrees with the value obtained by the TEM-derived results for histogram size (Figure 3d).

The synthesis of ZnO-NPs by precipitation is a method that requires a source of zinc and a precipitating agent. Different morphologies and sizes can be obtained by changing precursors, reaction temperatures, and calcination processes [34]. Sharma et al. [36] obtained similar results. The authors used the same precursors, zinc sulfate heptahydrate and sodium hydroxide, in aqueous solution, with a stirring time of 15 min and microwave treatment for 1 min, producing spherical ZnO-NPs with an average size of 2 to 28 nm. Similarly, Ni et al. [37] produced ZnO-NPs of nanorod morphology with peaks also indexed as the hexagonal wurtzite structure of ZnO (JCPDS n. 36-1451) [38].

These characterizations are important because the antimicrobial properties of nanoparticles can be altered according to their morphologies and average sizes. This can be attributed to the surface areas of particles, which affect antimicrobial efficacy [34]. The effect of particle size on antimicrobial activity was studied by Padmavathy and Vijayaraghavan [39], who studied *Escherichia coli* in suspensions of ZnO-NPs with sizes ranging from 12 to 45 nm. Via disk-diffusion plating on agar, starting from the radius of the zone of inhibition, the authors observed that a ZnO-NP suspension with 12 nm nanoparticles was more effective than suspensions with larger particles, suggesting that the size of the nanoparticles damaged cell membranes [34].

The bandgap is the energy distance between the valence band and the lowest empty conduction band [40]. Therefore, the bandgap determines the minimum energy required to excite electrons from one band to another, resulting in photoelectrons and holes [41].

Figure 5a shows the UV-visible absorption spectrum of ZnO-NPs dispersed in water, while Figure 5b shows a Tauc plot for bandgap energy.

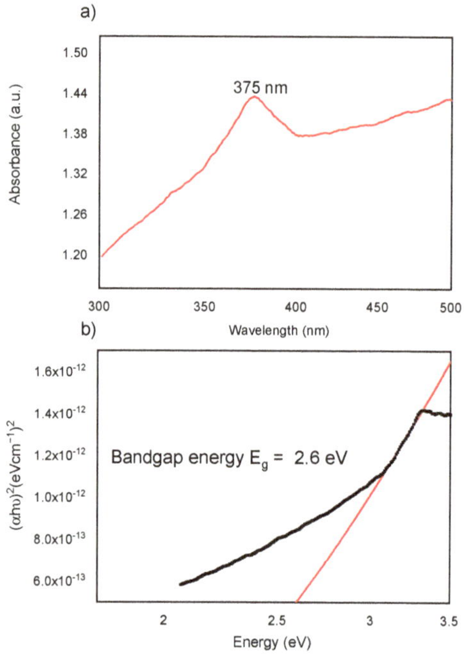

Figure 5. Optical absorption spectrum of colloidal ZnO-NPs: (**a**) centered at 375 nm and (**b**) Tauc plot for the energy gap of the ZnO-NPs. The straight line was extrapolated to determine the bandgap.

The broad absorption band centered at 375 nm is the characteristic peak for the hexagonal wurtzite ZnO structure [42]. The optical bandgap was calculated from the absorption spectrum based on the direct Tauc equation, Equation (6) [43]:

$$(\alpha hv)^2 = K(hv - Eg) \quad (6)$$

$$(2.303 \times A \times 1240/\lambda)^2 = k\,(1240/\lambda - Eg) \quad (7)$$

where A is the absorbance, λ is the wavelength, hv is the photon energy, Eg is the energy bandgap, k is a constant, and α is the absorption constant.

The results for the Tauc plot for the bandgap energy were extrapolated in the straight part of the graph by linear fitting to the energy (eV) axis, and the sample bandgap obtained was 2.6 eV.

The theoretical bandgap value for non-doped ZnO-NPs is around 3.37 eV; however, this value can change depending on the type of morphology analyzed and show values around 3.16 eV, 3.18 eV, and 2.72 eV [44]. Rusdi et al. [45] also found variation in energy bandgap values for different morphologies, for example, 3.35 eV for nanotubes, 3.29 eV for nanoroads, and 3.25 eV for spherical morphologies. This variation was associated with spacings in crystal structures, and smaller spacings resulted in higher bandgap values [45].

Compared with the values reported in the literature, the decrease in the bandgap value found in this study may be associated with high numbers of oxygen vacancies. Wang et al. [46] stated that a decrease may lead to a better efficacy of visible light absorption without affecting photocatalytic capacity. In this sense, the absorption of photons with energies equal to or greater than this bandgap leads to electron excitation from the valence band to the conduction band, thus generating holes in the valence band. The holes generate sites of oxidization or reduction of organic compounds, which are effective for effluent treatment [47].

3.2. Photocatalysis of Vinasse Using ZnO-NPs

Figure 6 shows a decrease in COD values after the treatments with photocatalysis during 4 h of exposure with intervals of 1 h.

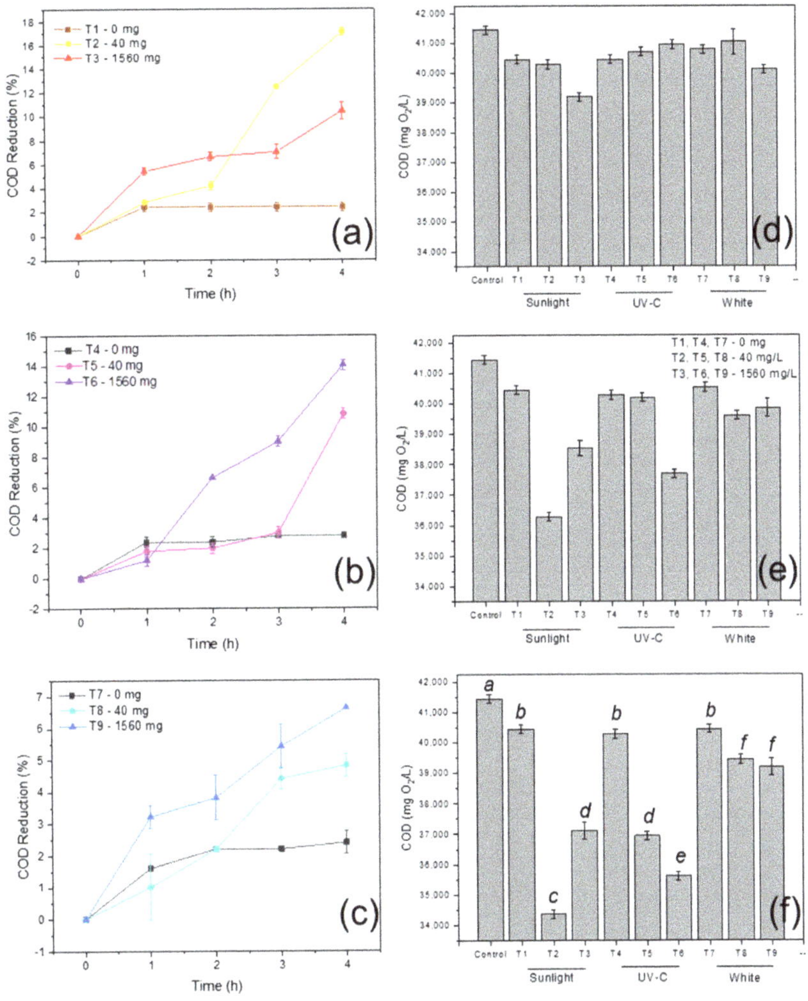

Figure 6. COD reduction (%) according to radiation source: (**a**) sunlight; (**b**) UV-C, and (**c**) white halogen light. Comparison of radiation sources after: (**d**) 1 h, (**e**) 3 h, and (**f**) 4 h. Different letters indicate significant statistical differences ($p < 0.05$, ANOVA, Tukey test).

The vinasse in natura sample without treatment presented a COD value of 41,441.7 ± 144 mg O_2/L, which was used as a parameter to determine the changes resulting from the treatment with ZnO-NPs. The exposure of the vinasse to different light sources without nanoparticles did not influence the COD values ($p > 0.05$).

After the first hour of exposure (Figure 6b), treatment T3 with sunlight 1560 mg/L resulted in the smallest COD value compared to the other light sources. This result was stable for up to 2 h. After 3 h (Figure 6d), treatment T3 resulted in the same COD value, while treatments T2 (40 mg/L ZnO-NPs and sunlight) and T6 (1560 mg/L ZnO and UV) resulted in COD reductions of 12.47% and 9.05%, respectively. After 4 h of exposure (Figure 6f), the comparison of different light sources showed that treatments T2 and T6

(40 and 1560 mg/L ZnO-NPs, sunlight and UV) showed greater efficacy in reducing COD values (17.09% and 14.08%) and that T2 was more efficient than T6 ($p < 0.05$). The samples did not present large variations in pH values at 21 °C, these remaining between 3.71 and 3.96. Sunlight has a higher power compared to artificial light, so it allows more transitions of electrons to the excited state. Such transitions result in recombination with the holes, thus promoting greater numbers of reactive oxygen species (ROS), which are responsible for photocatalysis [12]. As for NPS concentration, 40 mg/L presented a better result than 1560 when sunlight was used, which may be related to the optimal concentration. The application of a higher concentration of NPS may have saturated the sample. In addition, when exposed to the sun, greater numbers of nanoparticles can aggregate, and we did not use agitation in this experiment so as to simulate a real treatment application. When excited state transitions and recombination occur, the energy released by these nanoparticles can be self-absorbed around the NPs themselves (self-absorption) and not deposited in solution to provide reactive oxygen species (ROS). The results also indicate that further studies with varying concentrations are necessary to better elucidate the phenomenon.

Kee et al. [16] observed COD reductions for vinasse treated with 250, 500, 750, and 1000 mg/L ZnO-NPs/vinasse over 10 h of monitoring. The increase in the concentration of ZnO-NPs (1000 mg/L) affected the ·OH radicals on the catalyst surface [16]. The highest COD degradation of 42.85% was achieved with the lowest concentration (250 mg/L).

Similar to our results, and using an individual technique in the treatment, Apollo et al. [48] carried out photocatalysis with ultraviolet (UV) light for the treatment of molasses distillery wastewater, obtaining a COD reduction efficacy of <20%. Robles-González et al. [49] conducted vinasse treatment by ozonation and achieved a COD reduction between 4.5 and 11% (with contact times of up to 1.5 h). David, Arivazhagan, and Ibrahim [50] used photocatalysis with aluminum oxide nanoparticles (Al_2O_3) to treat distillery wastewater and observed that photocatalytic treatment efficacy depended mainly on the number of nanoparticles (0.5 to 2.5 g/L) and had better reduction rates at low pH levels (3) and 25 °C, while agitation had less effect on the treatment (50 to 250 rpm). Other studies on vinasse treatment from tequila production observed that the vinasse of blue *Agave tequilana* had a COD value of 343.30 mg O_2/L. To increase the quality of the treated effluent, different processes were applied, and combined experiments with Fenton reactions showed an efficacy of COD removal from 79 to 89% [5]. Guerreiro et al. [51] reported on the biodegradability efficacy of sugarcane vinasse. Vinasse samples were submitted to thermophilic anaerobic digestion (UASB) in series. For coagulation–flocculation analysis, a jar test apparatus (22–25 °C) was used, and the ferric chloride coagulant was added to vinasse with a pH of ~7. The dissolved iron salts that resulted from the coagulation–flocculation were used as a catalyst in Fenton oxidation. In sequence, BOD, COD, and dissolved iron levels were measured for clarified vinasse. The results showed a reduction of 45.7% for BOD, 69.2% for COD, and 270 mg/L for dissolved iron [51]. Some treatment methods, such as biodigestion and AOPs, require more complex infrastructures to carry out the processes; however, the processes have high efficacies. Studies in the literature have reported the influence of COD–sulfate ratios (12.0, 10.0, and 7.5) on COD removal and methane (CH_4) production from vinasse biodigestion [8]. At a COD–sulfate ratio of 7.5, CH_4 production was 35% lower compared to the 12.0 ratio. The diversion of electrons to sulfidogenesis was negligible at COD–sulfate ratios >25, considering the increase in CH_4 production [8]. The results showed that organic matter degradation was not affected by sulfidogenesis, with COD removal levels higher than 80%, regardless of the initial COD–sulfate ratio [8].

Afterward, BOD analysis was performed to complete the COD results for vinasse in natura (no treatment) and after 4 h of treatment with 40 and 1560 mg/L of ZnO-NPs (Table 2).

Table 2. BOD (mg/L) and reduction efficacy (%).

Sample	Description	BOD (mg/L)	% Reduction
Control	Vinasse in natura	17,666.7 ± 577	-
T2	40 mg/L (sunlight)	5000.0 ± 0	71.7
T6	1560 mg/L (UV-C)	6000.0 ± 0	66.0

After 5 days of analyses, treatment T2 showed the highest efficacy in reducing BOD (71.7%) and COD. Therefore, T2 was the most effective and the least costly treatment with the lowest concentration of nanoparticles (40 mg/L) and sunlight.

3.3. *Susceptibility of Microorganisms to ZnO-NPs*

The microorganisms identified via vinasse Gram staining were related to unicellular *fungi* or yeast of *Saccharomyces cerevisiae*, with an average size of 1 µm and oval and cylindrical morphologies. Yeast strains have phenotypes of high tolerance and resilience and are considered super-resistant microorganisms in the ethanol production process. *Saccharomyces cerevisiae* and other yeast strains, such as CAT-1 and PE-2, have been developed to tolerate different stress situations during ethanol fermentation, such as osmotic stress, oxidative stress, and high temperatures [52]. Possibly, the yeasts found in the vinasse samples analyzed in our work are organisms that resisted these processes and were active in the effluent.

Minimum inhibitory concentration (MIC) analysis provides quantitative results for antimicrobial effectiveness. MIC is the lowest antimicrobial concentration that inhibits visible microorganism growth in vitro [37]. The MICs for the ZnO-NPs against microorganisms in vinasse were determined through MIC experiments. Table 3 shows the ZnO-NP concentrations, and Figure 7 presents a plate photograph after resazurin staining.

Table 3. Minimum inhibitory concentrations (MICs) (ZnO-NPs mg/mL).

Sample	C+	C−	1	2	3	4	5	6	7	8	9	10
ZnO-NP concentration	0	50	25	12.5	6.25	3.125	1.562	0.781	0.390	0.195	0.097	0.049

C+: positive control, vinasse in natura (growth of viable cells); C−: ZnO-NP negative control with DMSO (absence of viable cells).

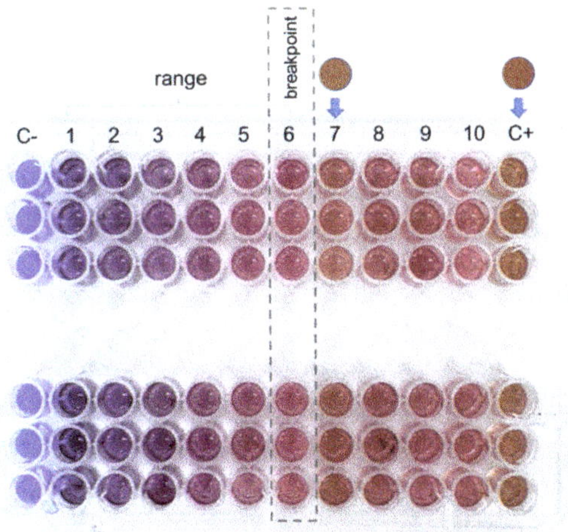

Figure 7. Photograph of 96-microwell plate after resazurin staining. Minimum inhibitory concentration (MIC) of ZnO-NPs is 0.781 mg/mL.

A gradual increase in inhibition was observed with increasing ZNO-NP concentrations (Table 3 and Figure 7), showing that the microorganisms were sensitive to ZnO-NPs between 25 and 0.781 mg/mL, with the breakpoint at 0.781 mg/mL. Low concentrations of ZnO-NPs showed no activity against microorganisms, and *Saccharomyces cerevisiae* strains were resistant between 0.390 and 0.049 mg/mL. MIC analysis allows the classification of microorganisms as sensitive or resistant to an antifungal when ranges and breakpoints are obtained [33]. Table 4 lists the number of CFUs in the vinasse plating experiment with microdilutions of ZnO-NPs.

Table 4. Number of colony-forming units (CFUs) according to the concentration of ZnO-NPs used in the vinasse treatment.

ZnO-NPs (mg/mL)	C+	C−	25	12.5	6.25	3.125	1.562	0.781	0.390	0.195	0.097	0.049
CFUs and SD	267	0	0	0	0	0	0	1 ± 0	3 ± 2	10 ± 3	11 ± 4	12 ± 5
Reduction in CFUs (%)	n/a	n/a	100^{a}	100^{a}	100^{a}	100^{a}	100^{a}	99.63^{b}	98.88^{b}	96.25^{c}	95.88^{c}	95.51^{c}

SD: standard deviation; n/a: not applied. Different letters indicate significant statistical differences ($p < 0.05$).

The non-parametric Kruskal–Wallis and the Student–Newman–Keuls tests showed that the vinasse treatment with ZnO-NPs was effective in eliminating/reducing the microbial load of vinasse up to a concentration of 1.562 mg/mL (Table 4). In the treatment with a concentration breakpoint visually determined, 0.781 mg/mL, viable cells were observed (Table 4). Thus, the MIC value is 1.562 mg/mL, which should be used for the treatment of microorganisms with vinasse in natura.

Other studies analyzed the in vitro toxicity of ZnO-NPs in *Saccharomyces cerevisiae* strains and reported that growth was inhibited by 80% with 250 mg/L [53]. Babele et al. [54] also investigated the toxic effects of ZnO-NPs using 5, 10, 15, and 20 $mg{\cdot}L^{-1}$ in *Saccharomyces cerevisiae*, as well as the underlying mechanisms. Cell walls were damaged, and the accumulation of reactive oxygen species (ROS) led to cell death at 10, 15, and 20 $mg{\cdot}L^{-1}$ of ZnO-NPs. Furthermore, exposure to ZnO-NPs caused cellular toxicity due to lipid disequilibrium and proteostasis. Galván-Marquéz et al. [26] assessed the effects of 1.5 mg/mL of ZnO-NPs in *Saccharomyces cerevisiae* and suggested that vinasse microorganisms showed greater sensitivity due to ZnO-NPs affecting cell wall integrity and/or ROS accumulation.

Other works have addressed promising and innovative techniques in agriculture, such as the application of metal oxide nanoparticles as plant fertilizers. As micronutrients in soils, ZnO-NP ions promote richness and diversity in soil microbial communities, increasing crop growth and yield by stimulating photosynthesis and respiration in plants and suppressing the growth of soil pathogens [55,56]. Therefore, the presence of ZnO-NPs in treated vinasse favors and does not impede its use as a fertilizer, nor is it a barrier to its sustainable reinsertion into the agro-industrial productivity cycle.

4. Conclusions

In this work, treatments with sugarcane vinasse were carried out by heterogeneous photocatalysis with different concentrations of zinc oxide nanoparticles (ZnO-NPs) and radiation sources, including sunlight, UV-C, and white halogen light. Our results showed reductions in COD and BOD and vinasse microorganism populations using a low concentration of nanoparticles (40 mg/L) and sunlight, demonstrating an alternative treatment which does not require complex infrastructures or artificial radiation sources. The addition of ZnO to vinasse added value to this effluent as a fertilizer, enabling its reinsertion into the ethanol production chain. Future studies should consider different concentrations and treatment durations to increase COD reduction efficiency.

5. Patents

Patent resulting from the work reported in the article: Innovation Privilege. Registration number: BR1020210201789; title: "Treatment of sugarcane vinasse with zinc oxide nanoparticles to reduce microbial load, chemical oxygen demand and biochemical oxygen demand"; institution of registration: INPI—Brazilian National Institute of Industrial Property; deposited: 10 July 2021.

Author Contributions: Conceptualization, J.R.T. and A.K.; methodology, J.R.T., I.S.L., F.F.G.d.P., L.H.P.S., D.V.M.d.O., A.K. and O.B.; validation, O.B. and A.K.; formal analysis, O.B. and A.K.; investigation, J.R.T., I.S.L., D.V.M.d.O., O.B. and A.K.; writing—original draft preparation, J.R.T., I.S.L., F.F.G.d.P., L.H.P.S., O.B. and A.K.; writing—review and editing, J.R.T., O.B. and A.K.; supervision, A.K. All authors have read and agreed to the published version of the manuscript.

Funding: This research was funded by Coordination for the Improvement of Higher Education Personnel (CAPES)—Brazil—Finance Code 001, grants 88882.365281/2019-01 and 88881.623268/2021-01; São Paulo Research Foundation (FAPESP) grants 2013/07699-0 and 304107/2019-0; and a National Council for Scientific and Technological Development (CNPq) grant (309186/2020–0).

Institutional Review Board Statement: Not applicable.

Informed Consent Statement: Not applicable.

Data Availability Statement: The data can be made available by the authors upon request.

Acknowledgments: The authors thanks Daniel Angelo Macena (Unoeste) for technical support.

Conflicts of Interest: The authors declare no conflict of interest.

References

1. Iltchenco, J.; Peruzzo, V.; Magrini, F.E.; Marconatto, L.; Torres, A.P.; Beal, L.L.; Paesi, S. Microbiota Profile in Mesophilic Biodigestion of Sugarcane Vinasse in Batch Reactors. *Water Sci. Technol.* **2021**, *84*, 2028–2039. [CrossRef]
2. Fernandes, J.M.C.; Sousa, R.M.O.F.; Fraga, I.; Sampaio, A.; Amaral, C.; Bezerra, R.M.F.; Dias, A.A. Chemosphere Fungal Biodegradation and Multi-Level Toxicity Assessment of Vinasse from Distillation of Winemaking by-Products. *Chemosphere* **2020**, *238*, 124572. [CrossRef]
3. Benke, M.B.; Mermut, A.R.; Shariatmadari, H. Retention of Dissolved Organic Carbon from Vinasse by a Tropical Soil, Kaolinite, and Fe-Oxides. *Geoderma* **1999**, *91*, 47–63. [CrossRef]
4. Carrilho, E.N.V.M.; Labuto, G.; Kamogawa, M.Y. Destination of Vinasse, a Residue from Alcohol Industry: Resource Recovery and Prevention of Pollution. In *Environmental Materials and Waste*; Elsevier Inc.: Amsterdam, The Netherlands, 2016; pp. 21–43. ISBN 978-0-12-803837-6.
5. Castillo-Monroy, J.; Godínez, L.A.; Robles, I.; Estrada-Vargas, A. Study of a Coupled Adsorption/Electro-Oxidation Process as a Tertiary Treatment for Tequila Industry Wastewater. *Environ. Sci. Pollut. Res.* **2021**, *28*, 23699–23706. [CrossRef]
6. Arreola, A.R.; Tizapa, M.S.; Zurita, F.; Morán-Lázaro, J.P.; Valderrama, R.C.; Rodríguez-López, J.L.; Carreon-Alvarez, A. Treatment of Tequila Vinasse and Elimination of Phenol by Coagulation–Flocculation Process Coupled with Heterogeneous Photocatalysis Using Titanium Dioxide Nanoparticles. *Environ. Technol.* **2018**, *41*, 1023–1033. [CrossRef]
7. Kusumaningtyas, R.D.; Hartanto, D.; Rohman, H.A.; Mitamaytawati; Qudus, N.; Daniyanto. Valorization of Sugarcane-Based Bioethanol Industry Waste (Vinasse) to Organic Fertilizer. In *Valorisation of Agro-industrial Residues–Volume II: Non-Biological Approaches*; Springer: Cham, Switzerland, 2020; pp. 203–223.
8. Buller, L.S.; Romero, C.W.D.S.; Lamparelli, R.A.C.; Ferreira, S.F.; Bortoleto, A.P.; Mussatto, S.I.; Forster-Carneiro, T. A Spatially Explicit Assessment of Sugarcane Vinasse as a Sustainable By-Product. *Sci. Total Environ.* **2021**, *765*, 142717. [CrossRef]
9. Hoarau, J.; Caro, Y.; Grondin, I.; Petit, T. Sugarcane Vinasse Processing: Toward a Status Shift from Waste to Valuable Resource. A Review. *J. Water Process Eng.* **2018**, *24*, 11–25. [CrossRef]
10. Coelho, M.P.M.; Correia, J.E.; Vasques, L.I.; Marcato, A.C.D.C.; de Guedes, T.A.; Soto, M.A.; Basso, J.B.; Kiang, C.; Fontanetti, C.S. Toxicity Evaluation of Leached of Sugarcane Vinasse: Histopathology and Immunostaining of Cellular Stress Protein. *Ecotoxicol. Environ. Saf.* **2018**, *165*, 367–375. [CrossRef]
11. Soto, M.F.; Diaz, C.A.; Zapata, A.M.; Higuita, J.C. BOD and COD Removal in Vinasses from Sugarcane Alcoholic Distillation by Chlorella Vulgaris: Environmental Evaluation. *Biochem. Eng. J.* **2021**, *176*, 108191. [CrossRef]
12. Fuess, L.T.; Garcia, M.L. Implications of Stillage Land Disposal: A Critical Review on the Impacts of Fertigation. *J. Environ. Manag.* **2014**, *145*, 210–229. [CrossRef]
13. Fuess, L.T.; Rodrigues, I.J.; Garcia, M.L. Fertirrigation with Sugarcane Vinasse: Foreseeing Potential Impacts on Soil and Water Resources through Vinasse Characterization. *J. Environ. Sci. Health Part A* **2017**, *52*, 1063–1072. [CrossRef] [PubMed]

14. Carpanez, T.G.; Moreira, V.R.; Assis, I.R.; Amaral, M.C.S. Sugarcane Vinasse as Organo-Mineral Fertilizers Feedstock: Opportunities and Environmental Risks. *Sci. Total Environ.* **2022**, *832*, 154998. [CrossRef] [PubMed]
15. Mahmoudabadi, Z.S.; Rashidi, A.; Maklavany, D.M. Optimizing Treatment of Alcohol Vinasse Using a Combination of Advanced Oxidation with Porous α-Fe2O3 Nanoparticles and Coagulation-Flocculation. *Ecotoxicol. Environ. Saf.* **2022**, *234*, 113354. [CrossRef]
16. Kee, W.-C.; Wong, Y.-S.; Ong, S.-A.; Lutpi, N.A.; Sam, S.-T.; Chai, A.; Eng, K.-M. Photocatalytic Degradation of Sugarcane Vinasse Using ZnO Photocatalyst: Operating Parameters, Kinetic Studies, Phytotoxicity Assessments, and Reusability. *Int. J. Environ. Res.* **2022**, *16*, 3. [CrossRef] [PubMed]
17. Kiyuna, L.S.M.; Fuess, L.T.; Zaiat, M. Unraveling the Influence of the COD/Sulfate Ratio on Organic Matter Removal and Methane Production from the Biodigestion of Sugarcane Vinasse. *Bioresour. Technol.* **2017**, *232*, 103–112. [CrossRef]
18. Misra, M.; Akansha, K.; Sachan, A.; Sachan, S.G. Removal of Dyes from Industrial Effluents by Application of Combined Biological and Physicochemical Treatment Approaches. In *Combined Application of Physico-Chemical & Microbiological Processes for Industrial Effluent Treatment Plant*; Springer: Singapore, 2020; pp. 365–407.
19. Kumar, V.; Singh, K.; Shah, M.P. Advanced Oxidation Processes for Complex Wastewater Treatment. In *Advanced Oxidation Processes for Effluent Treatment Plants*; Elsevier: Amsterdam, The Netherlands, 2021; pp. 1–31.
20. Sales, H.B.; Menezes, R.R.; Neves, G.A.; de Souza, J.J.N.; Ferreira, J.M.; Chantelle, L.; Oliveira, A.L.M.d.; de Lira, H.L. Development of Sustainable Heterogeneous Catalysts for the Photocatalytic Treatment of Effluents. *Sustainability* **2020**, *12*, 7393. [CrossRef]
21. Ameta, R.; Solanki, M.S.; Benjamin, S.; Ameta, S.C. Photocatalysis. In *Advanced Oxidation Processes for Wastewater Treatment: Emerging Green Chemical Technology*; Elsevier: Amsterdam, The Netherlands, 2018; pp. 135–175. ISBN 9780128105252.
22. Kwarciak-Kozlowska, A. *Removal of Pharmaceuticals and Personal Care Products by Ozonation, Advance Oxidation Processes, and Membrane Separation*; Elsevier Inc.: Amsterdam, The Netherlands, 2019; ISBN 9780128161890.
23. Yarahmadi, M.; Maleki-Ghaleh, H.; Mehr, M.E.; Dargahi, Z.; Rasouli, F.; Siadati, M.H. Synthesis and Characterization of Sr-Doped ZnO Nanoparticles for Photocatalytic Applications. *J. Alloys Compd.* **2021**, *853*, 157000. [CrossRef]
24. Venieri, D.; Mantzavinos, D.; Binas, V. Solar Photocatalysis for Emerging Micro-Pollutants Abatement and Water Disinfection: A Mini-Review. *Sustainability* **2020**, *12*, 10047. [CrossRef]
25. Bica, B.O.; de Melo, J.V.S. Concrete Blocks Nano-Modified with Zinc Oxide (ZnO) for Photocatalytic Paving: Performance Comparison with Titanium Dioxide (TiO_2). *Constr. Build. Mater.* **2020**, *252*, 119120. [CrossRef]
26. Márquez, I.G.; Ghiyasvand, M.; Massarsky, A.; Babu, M.; Samanfar, B.; Omidi, K.; Moon, T.W.; Smith, M.L.; Golshani, A. Zinc Oxide and Silver Nanoparticles Toxicity in the Baker's Yeast, Saccharomyces Cerevisiae. *PLoS ONE* **2018**, *13*, e0193111. [CrossRef]
27. Rasband, W.S. *Image J: Image Processing and Analysis in Java*; Astrophysics Source Code Library: Online, 1997.
28. APHA; AWWA; WEF. 5220-D Chemical Oxygen Demand (COD). In *Standard Methods for the Examination of Water and Wastewater*; Baid, R.B., Eaton, A.D., Rice, E.W., Eds.; American Public Health Association: Washington, DC, USA, 2018; pp. 5–14. ISBN 9780875530130.
29. APHA; AWWA; WEF. 5210 Biochemical Oxygen Demand (BOD). In *Standard Methods for the Examination of Water and Wastewater*; Baid, R.B., Eaton, A.D., Rice, E.W., Eds.; American Public Health Association: Washington, DC, USA, 2018; ISBN 9780875530130.
30. Unoeste. *Dados 2022*; Presidente Prudente: São Paulo, Brazil, 2022.
31. Otieno, B.O.; Apollo, S.O.; Naidoo, B.E.; Ochieng, A. Photodecolorisation of Melanoidins in Vinasse with Illuminated TiO_2-ZnO/Activated Carbon Composite. *J. Environ. Sci. Health Part A Toxic/Hazardous Subst. Environ. Eng.* **2017**, *52*, 616–623. [CrossRef] [PubMed]
32. National Health Surveillance Agency. *ANVISA Microbiologia Clínica Para o Controle de Infecção Relacionada à Assistência à Saúde-Detecção e Identificação Dos Fungos de Importância Médica*; National Health Surveillance Agency: Brasília, Brazil, 2020; Volume 10.
33. Arendrup, M.; Meletiadis, J.; Mouton, J.W.; Lagrou, K.; Hamal, P.; de Almeida Júnior, J.N.J.; Ishida, K. D.E.DEF. 7.3.2—Método Para Determinação de Concentração Inibitória Mínima Em Caldo Dos Agentes Antifúngicos Para Leveduras. *BrCAST Braz. Committe Antimicrob. Susceptibility Test* **2020**, 1–19. Available online: https://brcast.org.br/wp-content/uploads/2022/08/BrCAST-EUCAST-E-Def-732-CIM-levedura-20202.pdf (accessed on 5 May 2022).
34. Dimapilis, E.A.S.; Hsu, C.S.; Mendoza, R.M.O.; Lu, M.C. Zinc Oxide Nanoparticles for Water Disinfection. *Sustain. Environ. Res.* **2018**, *28*, 47–56. [CrossRef]
35. Rashidi, H.; Ahmadpour, A.; Bamoharram, F.F.; Zebarjad, S.M.; Heravi, M.M.; Tayari, F. Controllable One-Step Synthesis of ZnO Nanostructures Using Molybdophosphoric Acid. *Chem. Pap.* **2014**, *68*, 516–524. [CrossRef]
36. Sharma, D.; Rajput, J.; Kaith, B.S.; Kaur, M.; Sharma, S. Synthesis of ZnO Nanoparticles and Study of Their Antibacterial and Antifungal Properties. *Thin Solid Films* **2010**, *519*, 1224–1229. [CrossRef]
37. Ni, Y.H.; Wei, X.W.; Hong, J.M.; Ye, Y. Hydrothermal Preparation and Optical Properties of ZnO Nanorods. *Mater. Sci. Eng. B Solid-State Mater. Adv. Technol.* **2005**, *121*, 42–47. [CrossRef]
38. Chateigner, D.; Chen, X.; Ciriotti, M.; Downs, R.T.; Gražulis, S.; Kaminsky, W. Crystallography Open Database. Available online: http://crystallography.net/cod/index.php (accessed on 10 April 2022).
39. Padmavathy, N.; Vijayaraghavan, R. Enhanced Bioactivity of ZnO Nanoparticles—An Antimicrobial Study. *Sci. Technol. Adv. Mater.* **2008**, *9*, 035004. [CrossRef]

40. Ekennia, A.C.; Uduagwu, D.N.; Nwaji, N.N.; Oje, O.O.; Emma-Uba, C.O.; Mgbii, S.I.; Olowo, O.J.; Nwanji, O.L. Green Synthesis of Biogenic Zinc Oxide Nanoflower as Dual Agent for Photodegradation of an Organic Dye and Tyrosinase Inhibitor. *J. Inorg. Organomet. Polym. Mater.* **2021**, *31*, 886–897. [CrossRef]
41. Sze, S.M.; Kwok, K.N. *Physics of Semiconductor Devices*; John Wiley & Sons, Inc.: Hoboken, NJ, USA, 2006; ISBN 978-0-470-06832-8.
42. Wooten, A.J.; Werder, D.J.; Williams, D.J.; Casson, J.L.; Hollingsworth, J.A. Solution-Liquid-Solid Growth of Ternary Cu-in-Se Semiconductor Nanowires from Multiple- and Single-Source Precursors. *J. Am. Chem. Soc.* **2009**, *131*, 161–177. [CrossRef]
43. Obasuyi, A.R.; Alabi, A.B.; Kajewole, E.D.; Robert, O.; Abejide, F.; Adeleke, T.J.; Akomolafe, T. Capacitive and Bioenzymevoltaic Characteristics of Photosynthetic Chromolena Odorata—Nanostructured Zinc Oxide System. *Sri. Lankan. J. Phys.* **2017**, *18*, 17. [CrossRef]
44. Montero-Muñoz, M.; Ramos-Ibarra, J.E.; Rodríguez-Páez, J.E.; Teodoro, M.D.; Marques, G.E.; Sanabria, A.R.; Cajas, P.C.; Páez, C.A.; Heinrichs, B.; Coaquira, J.A.H. Role of Defects on the Enhancement of the Photocatalytic Response of ZnO Nanostructures. *Appl. Surf. Sci.* **2018**, *448*, 646–654. [CrossRef]
45. Rusdi, R.; Rahman, A.A.; Mohamed, N.S.; Kamarudin, N.; Kamarulzaman, N. Preparation and Band Gap Energies of ZnO Nanotubes, Nanorods and Spherical Nanostructures. *Powder Technol.* **2011**, *210*, 18–22. [CrossRef]
46. Wang, J.; Wang, Z.; Huang, B.; Ma, Y.; Liu, Y.; Qin, X.; Zhang, X.; Dai, Y. Oxygen Vacancy Induced Band-Gap Narrowing and Enhanced Visible Light Photocatalytic Activity of ZnO. *ACS Appl. Mater. Interfaces* **2012**, *4*, 4024–4030. [CrossRef] [PubMed]
47. Chavillon, B.; Cario, L.; Renaud, A.; Tessier, F.; Cheviré, F.; Boujtita, M.; Pellegrin, Y.; Blart, E.; Smeigh, A.; Hammarström, L.; et al. P-Type Nitrogen-Doped ZnO Nanoparticles Stable under Ambient Conditions. *J. Am. Chem. Soc.* **2012**, *134*, 464–470. [CrossRef] [PubMed]
48. Apollo, S.; Onyango, M.S.; Ochieng, A. An Integrated Anaerobic Digestion and UV Photocatalytic Treatment of Distillery Wastewater. *J. Hazard. Mater.* **2013**, *261*, 435–442. [CrossRef]
49. Robles-González, V.; Galíndez-Mayer, J.; Rinderknecht-Seijas, N.P.H. Treatment of Mezcal Vinasses: A Review. *J. Biotechnol.* **2012**, *157*, 521–546. [CrossRef]
50. David, C.; Arivazhagan, M.; Ibrahim, M. Spent Wash Decolourization Using Nano-Al2O3/Kaolin Photocatalyst: Taguchi and ANN Approach. *J. Saudi Chem. Soc.* **2015**, *19*, 537–548. [CrossRef]
51. Guerreiro, L.F.; Rodrigues, C.S.D.; Duda, R.M.; de Oliveira, R.A.; Boaventura, R.A.R.; Madeira, L.M. Treatment of Sugarcane Vinasse by Combination of Coagulation/Flocculation and Fenton's Oxidation. *J. Environ. Manag.* **2016**, *181*, 237–248. [CrossRef]
52. de Souza, J.P.; do Prado, C.D.; Eleutherio, E.C.A.; Bonatto, D.; Malavazi, I.; da Cunha, A.F. Improvement of Brazilian Bioethanol Production—Challenges and Perspectives on the Identification and Genetic Modification of New Strains of Saccharomyces Cerevisiae Yeasts Isolated during Ethanol Process. *Fungal. Biol.* **2018**, *122*, 583–591. [CrossRef]
53. Kasemets, K.; Ivask, A.; Dubourguier, H.C.; Kahru, A. Toxicity of Nanoparticles of ZnO, CuO and TiO_2 to Yeast Saccharomyces Cerevisiae. *Toxicol. Vitr.* **2009**, *23*, 1116–1122. [CrossRef]
54. Babele, P.K.; Thakre, P.K.; Kumawat, R.; Tomar, R.S. Zinc Oxide Nanoparticles Induce Toxicity by Affecting Cell Wall Integrity Pathway, Mitochondrial Function and Lipid Homeostasis in Saccharomyces Cerevisiae. *Chemosphere* **2018**, *213*, 65–75. [CrossRef] [PubMed]
55. Liu, Y.; Li, Y.; Pan, B.; Zhang, X.; Zhang, H.; Steinberg, C.E.W.; Qiu, H.; Vijver, M.G.; Peijnenburg, W.J.G.M. Application of Low Dosage of Copper Oxide and Zinc Oxide Nanoparticles Boosts Bacterial and Fungal Communities in Soil. *Sci. Total Environ.* **2021**, *757*, 143807. [CrossRef] [PubMed]
56. Rizwan, M.; Ali, S.; Qayyum, M.F.; Ok, Y.S.; Adrees, M.; Ibrahim, M.; Zia-ur-Rehman, M.; Farid, M.; Abbas, F. Effect of Metal and Metal Oxide Nanoparticles on Growth and Physiology of Globally Important Food Crops: A Critical Review. *J. Hazard. Mater.* **2017**, *322*, 2–16. [CrossRef] [PubMed]

sustainability

MDPI

Article

Biodegradation of Cyanide by a New Isolated *Aerococcus viridans* and Optimization of Degradation Conditions by Response Surface Methodology

Wenjin Jiang, Yang Lu, Zezhong Feng, Haixiao Yu, Ping Ma, Jinqi Zhu, Yingnan Wang and Jinfu Sun *

College of Life and Health Sciences, Northeastern University, Shenyang 110169, China
* Correspondence: sunjinfu@mail.neu.edu.cn; Tel.: +86-24-8365-6098

Abstract: Microbial treatment of cyanide pollution is an effective, economical, and environmentally friendly method compared with physical or chemical approaches. A cyanide-degrading bacterium was isolated from electroplating sludge and identified as *Aerococcus viridans* (termed *A. viridans* T1) through an analysis of the biochemical reaction and 16 S rDNA gene sequence. *A. viridans* T1 showed a maximum resistance to 550 mg L^{-1} CN^{-}. The effect of pH and temperature on cyanide degradation and bacterial growth was evaluated. The highest cyanide removal efficiency and bacterial growth occurred at pH 8 and pH7, respectively. The optimum temperature for cyanide degradation and bacterial growth was 34 °C. In addition, the carbon source and nitrogen source for cyanide degradation were optimized. The optimal carbon source and nitrogen source were glycerol and peptone, respectively. The cyanide degradation experiment indicated that *A. viridans* T1 was able to remove 84.1% of free cyanide at an initial concentration of 200 mg L^{-1} CN^{-} within 72 h and 86.7% of free cyanide at an initial concentration of 150 mg L^{-1} CN^{-} within 56 h. To improve the cyanide-degrading efficiency of *A. viridans* T1, eight process variables were further optimized using a response surface methodology. Three significant variables (soybean meal, corn flour, and L-cysteine) were identified using a Plackett–Burman design, and the variable levels were optimized using a central composite design. The optimal values of soybean meal, corn flour, and L-cysteine were 1.11%, 1.5%, and 1.2%, respectively. Under these optimal conditions, the confirmatory experiments showed that the actual degradation rate was 97.3%, which was similar to the predicted degradation rate of 98.87%. Its strong resistance to cyanide and cyanide-degrading activity may allow *A. viridans* T1 to be a candidate for the bioremediation of cyanide-contaminated environments.

Keywords: cyanide; biodegradation; *Aerococcus viridans*; response surface methodology

Citation: Jiang, W.; Lu, Y.; Feng, Z.; Yu, H.; Ma, P.; Zhu, J.; Wang, Y.; Sun, J. Biodegradation of Cyanide by a New Isolated *Aerococcus viridans* and Optimization of Degradation Conditions by Response Surface Methodology. *Sustainability* **2022**, *14*, 15560. https://doi.org/10.3390/su142315560

Academic Editors: Sunil Kumar, Pooja Sharma, Deblina Dutta and Francesco Ferella

Received: 16 September 2022
Accepted: 18 November 2022
Published: 23 November 2022

Publisher's Note: MDPI stays neutral with regard to jurisdictional claims in published maps and institutional affiliations.

1. Introduction

Cyanide is a highly toxic chemical that is widely used in many industrial processes, including gold and silver extraction, the jewelry industries, electroplating, and plastic production. Discharged cyanide from these industries may cause serious environmental pollution. In addition, some kinds of bacteria, fungi, plants, and animals synthesize cyanide for their self-defense [1]. Cyanide exists in the natural environment in three forms; namely, free cyanide, weak acid-dissociable cyanide, and strong acid-dissociable cyanide. HCN and CN^{-} are extremely toxic to almost all living organisms [2]. Therefore, cyanide in drinking water—in particular, class I, II, III, IV, and V surface water—has been limited to no more than 0.05, 0.005, 0.05, 0.2, 0.2, and 0.2 mg L^{-1} in China.

Cyanide pollution seriously threatens human health, the environment, and the sustainable development of human society. Therefore, wastes or wastewater containing cyanide must be treated to remove it. Various chemical or physical methods, such as oxidation, adsorption, electrolysis, electrocoagulation (EC), electrochemical oxidation, and photoelectrochemical degradation, are applied to eliminate cyanide from contaminated environments or industrial cyanide-containing wastes [3]. However, these methods are often expensive and generate harmful by-products, causing secondary environmental pollution. In contrast, microbial degradation of cyanide is economical, effective, and environmentally friendly [4]. Cyanide-degrading microorganisms (CDMs) can completely degrade cyanide into less or non-toxic products, such as formic acid, ammonia, formamide, carbon dioxide, and methane, through four types of enzymatic reactions; namely, hydrolysis, oxidation, reduction, and substitution/transfer. The hydrolytic reactions are catalyzed by cyanidase or cyanide hydratase, cyanidase degrades cyanide into formic acid and ammonia, and cyanide hydratase processes cyanide into formamide [5]. Cyanide monoxygenase catalyzes oxidative reactions to produce cyanate, which is then transformed into ammonia and carbon dioxide by cyanase. Cyanide dioxygenase directly catalyzes cyanide to carbon dioxide and ammonia [5]. Cyanide can also be degraded into ammonia and methane by nitrogenase in a reductive pathway [1]. Cyanoalanine synthase and cyanide sulfurtransferase catalyze cyanide assimilation in substitution/transfer reactions [6]. In addition, cyanide degradation by *Pseudomonas pseudoalcaligenes* CECT5344 involves quinone oxidoreductase and an associated cyanide-insensitive electron transfer chain [7].

Some bacteria, fungi, and algal have been reported to be able to biodegrade cyanide, such as Klebsiella pneumoniae [8], Pseudomonas pseudoalcaligenes [9], Pseudomonas putida and Pseudomonas stutzeri [10], Bacillus pumilus [11], Serratia marcescens [12], Exiguobacterium acetylicum, and Bacillus marisflavi [13], Escherichia coli [14], and Scenedesmus obliquus [15]. Cyanide-degrading microbes use their unique enzymic systems and metabolic pathways to degrade cyanide [16]. Therefore, the conditions for microbial degradation of cyanide vary with different species or strains of CDM. Some CDMs can degrade cyanide in neutral or acidic conditions, and some can grow and remove cyanide in alkaline conditions. For example, it was reported that Klebsiella pneumoniae [8], Serretia marcescens RL2b [17], and Rhodococcus UKMP-5M [18] degraded cyanide at pH 7, pH 6, and pH 6.3, respectively. Pseudomonas pseudoalcaligenes CECT5344 [9] and an isolated strain [19] have been shown to be able to grow and degrade cyanide at pH 9.5 and pH 10.3, respectively.

Given that CDMs possess specific cyanide-degrading enzyme systems and distinct cyanide degradation capacities, isolation of new cyanide-degrading strains and exploration of the resources of CDMs are needed for bioremediation of cyanide-contaminated environments and sustainable development of the ecological environment. The main aims of this study were to (1) isolate cyanide-degrading bacteria from electroplating sludge and analyze bacterial cyanide-degrading ability; (2) optimize physical and chemical conditions for cyanide degradation, including pH, temperature, carbon source, and nitrogen source; and (3) identify significant variables and optimize the levels of variables for cyanide degradation by isolated strains using a response surface methodology (RSM).

2. Materials and Methods

2.1. Reagents and Kits

A bacterial genomic DNA extraction kit with a pre-stained 1 kb DNA ladder was obtained from BioTeke Corporation (Beijing, China). The Platinum™ SuperFi II PCR premix GeneJET gel DNA recovery kit was obtained from Thermo Fisher Scientific Inc. (Waltham, MA, USA). Bacterial 16S rDNA primers were synthesized by GENEWIZ Inc. (Tianjin, China). A micro-biochemical identification tube for bacteria was obtained from Hope Bio-Technology Co., Ltd (Qingdao, China).

2.2. *Isolation and Screening of* Cyanide-*Degrading Bacteria*

The electroplating sludge used in this study was collected from an electroplating factory in Shenyang, Liaoning province, China. A total of 20 g electroplating sludge was suspended in 100 mL of nutrient broth medium containing 100 mg L^{-1} CN^- at pH 9.0 and incubated for 24 h in an incubator shaker at 160 rpm and 30 °C. Afterward, 10 mL of the liquid culture was inoculated into 100 mL of the same fresh medium and incubated under the same condition for another 24 h. Then, 0.2 mL of the liquid culture was spread onto beef extract peptone medium plates containing concentrations of 100 mg L^{-1} CN^-. Five plates were incubated at 30 °C for

24 h. A single colony with a different morphology growing on plates was selected for pure culture isolation and evaluation of resistance to cyanide.

2.3. Determination of the Minimal Inhibitory Concentration of Cyanide

The four isolates were inoculated into nutrient broth medium containing different concentrations of CN^- (300, 350, 400, 450, 500, 550, 600, 650, and 700 mg L^{-1}) and incubated at 30 °C for 24 h. Then, 0.2 mL of the liquid culture was plated onto beef extract peptone medium plates. These plates were incubated at 30 °C for 48 h. The MIC was the lowest concentration of CN^- at which bacterial growth was completely inhibited. The isolates with high resistance to CN^- were selected for further study.

2.4. Identification of the Isolated Strain

Among the isolates, the T1 strain was selected for identification through an analysis of the biochemical reaction and 16 S rDNA gene sequence. For the biochemical identification of the T1 strain, a micro-biochemical identification tube for bacteria (Hopebio, China) was used, following the manufacturer's instructions.

Bacterial genomic DNA was extracted using a bacterial genomic DNA extraction kit according to the instructions and reports published by Nasution et al. [20]. The 16S rDNA gene was amplified via PCR by using universal primers (F: 5′-GAGTTTGATCMTGGCTC AG-3′; R: 5′- ACGGCTACCTTGTTACGACTT-3′). The PCR program was as follows: 94 °C for 3 min, 30 cycles of denaturation at 94 °C for 40 s, annealing at 55 °C for 40 s, extension at 72 °C for 40 s, and, finally, extension at 72 °C for 7 min. PCR products were purified using the GeneJET gel DNA recovery kit (BioTeke, Beijing, China) and sequenced by GENEWIZ Inc. (Beijing, China). Highly similar sequences were searched using NCBI BLAST (http://blast.ncbi.nlm.nih.gov/Blast.cgi (accessed on 15 September 2022)). The phylogenetic tree was constructed using Clustalx and MEGA-7.0 software by using the maximum likelihood and neighbor-joining method.

2.5. Biodegradation of Cyanide by T1 Strain and Measurement of Concentration of Cyanide

Isolated strain T1 was cultured in 30 mL nutrient broth medium at an initial CN^- concentration of 200 mg L^{-1} and 30 °C, pH 9, 160 rpm, and 10% (*v/v*) inoculum or under the indicated degradation conditions. Uninoculated cyanide-containing medium served as control. After incubation for the indicated time, the residual concentration of CN^- was measured via isonicotinic acid-pyrazolone spectrophotometry. The absorbance of the developed color was measured colorimetrically at 638 nm by using an Lu-T3 ultraviolet visible spectrophotometer (LABOAO, Zhengzhou, China). The amount of cyanide removal by the T1 strain was quantified by calculating the residual concentration of CN^- difference between the culture solution inoculated and uninoculated with T1 strain. The cyanide removal efficiency was the ratio of the amount of cyanide removal to the initial cyanide content [21]. The experiments were performed in triplicate.

2.6. Effects of Physical and Chemical Conditions on Biodegradation of Cyanide by T1 Strain

The effects of pH value (7 to 11), temperature (22 °C to 38 °C), carbon sources (glucose, sucrose, corn flour, glycerol), and nitrogen sources (peptone, ammonium chloride, ammonium sulphate, soybean flour, soybean meal) were evaluated to optimize cyanide degrading conditions. The buffer solution (each liter of buffer solution contained 2.5 g Na_2HPO_4, 2.0 g KH_2PO_4, 0.5 g $MgSO_4{\cdot}7H_2O$, 30 mg $FeSO_4{\cdot}7H_2O$, and 60 mg $CaCl_2$, pH 9) [17] contained 1% glucose or 1% beef extract and was used for optimizing nitrogen or carbon sources, respectively.

2.7. Optimization of Process Variables with Response Surface Methodology

2.7.1. Plackett–Burman Design

A Plackett–Burman design (PBD) was used to identify the important variables for cyanide degradation [22]. A total of 30 mL of nutrient broth medium with a cyanide ion concentration of 200 mg L^{-1} was used in the experiments. Eight independent factors (i.e., pH, temperature, inoculation amount, corn flour, soybean meal, L-methionine, L-cysteine, and glucose) were used in the experiments. Each factor was set at two levels: high (+1) and low (−1). Design-Expert 8.0.5 was used to develop the design matrix and perform statistical analysis.

2.7.2. Central Composite Design

The important variables identified by the PBD were selected and further optimized using a central composite design (CCD) method to obtain the optimum levels of these variables [22]. Design-Expert 8.0.5 was used for statistical analysis. A total of 30 mL of nutrient broth medium with cyanide ion concentration of 200 mg L^{-1} was used in the experiments. Each variable was studied at five different levels. Twenty experiments were performed. The behavior of the system was explained using the following equation:

$$Y = \beta_0 + \sum \beta_i X_i + \sum \beta_{ii} X_i^2 + \sum \beta_{ij} X_i X_j$$

where Y represents the response variables; β_0 is the intercept coefficient; and βi, βii, and βij are coefficients of the linear, quadratic, and interaction effects, respectively.

2.8. *Statistical Analysis*

Design-Expert 8.0.5 was used for ANOVA of PBD and CCD data. Excel 2010 was used for data statistics and Origin 2019 software was used for graphing analysis. A p value of ≤ 0.05 was considered to be statistically significant.

3. Results

3.1. *Isolation of Cyanide-Degrading Bacteria*

Four bacteria strains (named T1–T4) were isolated based on the differences in their colony morphologies from electroplating sludge on beef extract peptone medium plates containing concentrations of 100 mg L^{-1} CN^-. Among the four isolates, the T1 strain showed the highest resistance to cyanide with the minimal inhibitory concentration (MIC) of CN^- of up to 550 mg L^{-1}. Therefore, the T1 strain was selected for further study.

3.2. *Identification of Cyanide-Degrading Bacteria Strain T1*

To identify the isolated bacteria T1, we performed Gram staining, morphological observation, and biochemical identification. The results showed that the isolated bacteria T1 was Gram-positive and spherical, and the biochemical characters of T1 (Table 1) were consistent with *Aerococcus viridans*. Therefore, the T1 strain was initially identified as *Aerococcus viridans* according to Bergey's Manual of Systematic Bacteriology.

Table 1. Biochemical reactions of isolate T1.

Characteristics	*Properties*	*Characteristics*	*Properties*
Glucose semi-solid agar	+	*Melibiose*	−
Ornithine decarboxylase	−	*Raffinose*	−
Lysine decarboxylase	−	*Glucose*	+
Urea enzyme	+	*Sucrose*	+
Hydrogen sulfide	−	*Xylose*	+
V-P test	+	*Escin*	+
Phenylalanine	−	*Rhamnose*	+
Mannitol	+	*Arginine dihydrolysis*	−
Sorbose	+	*Nitrate reduction*	−
Sorbitol	+	*Gram staining*	+

The 16S rDNA gene sequence of the T1 strain contained 1547 bp. The analysis of sequence similarity showed that the T1 strain presented 99% sequence identity with three strains of *A. viridans* (sequence IDs: NR118723.2, NR104708.1, and NR043443.1). The phylogenetic tree was constructed based on 16S rDNA sequence. Combined with the above biochemical characteristics (Table 1), the isolated strain T1 was identified as *A. viridans* (termed *A. viridans* T1).

3.3. *Cyanide-Degrading Experiment*

3.3.1. Cyanide Removal Efficiency of *A. viridans* T1

The cyanide-degrading ability of the isolated strain *A. viridans* T1 was investigated at pH 9, 30 °C, 160 rpm, and initial CN^- concentrations of 100, 150, and 200 mg/L, respectively. The residual CN^- concentrations in culture solution inoculated or uninoculated with the *A. viridans* T1 strain were monitored at indicated time points, and the removal efficiency of CN^- was calculated following the protocol described in the Materials and Methods" section (Section 2.5). The results indicated that *A. viridans* T1 was able to remove 84.1% of free cyanide at an initial CN^- concentration of 200 mg L^{-1} in 72 h and 86.7% of free cyanide at an initial CN^- concentration of 150 mg L^{-1} in 56 h (Figure 1).

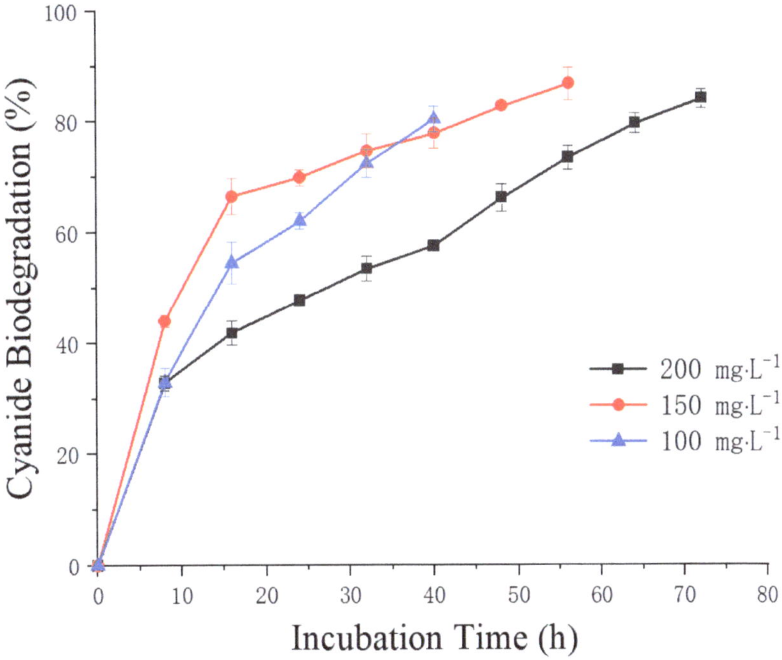

Figure 1. Cyanide biodegradation efficiency of *A. viridans* T1 at different initial concentrations (initial CN^- concentration of 200 mg L^{-1}, 30 °C, pH 9, 160 rpm rotation speed).

3.3.2. Effects of Temperature and pH on Cyanide Biodegradation and Growth of *A. viridans* T1

After incubation for 16 h, the cyanide removal efficiency of *A. viridans* T1 and bacterial growth were measured at the indicated pH and temperature. The highest cyanide removal efficiency (Figure 2A) and bacterial growth (Figure 2B) were observed at 34 °C. The optimum pH for cyanide degradation (Figure 3A) and bacterial growth (Figure 3B) were 8.0 and 7, respectively.

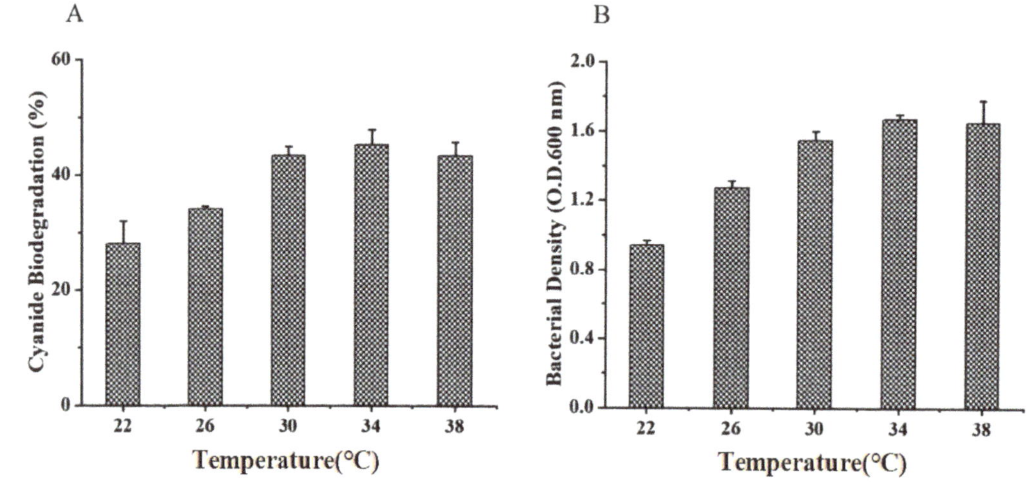

Figure 2. Effects of temperature on cyanide biodegradation (**A**) and bacterial growth (**B**) (initial CN^- concentration of 200 mg L^{-1}, pH 9, 16 h incubation time, 160 rpm rotation speed).

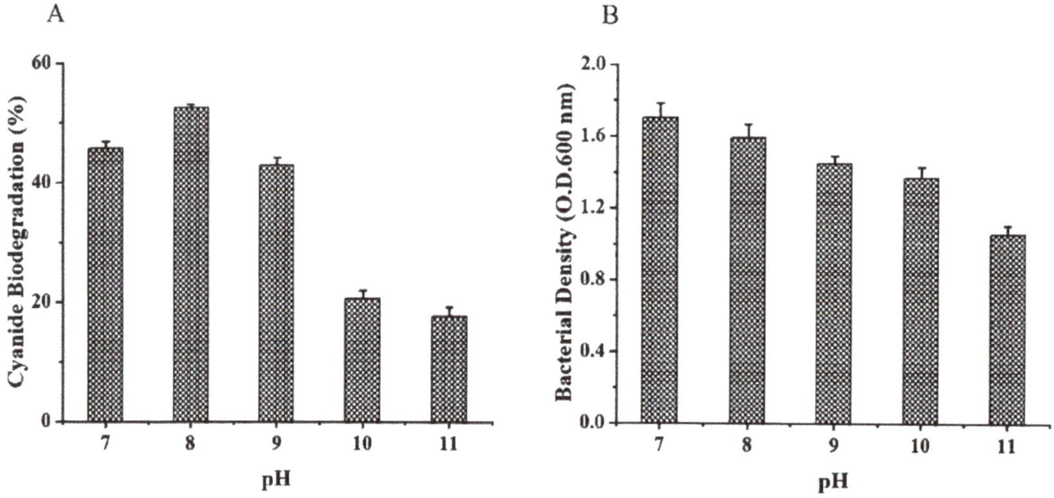

Figure 3. Effects of pH on cyanide biodegradation (**A**) and bacterial growth (**B**) (initial CN^- concentration of 200 mg L^{-1}, 30 °C, 16 h incubation time, 160 rpm rotation speed).

3.3.3. Effects of Carbon Source and Nitrogen Source on Cyanide Biodegradation by *A. viridans*

As shown in Figure 4, the carbon sources glycerol, glucose, corn flour, and sucrose presented cyanide degradation efficiencies up to 71.78%, 68.04%, 63.77%, and 47.46% in 48 h, respectively. Figure 5 shows that the nitrogen sources peptone, ammonium chloride, ammonium sulphate, soybean flour, and soybean meal demonstrated 67.96%, 54.32%, 43.84%, 45.97%, and 55.81% cyanide removal efficiencies in 48 h, respectively. Considering the economic costs, corn flour and soybean meal were selected as the respective carbon source and nitrogen source for further optimization.

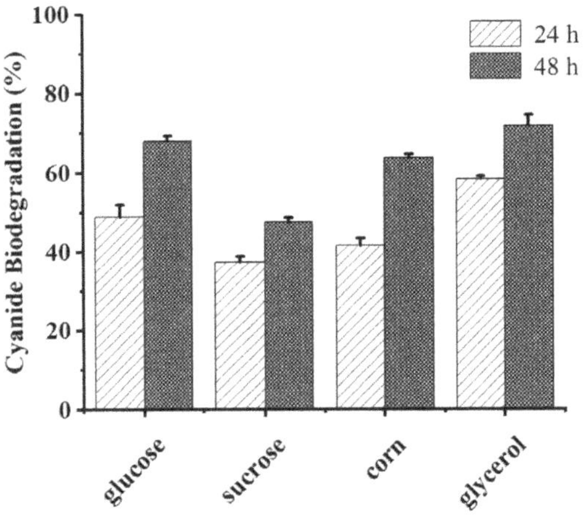

Figure 4. Effects of carbon sources on cyanide removal (initial CN^- concentration of 200 mg L^{-1}, 30 °C, pH 9, 160 rpm rotation speed).

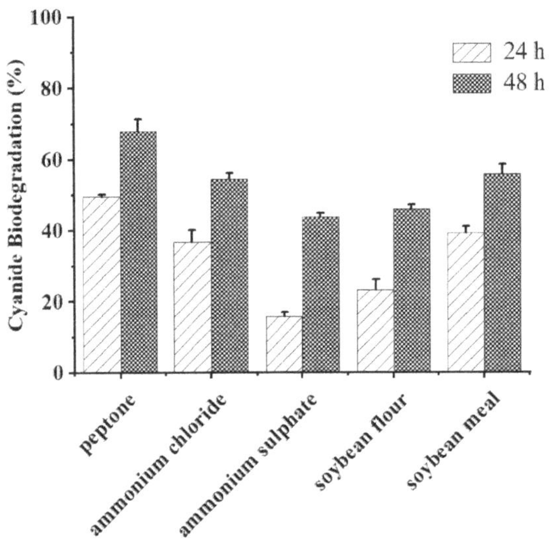

Figure 5. Effects of nitrogen sources on cyanide removal (initial CN^- concentration of 200 mg L^{-1}, 30 °C, pH 9, 160 rpm rotation speed).

3.4. Significant Factors Identified by PBD

The significant factors for the biodegradation of cyanide were identified through a 12 run PBD. The CN^- degradability is shown in Table 2. The data were statistically analyzed using Design-Expert 8.0 software and are shown in Table 3. According to the synergies of these eight factors for CN^- degradability and the related significant levels, three factors (soybean meal ($p = 0.0385$), corn flour ($p = 0.0147$), and L-cysteine ($p = 0.0079$)) were selected for further optimization to determine their optimal levels. The ranges of these variables are shown in Table 4.

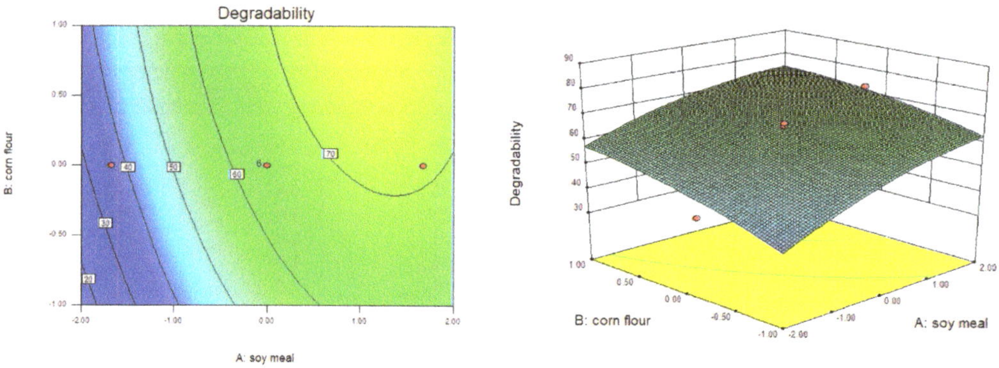

(**A**) (**B**)

Figure 6. Contour plot (**A**) and response surface plots (**B**) showing the effects of the interaction between soy meal and corn flour on CN^- degradation.

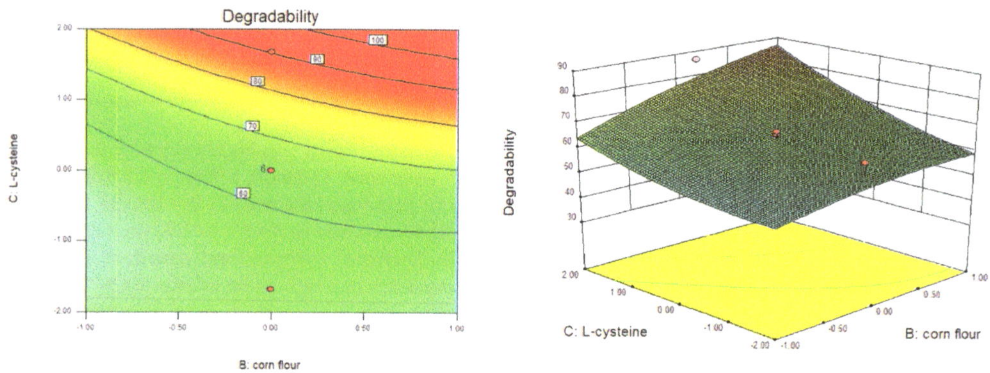

(**A**) (**B**)

Figure 7. Contour plot (**A**) and response surface plots (**B**) showing the effects of the interaction between soy meal and L-cysteine on CN^- degradation.

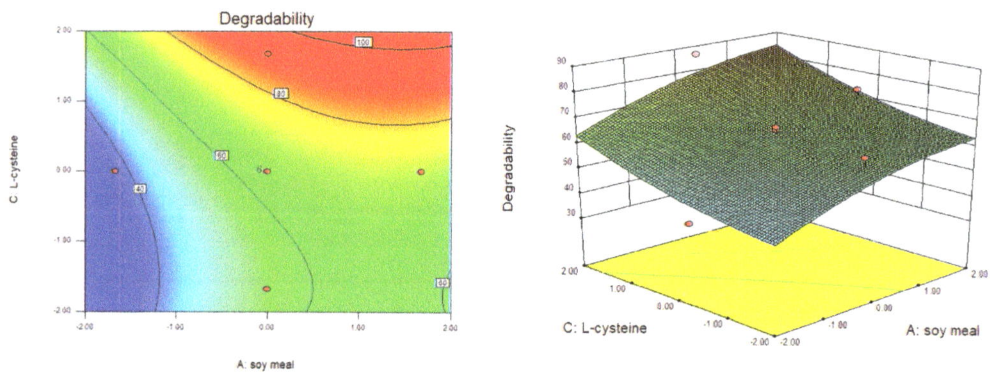

(**A**) (**B**)

Figure 8. Contour plot (**A**) and response surface plots (**B**) showing the effects of the interaction between corn flour and L-cysteine on CN^- degradation.

Table 7. CN^- degradation efficiency of *A. viridans* T1 under the optimal conditions.

Soybean Meal (%)	Corn Flour (%)	L-Cysteine (%)	CN Degradation (%)
1.11	1.5	1.2	97.3 ± 2.56

4. Discussion

Microbial degradation of cyanide is an attractive method. Various bacteria, fungi, and algae have been isolated and shown to be able to biodegrade cyanide. However, *A. viridans* has not been reported to be able to degrade cyanide. In this study, *A. viridans* T1 with potent cyanide-degrading ability was isolated from electroplating sludge. *A. viridans* is a Gram-positive, oxidase, catalase-negative, micro-aerophylic, and non-motile coccus that belongs to the family *Aerococcaceae* [23]. *A. viridans* is widely distributed in the environment, such as in air, water, and soil.

The cyanide-degrading ability of *A. viridans* T1 was investigated. As shown in Figure 1, *A. viridans* T1 was able to remove 84.1% of 200 mg L^{-1} free cyanide in 72 h and 86.7% of 150 mg L^{-1} free cyanide in 56 h. A higher degradation rate occurred in hours 0–16, with 41.75% removal of 200 mg L^{-1} CN^-, 66.5% removal of 150 mg L^{-1} CN^-, and 54.5% removal of 100 mg L^{-1} CN^- (Figure 2), showing higher cyanide removal efficiency compared to some previously reported CDMs. A previous study showed that *K. pneumoniae* was able to completely degrade 0.5 mM potassium cyanide within three days [17]. *Rhodococcus* UKMP-5M isolated from petroleum-contaminated soils was able to completely degrade 0.1 mM KCN within 24 h [18]. A strain of *pseudomonas putida* isolated from goldmine soil was able to remove 93.5% of 200 ppm cyanide within 13 days [24].

In this study, the cyanide degradation conditions of the isolated strain *A. viridans* T1 were optimized. Similar cyanide removal (43.3–44.25%) and bacterial growth (1.595–1.67 O.D. 600 nm) were observed at 30 °C–38 °C, and the highest removal of cyanide and bacterial growth occurred at 34 °C (Figure 3). Reports have indicated that *Klebsiella pneumoniae* [8] and *Serretia marcescens* RL2b [17] present maximum degradation of cyanide at 25 °C and 35 °C, respectively, while the optimum temperature for *Rhodococcus* UKMP-5M [18] and *pseudomonas putida* [24] was 30 °C. Isolated *A. viridans* T1 was able to grow in the pH range of 7–11 and the highest growth yield occurred at pH 7 (Figure 3B), while the maximum degradation of cyanide occurred at pH 8 (Figure 3A). In accordance with this study, pH 6.0–7.8 was determined as the optimal pH range for bacterial or enzymic degradation of cyanide [8,18,25]. In addition, some CDMs can grow and degrade cyanide under alkaline conditions. For example, *Pseudomonas aeruginosa* STK 03 can grow and degrade free cyanate at pH 10 [16]. Moradkhani et al. showed that the optimum pH for cyanide degradation by *Pseudomonas parafulva* C3 is 9.95 [24]. These studies indicate that temperature and pH are critical factors for the cyanide removal efficiency of CDMs.

Consistent with our study, glycerol carbon sources were found to result in the highest cyanide removal by *Serretia marcescens* RL2b [17]. However, Adjei et al. [26] indicated that *B. cepacia* C-3 could not proliferate and degrade cyanide with glycerol as a carbon source, and the optimal carbon source was fructose. In another study, glucose resulted in the highest cyanide removal by a new isolated strain, followed by fructose [19]. Among the five nitrogen sources tested in this study, peptone was determined as the optimal nitrogen source for cyanide degradation by *A. viridans* T1, followed by ammonium sulfate and soybean meal (Figure 6). Mirizadeh et al. [19] showed that ammonium sulfate, ammonium nitrate, or urea nitrogen sources inhibited cyanide degradation by a new isolated strain. Similar to this study, tryptone was shown to be the optimum nitrogen source for cyanide degradation by *Serretia marcescens* [17]. Adjei and Ohta [26] reported that cyanide biodegradation by *B.cepacia* C-3 increased by about twofold with casein and urea as organic nitrogen compared to that with KCN as sole nitrogen source, while ammonium sulfate and KNO_3 as inorganic nitrogen improved the cyanide degradation by about 3–3.5 fold. Maniyam et al. [18] evaluated the effect of glucose and yeast extract on cyanide degradation by *Rhodococcus* UKMP-5 M. When glucose and yeast were supplemented simultaneously, cyanide removal efficiency was four times greater than that in the presence of yeast extract alone, indicating a significantly higher effect from glucose compared to yeast extract on cyanide degradation. These studies show the difference in the utilization of carbon sources and nitrogen sources by CDMs and indicate that the carbon and nitrogen sources need to be optimized to improve the degradation efficiency of cyanide. In the present study, with the economic costs of practical utilization taken into consideration, corn flour and soybean meal were selected as the carbon source and nitrogen source, respectively, for further optimization.

The RSM is a powerful statistical tool and is often used to explore the effects of several independent variables [27]. In this study, three process variables (corn flour, soybean meal, and L-cysteine) were identified as significant variables for cyanide degradation by *A. viridans* T1, and the optimal levels of these variables were determined with an RSM. Wu et al. identified inoculum amount, rotary speed, and temperature as significant factors for cyanide biodegradation by *Bacillus sp.* CN-22 using an RSM, and the optimal variable levels were a rotary speed of 193 rpm, inoculum amount of 2.38%, and temperature of 31 °C [22]. Another study determined temperature, pH, and glucose concentration to be significant variables for cyanide biodegradation by *Pseudomonas parafulva*, and the optimal variable levels were determined to be a temperature of 32.23 °C, pH 9.95, and glucose concentration of 0.73 gL^{-1} [24]. These data indicate that the significant variables and variable levels that affect cyanide biodegradation are different for various CDMs.

5. Conclusions

In this study, a new cyanide-degrading bacteria was isolated and identified as *A. viridans*, its resistance to cyanide and cyanide degradation ability was evaluated, and its cyanide degradation conditions were optimized. *A. viridans* T1 was able to tolerate CN^- concentration of 550 mg L^{-1} and exhibited strong cyanide-degrading ability, with 84.1% cyanide removal at an initial concentration of 200 mg L^{-1} CN^- within 72 h. The result showed that the cyanide removal efficiency was highest at pH 8 and 34 °C. Under the conditions optimized by the RSM, the predicted cyanide degradation rate was 98.87%, and the confirmatory experiment showed a cyanide degradation rate of 97.3%. The strong cyanide degradation ability of *A. viridans* T1 makes it a good candidate for the bioremediation of cyanide-contaminated environments. However, further studies on its cyanide-degrading enzyme, genetic characteristics, and toxicity are needed.

Author Contributions: Conceptualization, W.J., Y.L. and J.S.; Methodology, W.J. and Y.L.; Investigation, W.J., Y.L., P.M. and J.Z.; Formal analysis, H.Y., Z.F. and Y.W.; Writing—original draft preparation, H.Y., Z.F. and Y.W.; Writing—review and editing, J.S.; Supervision, J.S. All authors have read and agreed to the published version of the manuscript.

Funding: This work was supported by the National Key Research and Development Program of China (no. 2018YFC1902002).

Institutional Review Board Statement: Not applicable.

Informed Consent Statement: Not applicable.

Data Availability Statement: Not applicable.

Conflicts of Interest: The authors declare that they have no conflict of interest.

References

1. Luque-Almagro, V.M.; Cabello, P.; Sáez, L.P.; Olaya-Abril, A.; Moreno-Vivián, C.; Roldán, M.D. Exploring anaerobic environments for cyanide and cyano-derivatives microbial degradation. *Appl. Microbiol. Biotechnol.* **2018**, *102*, 1067–1074. [CrossRef]
2. Razanamahandry, L.C.; Andrianisa, H.A.; Karoui, H.; Kouakou, K.M.; Yacouba, H. Biodegradation of free cyanide by bacterial species isolated from cyanide-contaminated artisanal gold mining catchment area in Burkina Faso. *Chemosphere* **2016**, *157*, 71–78. [CrossRef]
3. Ojaghi, A.; Shafaie Tonkaboni, S.Z.; Shariati, P.; Doulati Ardejani, F. Novel cyanide electro-biodegradation using Bacillus pumilus ATCC 7061 in aqueous solution. *J. Environ. Health Sci. Eng.* **2018**, *16*, 99–108. [CrossRef] [PubMed]

4. Akcil, A.; Karahan, A.G.; Ciftci, H.; Sagdic, O. Biological treatment of cyanide by natural isolated bacteria. *Miner. Eng.* **2003**, *16*, 643–649. [CrossRef]
5. Ebbs, S. Biological degradation of cyanide compounds. *Curr. Opin. Biotechnol.* **2004**, *15*, 231–236. [CrossRef]
6. Gupta, N.; Balomajumder, C.; Agarwal, V.K. Enzymatic mechanism and biochemistry for cyanide degradation: A review. *J. Hazard. Mater.* **2010**, *176*, 1–13. [CrossRef] [PubMed]
7. Luque-Almagro, V.M.; Merchán, F.; Blasco, R.; Igeño, M.I.; Martínez-Luque, M.; Moreno-Vivián, C.; Castillo, F.; Roldán, M.D. Cyanide degradation by *Pseudomonas pseudoalcaligenes* CECT5344 involves a malate: Quinone oxidoreductase and an associated cyanide-insensitive electron transfer chain. *Microbiology* **2011**, *157*, 739–746. [CrossRef]
8. Avcioglu, N.H.; Bilkay, I.S. Biological Treatment of Cyanide by Using Klebsiella pneumoniae Species. *Food. Technol. Biotechnol.* **2016**, *54*, 450–454. [CrossRef]
9. Huertas, M.J.; Sáez, L.P.; Roldán, M.D.; Luque-Almagro, V.M.; Martínez-Luque, M.; Blasco, R.; Castillo, F.; Moreno-Vivián, C.; García-García, I. Alkaline cyanide degradation by *Pseudomonas pseudoalcaligenes* CECT5344 in a batch reactor. Influence of pH. *J. Hazard. Mater.* **2010**, *179*, 72–78. [CrossRef]
10. Singh, U.; Arora, N.K.; Sachan, P. Simultaneous biodegradation of phenol and cyanide present in coke-oven effluent using immobilized *Pseudomonas putida* and *Pseudomonas stutzeri*. *Braz. J. Microbiol.* **2018**, *49*, 38–44. [CrossRef]
11. Meyers, P.R.; Gokool, P.; Rawlings, D.E.; Woods, D.R. An efficient cyanide-degrading *Bacillus pumilus* strain. *J. Gen. Microbiol.* **1991**, *137*, 1397–1400. [CrossRef] [PubMed]
12. Karamba, K.I.; Ahmad, S.A.; Zulkharnain, A.; Yasid, N.A.; Ibrahim, S.; Shukor, M.Y. Batch growth kinetic studies of locally isolated cyanide-degrading *Serratia marcescens* strain AQ07. *3 Biotech* **2018**, *8*, 11. [CrossRef] [PubMed]
13. Mekuto, L.; Alegbeleye, O.O.; Ntwampe, S.K.; Ngongang, M.M.; Mudumbi, J.B.; Akinpelu, E.A. Co-metabolism of thiocyanate and free cyanide by *Exiguobacterium acetylicum* and *Bacillus marisflavi* under alkaline conditions. *3 Biotech* **2016**, *6*, 173. [CrossRef] [PubMed]
14. Figueira, M.M.; Ciminelli, V.S.; de Andrade, M.C.; Linardi, V.R. Cyanide degradation by an *Escherichia coli* strain. *Can. J. Microbiol.* **1996**, *42*, 519–523. [CrossRef] [PubMed]
15. Gurbuz, F.; Ciftci, H.; Akcil, A. Biodegradation of cyanide containing effluents by *Scenedesmus obliquus*. *J. Hazard. Mater.* **2009**, *162*, 74–79. [CrossRef] [PubMed]
16. Mahendran, R.; Bs, S.; Thandeeswaran, M.; kG, K.; Vijayasarathy, M.; Angayarkanni, J.; Muthusamy, G. Microbial (Enzymatic) degradation of cyanide to produce pterins as cofactors. *Curr. Microbiol.* **2020**, *77*, 578–587. [CrossRef]
17. Kumar, V.; Kumar, V.; Bhalla, T.C. In vitro cyanide degradation by *Serretia marcescens* RL2b. *Int. J. Environ. Sci.* **2013**, *3*, 1969–1979.
18. Maniyam, M.N.; Sjahrir, F.; Ibrahim, A.; Cass, A.G. Biodegradation of cyanide by *Rhodococcus* UKMP-5 M. *Biologia* **2013**, *68*, 177–185. [CrossRef]
19. Mirizadeh, S.; Yaghmaei, S.; Ghobadi Nejad, Z. Biodegradation of cyanide by a new isolated strain under alkaline conditions and optimization by response surface methodology. *J. Environ. Health Sci. Eng.* **2014**, *12*, 85. [CrossRef]
20. Nasution, F.; Theanhom, A.A.; Sukartini; Bhuyar, P.; Chumpookam, J. Genetic diversity evaluation in wild *Muntingia calabura* L. based on Random Amplified Polymorphic DNA (RAPD) markers. *Gene Rep.* **2021**, *25*, 101335. [CrossRef]
21. Mekuto, L.; Ntwampe, S.K.O.; Kena, M.; Golela, M.T.; Amodu, O.S. Free cyanide and thiocyanate biodegradation by Pseudomonas aeruginosa STK 03 capable of heterotrophic nitrification under alkaline conditions. *3 Biotech* **2016**, *6*, 6. [CrossRef] [PubMed]
22. Wu, C.F.; Xu, X.M.; Zhu, Q.; Deng, M.C.; Feng, L.; Peng, J.; Yuan, J.P.; Wang, J.H. An effective method for the detoxification of cyanide-rich wastewater by Bacillus sp. CN-22. *Appl. Microbiol. Biotechnol.* **2014**, *98*, 3801–3807. [CrossRef] [PubMed]
23. Williams, R.E.; Hirch, A.; Cowan, S.T. Aerococcus, a new bacterial genus. *J. Gen. Microbiol.* **1953**, *8*, 475–480. [CrossRef] [PubMed]
24. Moradkhani, M.; Yaghmaei, S.; Nejad, Z.G. Biodegradation of cyanide under alkaline conditions by a srain of *Pseudomonas putida* isolated from gold mine soil and optimization of process variables through response surface methodology (RSM). *Period Polytech-Chem.* **2018**, *63*, 265–273.
25. Jandhyala, D.M.; Willson, R.C.; Sewell, B.T.; Benedik, M.J. Comparison of cyanide-degrading nitrilases. *Appl. Microbiol. Biotechnol.* **2005**, *68*, 327–335. [CrossRef]
26. Adjei, M.D.; Ohta, Y. Factors affecting the biodegradation of cyanide by *Burkholderia cepacia* strain C-3. *J. Biosci. Bioeng.* **2000**, *89*, 274–277. [CrossRef]
27. Myers, R.H.; Montgomery, D.C.; Adweson-Cook, C.M. *Response Surface Methodology: Process and Product Optimization Using Designed Experiments*; John Wiley and Sons, Inc.: New York, NY, USA, 2008.

sustainability

MDPI

Article

"Nature-like" Cryoimmobilization of Phototrophic Microorganisms: New Opportunities for Their Long-Term Storage and Sustainable Use

Olga Senko [1,2], Nikolay Stepanov [1,2], Olga Maslova [1] and Elena Efremenko [1,2,*]

1 Faculty of Chemistry, Lomonosov Moscow State University, 119991 Moscow, Russia; senkoov@gmail.com (O.S.); na.stepanov@gmail.com (N.S.); olga-still@mail.ru (O.M.)
2 N.M. Emanuel Institute of Biochemical Physics, Russian Academy of Sciences, 119334 Moscow, Russia
* Correspondence: elena_efremenko@list.ru; Tel.: +7-495-939-3170; Fax: +7-495-939-5417

Abstract: It was found that immobilization of cells in poly(vinyl alcohol) (PVA) cryogel can be successfully applied for concurrent cryoimmobilization, cryoconservation and long-term storage of the cells of various phototrophic microorganisms (green and red microalgae, diatoms and cyanobacteria). For the first time, it was shown for 12 different immobilized microalgal cells that they can be stored frozen for at least 18 months while retaining a high level of viability (90%), and can further be used as an inoculum upon defrosting for cell-free biomass accumulation. Application of cryoimmobilized *Chlorella vulgaris* cells as inocula allowed the loading of a high concentration of the microalgal cells into the media for free biomass accumulation, thus increasing the rate of the process. It was shown that as minimum of 5 cycles of reuse of the same immobilized cells as inocula for cell accumulation could be realized when various real wastewater samples were applied as media for simultaneous microalgae cultivation and water purification.

Keywords: phototrophic microorganisms; cryoimmobilization; poly(vinylalcohol) cryogel; cryoconservation; wastewater treatment; microalgal inoculum; long-term storage

Citation: Senko, O.; Stepanov, N.; Maslova, O.; Efremenko, E. "Nature-like" Cryoimmobilization of Phototrophic Microorganisms: New Opportunities for Their Long-Term Storage and Sustainable Use. *Sustainability* **2022**, *14*, 661. https://doi.org/10.3390/su14020661

Academic Editors: Sunil Kumar, Pooja Sharma and Deblina Dutta

Received: 26 November 2021
Accepted: 5 January 2022
Published: 7 January 2022

Publisher's Note: MDPI stays neutral with regard to jurisdictional claims in published maps and institutional affiliations.

1. Introduction

Interest in biotechnological developments based on the use of microalgae and cyanobacteria cells has not decreased in recent years, but has even increased [1,2], since these cells are attractive for solving many problems of ecology and biocatalysis in the interests of sustainable ecosystem development. Cyanobacteria and microalgae are both used as a valuable source of proteins, vitamins and polyunsaturated fatty acids for humans or for the purpose of enriching animal feed with protein, vitamins and trace elements [3]. The effective use of cyanobacteria and microalgae as a source of omega-3 fatty acids, which are used in medicine for the treatment of cardiovascular diseases, asthma, migraines, arthritis, psoriasis, etc. has been shown. The biomass of these phototrophic microorganisms can also be used as a raw material for the production of fuel and additives to it (ethanol, butanol, biogas, biodiesel, hydrogen) [4,5]. Some microalgae and cyanobacteria can be used in bioindication to assess the level of environmental pollution by aquaculture wastewater [6].

Of particular interest are processes based on the use of microalgae and cyanobacteria for the purification of industrial, agricultural and domestic wastewater, with further processing of the accumulated biomass of phototrophic microorganisms acting as a source of bio-renewable raw materials [7]. At the same time, such strings serve as an excellent medium for the cultivation of phototrophic cells, allowing the maintenance of mixotrophic conditions, enhancing the growth rate of the microalgal cells [8] This allows both a reduction in the cost of biomass production and the solving of several ecological issues of wastewater treatment, which is, in turn, a part of the successful solving of the general tasks of sustainable environmental development. However, the cultivation of microalgae under such conditions is complicated by the inconstancy of the wastewater composition, and by the presence of toxicants which can inhibit microalgal cell growth [9]. Therefore, using immobilized microalgal cells as inocula looks like an attractive approach to improving the characteristics of microalgae wastewater treatment [10]. The advantages of this approach are as follows: the immobilization stabilizes the metabolic activity of the cells and allows introduction into the medium of an inoculum with a high concentration of cells uniformly distributed within the carrier; this ensures favorable mass transfer conditions for all the cells and causes an essential increase in the accumulation rate of the biomass of the free cells of filial generations [11,12].

Interestingly, the self-immobilization of phototrophic cells is a preferable natural state of the microorganisms. Microalgae assume the form of polymicrobial aggregates at the interface [13]. The materials that hold these cells together to form a heterogeneous matrix are synthesized and excreted by microalgae themselves into their immediate environment [14]. The lifestyle of microalgae cells in the frame of aggregates is completely different from the lifestyle of suspended cells. For the active use of stably functioning microalgae and cyanobacteria cells in immobilized form, the use of various carriers and methods of immobilization is presently in practice, as follows: sorption, inclusion of gel structures in pores, chemical stitching to carriers, etc. [15–17]. Analysis of known technical solutions as part of the most successful approaches to immobilized microalgae development allows the formulation of the following main requirements for the carrier of the cells to be used as an inoculum for biomass accumulation in wastewater: firstly, the carrier should ensure favorable conditions for mass transfer processes while preserving the high metabolic activity of the phototrophic cells; secondly, it should not hinder the microalgal cells of filial generations from leaving the polymeric matrix and accumulating in a suspension form in the culture medium outside the carrier.

In the view of all the mentioned characteristics, the application of poly(vinyl alcohol) (PVA) cryogel in microalgal cell immobilization seems to us to be prospective for a variety of reasons. These include the (typically high) pore size of this carrier, which are formed in the course of the "nature-like" structuring of high-molecular compounds via a simple freeze-thaw treatment, and which can be controlled depending on the requirements of the target process. The cryogel itself has a high elasticity and mechanical strength, and the pore structure is quite regular due to the use of a synthetic polymer with controllable characteristics [18,19]. The PVA cryogel has been successfully used for the immobilization of microorganism cells of various types (yeast, bacteria and fungi) [4,11,20–24], ensuring the high viability of these cells even upon their long-term (several years) frozen storage in an immobilized form [20,23]. The possible long-term storage of cells immobilized in PVA cryogel makes another aspect of the application of this carrier for phototrophic microorganisms interesting as an alternative solution used in collections to preserve the biological diversity of microalgae cells.

Generally, the possible negative influence of cryoconservation on phototrophic cells is well known, however the techniques were reported [25] to successfully preserve the viability and productivity of microalgal cells, involving freezing of the cells with the application of various cryoprotectants (dimethyl sulfoxide, methanol, glycerol, etc.) and very low temperatures (down to temperature of liquid nitrogen) [26]. However, some of the cryoprotectants possess cytotoxicity or non-effectiveness, and the storage of cells at the temperature of liquid nitrogen requires power-consuming and expensive cryogenic equipment.

Therefore, a more practicable approach to cryoconservation can be realized via cryoimmobilization of the phototrophic cells in appropriate polymeric gel carriers. The polymer in such systems should play both the role of cryoprotectant and the base of the matrix containing the cells. In order to produce such a matrix with this technique, the cell suspension in the polymer solution can be frozen at a temperature within the range of -15–-70 °C, which is much higher than that of liquid nitrogen.

Immobilization of the microalgal cells into the PVA cryogel matrix appears more prospective due to the following reasons. First, PVA is known for its cryoprotective properties [27–29], and second, the genuinely successful storage of various microbial cells immobilized in PVA cryogel for two years at −20 °C without an essential loss of their metabolic and catalytic activity has been previously established [20,23,30].

Thus, the main tasks of this work were the following: firstly, to study the possibility of the obtainment and long-term storage of various phototrophic microorganisms immobilized in PVA cryogel; secondly, to evaluate the biochemical composition of free cells accumulated using an immobilized inoculum. Another goal was to analyze the subsequent use of such immobilized cells as an inoculum for cultivation in various types of wastewater for the accumulation of free biomass of microalgal cells (using the example of *Chlorella vulgaris* culture). In this work, a fairly large variety of cultures were used as objects for cryopreservation. It is known that eukaryotes are more complex organisms than prokaryotes, so it seemed to us to be most interesting and important to test the investigated approach based on cell immobilization in a PVA cryogel with a wider use of eukaryotes [31–34]. The choice in favor of cyanobacteria of the filamentous type was made by us, based on similar considerations. It is also known that cyanobacteria of the filamentous type have a lower survival rate than many other cyanobacteria during cryoconservation, including in situations when they are frozen in liquid nitrogen [32,35,36].

2. Materials and Methods

2.1. Strains and Media

The following cell cultures were used in this research: *C. vulgaris* [Beier.] rsemsu Chv-20/11, *Dunaliella salina rsemsu* Dns-26/11, *Nannochloropsis* sp. rsemsu N-1/11-B, *Chlamydomonas* sp. rsemsu Chlam-10/11, *Chlorococcum* sp. rsemsu Ccc-24/11, *Cosmarium* sp. rsemsu Cos-19/11, *Galdieria partita* rsemsu Gp-17/11, *Haematococcus pluvialis Flotow* em. Wille rsemsu Hp-1/11, *Thalassiosira weissflogii* rsemsu Twl-11/11, *Nostoc* sp. rsemsu Nss-14/11, *Arthrospira platensis* (Nordst.) Geitl. rsemsu 1/02 and *Gloeotrichia echinulata* rsemsu Ge-15/11, all obtained from the IBCP RAS collections.

Cultivation of the microalgal cells was performed in the corresponding typical media recommended for the cells' specific conditions [4] at 25 °C and under round-the-clock lighting with Osram Fluora 77 luminescent lamps (30 W, Munich, Germany). The OD_{540} of the cell suspensions was controlled to investigate the kinetic curve of growth.

The biomass of microalgal cells was separated from the culture media by centrifugation (4000× *g*, 15 min), using an Avanti J 25 centrifuge (Beckman Coulter, Indianapolis, IN, USA). Immobilization of cells was conducted using the patented procedure [37]. Poly(vinyl alcohol) 16/1 (84 kDa) was purchased from Sinopec Corp. (Beijing, China). The biomass of phototrophic microorganisms immobilized in PVA cryogel was stored at −70 °C using a DS 78 compact freezer (Dairei Asia Sdn. Bhd, Kuala Lumpur, Malaysia) over 18 months, and was then slowly defrosted via the following two stages: the first one, at −20 °C, using a GN 3613 freezer (Liebherr, Biberach, Germany) for 3 h, and the second one, at 8 ± 2 °C in a 2201 Combicoldrac II refrigerator (LKB Instruments Haglund, Saffle, Sweden).

For multiple uses of immobilized *C. vulgaris* cells under batch conditions after each working cycle, the granules were washed with a sterile 0.9% sodium chloride solution and loaded in the reactor with a new portion of the culture medium (horticultural, municipal or milk plant wastewater).

In order to accumulate the *C. vulgaris* biomass we used municipal wastewater (Moszelenkhoz, Moscow, Russia), wastewater from a milk processing plant (OOO Ostankinsky Molochny Kombinat, Moscow, Russia) and the horticultural water from a gardening facility (Ostankinsky sovkhoz dekorativnogo sadovodstva, Moscow, Russia). The chemical content (lipids/proteins/carbohydrates, g/L) and pH of the wastewater samples were as follows: milk plant—0.35/0.28/0.56 (pH 6.8); horticultural water—0.17/0.13/0.27 (pH 6.5); municipal water—0.08/0.06/0.04 (pH 7.1). The cultivation was performed in 750 mL Erlenmeyer flasks with 100 mL of the cultivation medium. The cells were harvested after cultivation using an Avanti J 25 centrifuge (Beckman Coulter, Indianapolis, IN, USA) at 8000 rpm for 10 min.

2.2. Analytical Methods

The biochemical composition of the cell biomass (content of lipids, proteins, hydrocarbons) was analyzed using the standard methods described previously [4]. For the thermal treatment (121 °C, 0.5 h), the wet biomass (100 g cell dry weight (CDW)/L) was suspended in the 0.1 M citrate buffer with pH 5.0.

The lipid extraction was carried out by the treating of wet concentrated biomass with *n*-hexane via the known method [38]. Upon the extraction of the lipids, the microalgal cell debris was removed from the medium via centrifugation (8000 rpm during 10 min), and was used for the detection of protein and hydrocarbon content by well-known methods [4].

The dry weight of cell biomass (CDW) was determined via a standard gravimetrical method by drying a sample at 105 °C to a constant weight.

The concentration of intracellular adenosine triphosphate (ATP) was determined by the bioluminescent luciferin–luciferase method as described previously [30].

Chemical oxygen demand (COD), total suspended solids (TSS) content, total nitrogen (TN) content and total phosphorus (TP) content were measured by standard methods [39–42].

To determine the fluorescent characteristics of microalgal cell pigments, the obtained granules with immobilized cells were placed in a nutrient medium after thawing for a long period of storage. Further, appropriate measurements were carried out to determine the relative fluorescence variable (Fv/Fm) of the pigment of the cells according to a known technique [43]. Fluorescence excitation was performed at 455 nm with an electronic photomultiplier through a KS-18 filter. The intensity of chlorophyll fluorescence under the conditions of open reaction centers of the Photosystem II (F_0) and the maximum chlorophyll-fluorescence intensity under the conditions of completely closed reaction centers of Photosystem II (Fm) were measured at the intensity of an excitation light with a density of 0.8 and 6000 μmol quanta/(m^2/s), respectively. The potential efficiency of the primary photosynthesis processes (maximum photochemical quantum yield of Photosystem II) $Fv/Fm = (Fm - F_0)/Fm$, $Fv = Fm - F_0$ is the variable fluorescence. The fluorescence values are presented in relative units.

The pH value of the media and buffers prepared was measured potentiometrically using a Corning Pinnacle 530 pH meter (Corning, Root, Switzerland).

The Reducing Power (RP) of microalgal biomass samples was determined in triplicate according to the described method [44]. Absorbance was measured at 700 nm. Ascorbic acid (vitamin C) was used as a positive control. Results were expressed as mg vit C/g CDW.

Unsaturated fatty acids in lipids of microalgal biomass was analyzed using the published procedure [45] of gas chromatography–mass spectrometry on a 7820A gas chromatograph with a 5977B MSD Bundle with a 7820 GC mass detector (Agilent Technologies, Waldbronn, Germany).

To determine the standard deviation (±SD) of the results, data were obtained in at least three independent experiments. Statistical analysis was realized using SigmaPlot (ver. 12.5, Systat Software Inc., San Jose, CA, USA). The one-way analysis of variance (abbreviated one-way ANOVA) was used. All pairwise multiple comparison procedures were undertaken using the Holm-Sidak method: overall significance level = 0.05.

3. Results

3.1. Cryoimmobilization of Phototrophic Microbial Cells in PVA Cryogel and Their Long-Term Storage

Initially, the effect of immobilization in PVA cryogel was tested on the surviving chlorella cells, and if successful, it was decided to switch to using the same cryopreservation method for other phototrophs.

To estimate the potential applicability of the cryoimmobilization technique to the different microalgal species, this approach was applied to 12 phototrophic microorganisms (Table 1). The samples of the immobilized cells were prepared at −70 °C and stored at the same temperature for 18 months. Periodically, the samples of polymeric granules with entrapped cells were slowly thawed and were subsequently used as immobilized inocula for biomass accumulation in the typical media for autotrophic cell growth (Figure 1). The cell survival level was analyzed by the detection of intracellular ATP concentration in immobilized cells upon their thawing, and their opportunity to reproduce was controlled by the determination of the concentration of accumulated free cells in the medium. Percentage ratio (PRB) between biomass accumulated for 72 h in the media with immobilized inoculum after and before the cell storage for 18 months was specially calculated (Table 1).

Table 1. Characteristics of the immobilization process in PVA cryogel (−70 °C) optimal for various phototrophic microorganisms.

Phototrophic Microorganism Species	Concentration of PVA Solution Used for Cell Immobilization, %	Cell Biomass Concentration (CDW) in the Immobilized Sample, %	PRB *, %
	Green microalgae		
C. vulgaris	7.0	4.0	95 ± 3
Dunaliella salina	8.0	4.3	94 ± 4
Nannochloropsis sp.	7.5	3.8	93 ± 4
Chlamydomonas sp.	6.5	4.3	90 ± 3
Chlorococcum sp.	7.0	4.2	92 ± 3
Cosmarium sp.	8.0	3.7	94 ± 4
	Red microalgae		
Galdieria partita	8.0	4.1	91 ± 3
Haematococcus pluvialis	7.0	4.3	92 ± 3
	Diatoms		
Thalassiosira weissflogii	8.7	4.0	95 ± 4
	Cyanobacteria		
Nostoc sp.	6.5	4.0	91 ± 3
Arthrospira platensis	6.0	3.8	90 ± 3
Gloeotrichia echinulata	8.5	3.7	93 ± 4

* PRB is Percentage Ratio between Biomass accumulated in the media with immobilized inoculum after (Cs) and before (C_0) its storage for 18 months, calculated as $L = Cs/C_0 \times 100\%$. Biomass accumulation was performed for 72 h.

The variations in the concentrations of PVA solution and cell biomass taken for mixing in the obtaining of granule samples with immobilized cells revealed the influence of these parameters on the final successful results, namely, values of intracellular ATP and PRB. Table 1 contains the experimentally established optimal conditions for the immobilization of the microorganisms in PVA cryogel. It was demonstrated that the suggested carrier and the applied immobilization technique ensure a high viability level of different microalgal cells (the average viability level was 90–95%).

The high-enough levels of residual intracellular ATP concentrations in all the performed phototrophic cultures after their different periods of storage at −70 °C in immobilized form as compared to the initial levels of same parameter obtained for immobilized cells without their storage are shown in Figure 1.

Additionally the levels of fluorescent characteristics of phototrophic cell pigments were analyzed for immobilized cells (upon long-term storage), in comparison with the same cells in a free form (Figure 2). It was shown that cryoimmobilization and storage result in a temporary decrease in the relative fluorescence values of most of the cells except for the *Thallasiosira*. However, the analyzed characteristics of immobilized cells restored up to the level of free cells during the further cultivation of the microalgae for 3 days (Figure 3).

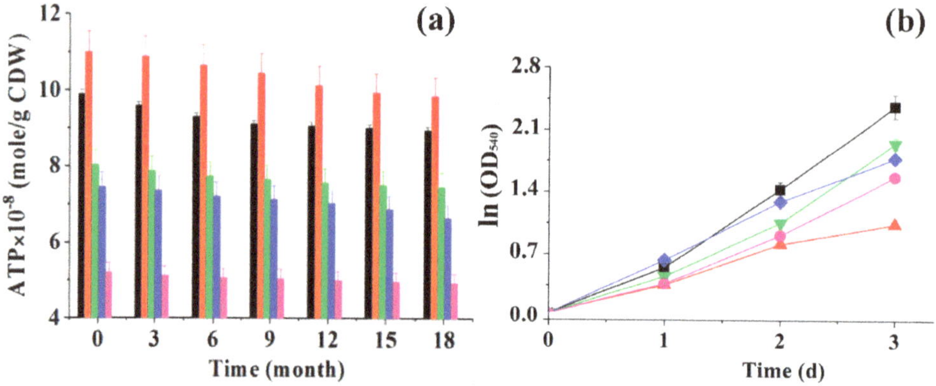

Figure 1. Changes in the concentration of ATP in microalgal cells immobilized in PVA cryogel during their storage for 18 months (**a**) and accumulation of free cell biomass during cultivation of immobilized cells as inoculum after storage for 18 months in media appropriate for certain type of culture (**b**), where *Chlorella* (black ■), *Nostoc* (red ▲), *Nannocnloropsis* (green ▼), *Arthrospira* (blue ◆), *Thallasiosira* (magenta ●).

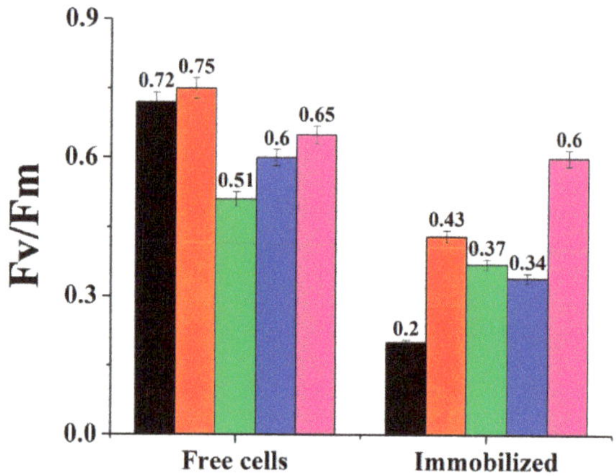

Figure 2. Relative variable fluorescence values of free and immobilized microalgal cells immediately after their defrosting upon 18 months of storage, where *Chlorella* (black ■), *Nostoc* (red ▲), *Nannocnloropsis* (green ▼), *Arthrospira* (blue ◆), *Thallasiosira* (magenta ●).

The cellular biochemical composition of free phototrophic microbial cell biomass accumulated as result of the cultivation of immobilized inoculum, used after its storage for 18 months, was specially analyzed (Table 2).

In general, it was noted that the content of the main components of suspended cells obtained by cultivating the immobilized inoculum corresponded to similar characteristics typical for cells usually accumulated in the case of using a traditional free cell inoculum. Antioxidant activity in the microalgal free cell biomass was at a regular level.

Thus, it was shown for the first time that cryoimmobilization and cryoconservation can be united in the same stage of the applied approach to the long-term storage of cells of various microalgae, which can be further successively implemented as inocula to grow and accumulate the free cell biomass of the phototrophic microorganisms useful for treatment in various processes owing mainly to biochemical composition, which are similar to common suspended cells.

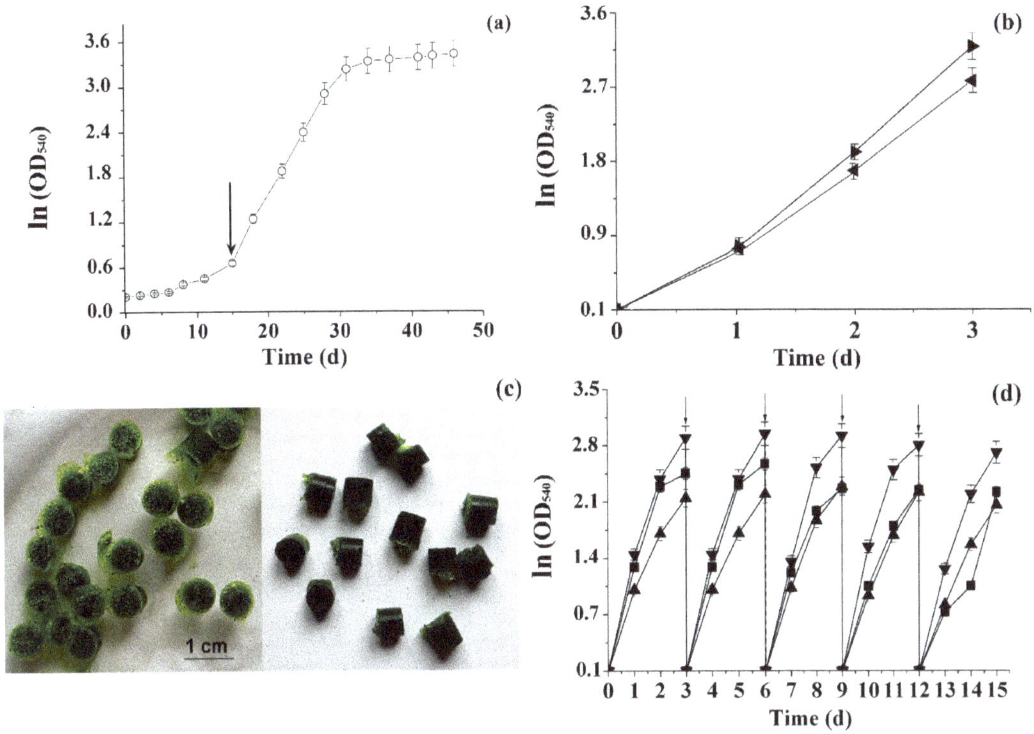

Figure 3. The kinetic curve of the autotrophic growth of *C. vulgaris* in the Tamiya medium in semi-logarithmic coordinates, the arrow marking the beginning of the most intensive biomass growth. (**a**). Free biomass accumulation during the cultivation of immobilized *C. vulgaris* cells in semi-logarithmic coordinates: in a Tamiya medium with addition of 2 g/L of glucose (▶) or 2 g/L of glycerol (◀) (**b**). The PVA cryogel granules with the immobilized *C. vulgaris* cells before (left) and after (right) their use as inocula (**c**). Free biomass accumulation upon multiple uses of the immobilized inoculum in horticultural (■), municipal (▲) and milk plant (▼) wastewater in semi-logarithmic coordinates. The arrows mark the time of inoculum transfer to the fresh medium. (**d**). The concentration of the cells in the initial inoculum was: 0.05 ± 0.002 g CDW/L (○), 1.17 ± 0.08 g CDW/L (◀), and 1.1 ± 0.05 g CDW/L (▶, ■, ▲, ▼).

Table 2. Biochemical composition of biomass of phototrophic microorganisms obtained using immobilized inoculates.

Phototrophic Microorganism Species	Lipids/ Proteins/ Hydrocarbons, %	Unsaturated Fatty Acids, % from Total Fatty Acids in Lipid Content	RP, mg vit C/g CDW
	Green microalgae		
C. vulgaris	16.5 ± 0.8/7.0 ± 0.3/55.6 ± 2.7	53.6 ± 2.6	0.01 ± 0.0004
Dunaliella salina	9.1 ± 0.4/5.4 ± 0.2/69.7 ± 3.4	26.3 ± 1.3	0.02 ± 0.001
Nannochloropsis sp.	25.0 ± 1.2/7.0 ± 0.3/54.9 ± 2.7	24.4 ± 1.2	1.21 ± 0.06
Chlamydomonas sp.	48.2 ± 2.4/17.0 ± 0.8/21.3 ± 0.9	43.6 ± 2.1	0.42 ± 0.02
Chlorococcum sp.	19.3 ± 0.9/52.7 ± 2.6/17.2 ± 0.8	47.4 ± 2.3	1.09 ± 0.05
Cosmarium sp.	6.7 ± 0.3/19.9 ± 0.9/58.4 ± 2.9	18.4 ± 0.9	0.05 ± 0.002

Table 2. *Cont.*

Phototrophic Microorganism Species	Lipids/ Proteins/ Hydrocarbons, %	Unsaturated Fatty Acids, % from Total Fatty Acids in Lipid Content	RP, mg vit C/g CDW
Red microalgae			
Galdieria partita	7.5 ± 0.3/41.8 ± 2.0/50.0 ± 2.4	17.5 ± 0.8	1.05 ± 0.05
Haematococcus pluvialis	15.6 ± 0.7/22.2 ± 1.1/52.7 ± 2.6	30.6 ± 1.5	3.6 ± 0.1
Diatoms			
Thalassiosira weissflogii	42.1 ± 2.1/30.6 ± 1.5/22.3 ± 1.2	9.4 ± 0.4	-
Cyanobacteria			
Nostoc sp.	9.2 ± 0.4/20.4 ± 0.9/52.3 ± 2.6	15.6 ± 0.7	0.83 ± 0.04
Arthrospira platensis	19.0 ± 0.9/40.9 ± 1.9/40.8 ± 1.9	22.9 ± 1.1	1.46 ± 0.07
Gloeotrichia echinulata	15.8 ± 0.7/32.7 ± 1.6/48.5 ± 2.4	33.4 ± 1.6	-

3.2. Accelerating the Accumulation of Biomass of Chlorella Vulgaris via the Use of Immobilized Inoculum

The possible effect of microalgae immobilization on the growth rate of progeny cells was studied using the example of *C. vulgaris*. Initially, the kinetic curve of autotrophic growth of the *C. vulgaris* cells was studied in a Tamiya medium typical for chlorella cell growth, with an initial free cell inoculum concentration of 0.05 ± 0.002 g CDW/L (Figure 3a).

It was shown that the rate of free cell biomass growth reached its maximum when the concentration of 0.35 ± 0.01 g CDW/L was reached in the medium with an initial free inoculum. It appeared that the biomass growth rate depended on the cell concentration in the medium (Figure 3a). This effect was also noted by other researchers [46,47], and it was observed, most probably, due to the intracellular interactions in the frame of the quorum sensing (QS) mechanism in microalgal cells. The increase in cell concentration within limited volume results in a reduction of the intracellular distance and an increase in the local concentration of the inductor molecules of the QS mechanism [48].

It is known that the introduction of additional components into the mineral medium can cause an essential increase in the biomass growth rate [49]. In order to estimate the accumulation rate of the *C. vulgaris* free cell biomass in the presence of a certain concentration of inocula, the immobilized inoculum with a concentration of cells higher than that in the experiment with free cell inoculum (Figure 3a) was applied in a further investigation (Figure 3b). For this purpose, the PVA cryogel granules containing the immobilized *C. vulgaris* cells were introduced into the Tamiya medium with additional organic substrates (2 g/L of glucose or 2 g/L of glycerol) (Figure 3b), and the comparative estimation of similar growth in a Tamiya medium under autotrophic conditions (Figure 1) was conducted.

It was established that the use of the immobilized inocula under the mixotrophic conditions allowed a notable increase in the cell-free biomass accumulation within 72 h as compared both to free-cell inocula (Figure 3a) and to the immobilized inoculum applied in the Tamiya medium without special organic additives (Figure 1). It was found that after their use, the granules containing the cryoimmobilized inoculum were practically unchanged in appearance; they could be easily separated from the medium and multiply reused for further biomass accumulation (Figure 3c).

It was reasonable to suppose that the observed effects can be achieved in case of biomass growth in other media, e.g., wastewater samples. In the present, the possibility of multiple uses of *C. vulgaris* cells immobilized in PVA cryogel as inocula for biomass accumulation was demonstrated in wastewater of different types (Figure 3d). The most intensive biomass accumulation was observed within the first 48 h of cultivation. The greatest biomass amount was produced in milk plant wastewater. The biomass accumulation was accompanied with an efficient purification of wastewater (Table 3).

Table 3. Wastewater baseline characteristics and removal efficiency of components after 72 h of *Chlorella* biomass cultivation with the use of immobilized inoculum.

Wastewater	Baseline Characteristics (mg/L)				Removal Efficiencies after 72 h (%)			
	COD	TSS	TN	TP	COD	TSS	TN	TP
Horticultural	780 ± 28	70 ± 3	47.6 ± 2.1	7.3 ± 0.2	81 ± 3	79 ± 3	84 ± 3	93 ± 3
Municipal	230 ± 9	50 ± 2	38.4 ± 1.6	4.6 ± 0.2	88 ± 4	94 ± 4	91 ± 4	95 ± 4
Milk plant	1400 ± 61	350 ± 15	60.0 ± 2.7	8.0 ± 0.3	92 ± 4	76 ± 3	78 ± 4	88 ± 3

It was revealed that upon five cycles of cryoimmobilized inoculum reuse, the amount of *C. vulgaris* biomass could be in the range of 4.7 ± 0.2 up to 6.1 ± 0.3 g CDW/L, depending on the wastewater type. Note that the productivity (in terms of accumulated biomass amount) of the cryoimmobilized inoculum decreased by no more than 10% after 5 reuse cycles.

4. Discussion

Taking into account the obtained results, it was found that the cryoimmobilization of phototrophic microbial cells (green and red microalgae, diatoms and cyanobacteria) in PVA cryogel is suitable for cryoconservation of the cells (Table 1). The cryoconservation of the phototrophic microorganism cells in the suggested approach occurs simultaneously with their cryoimmobilization in PVA cryogel, and ensures long-term storage of the cells for at least 18 months while preserving their high viability. This immobilization technique can be successfully applied in the cryoconservation of a wide-enough range of phototrophic microorganism cells, and thus has a claim to a certain universality. Thus, it may be attractive for microalgae collections existing in different countries, since the method does not require the use of an additional cryoprotectant, for example DMSO, penetrating cells and acting as a toxicant, nor does it require the use of super-low temperatures for cell storage.

The results showed that the previously well-studied porosity of the PVA cryogels [18,19,29] is such that the mass transfer of microalgae cells has no limitations, so that the cells cryoimmobilized are well supplied with substances necessary for their growth and propagation, and the daughter cells can exit from the matrix into the cultivation medium. Thus, using the inoculum of cells cryoimmobilized in the PVA cryogel under used conditions is prospective for microalgae biomass accumulation (Figures 1 and 3).

The cell biomass accumulated using inocula such as those immobilized in PVA cryogel possessed a biochemical composition (Table 2) that can be applied for producing various commercially valuable substances. The performed studies have shown that the application of cryoimmobilized inocula can allow an improvement of the accumulation rate of *C. vulgaris* cell biomass when using wastewater of various origins as media for microalgae cultivation (Table 3). The obtained results can be further used for creating biotechnological complexes involving efficient biocatalytic processes aimed at microalgae biomass accumulation accompanied by treatment of wastewater of various origins, with a further transformation of this biomass into various target products, while providing a possibility of the long-term storage of viable immobilized cells.

It is known that when certain concentrations are reached in a population, the cells of microorganisms can change the nature of their behavior due to the manifestation of the QS effect [50]. Microalgae and cyanobacteria cells are no exception to this rule [51,52]. The immobilization of microalgae cells in PVA cryogel allows the creation and cultivation of populations with a high density of cells with their simultaneous uniform distribution over the volume of the nutrient medium. In this regard, the kinetics of cell growth in the case of using an inoculate loaded into the nutrient medium in an immobilized form can be regulated and can reach a level (Figure 3b) that is characteristic of the growth curves of free cells at their increased concentration in the medium (Figure 3a).

The demonstrated approach to cryoconservation of microalgae cells does not require the use of special low-temperature equipment and thermal insulation, as is the case when liquid nitrogen is applied for the long-term cryoconservation of analogue cell samples. Therefore this approach to storing large quantities of frozen samples under easily reproducible conditions allows the use of equipment routinely applied in biotechnology. The demonstrated approach is a clear winner in comparison with another example (Table 4), involving the immobilization of the diatoms (*Haslea ostrearia* cells) in Ca-alginate gel, followed by its impregnation with a 0.7 M solution of sucrose (used as a cryoprotectant), freezing at −80 °C, and subsequent storage in liquid nitrogen [53].

Our approach also stands well in comparison with another example [54], which involves the initial immobilization of the cells of *Rivularia aquatic* and *Gloeotrichia echinulata* cyanobacteria in Ca-alginate gel followed by freezing and storage at −20 °C. It was shown that upon the storage of such an immobilized inoculum for 1 year at the mentioned temperature, the resulting viability of these two cell cultures was 47% and 49%, respectively. Such a low level of cell viability shows that the polymer matrix used in this technique did not ensure the desired level of cell cryoprotection.

Table 4. Various cryopreservation methods of microalgal biomass and cell viability after long-term storage.

Strain [Reference]	Cryopreservation of Biomass *	Storage Time	Cell Viability, %
C. vulgaris 211-11b [55]	Freezing and storage at –80 °C using a freezing container or a simple polystyrene box. Cryoprotectant mixture: 10% (*v*/*v*) DMSO, 10% (*v*/*v*) EG, and 10% (*w*/*v*) L-proline.	24 h	63 ± 2
C. vulgaris C-27 [56]	Freezing at –40 °C for 4 h and then storage in liquid nitrogen Cryoprotectant mixture: 5% (*v*/*v*) DMSO, 5% (*v*/*v*) EG, and 5% (*w*/*v*) L-proline.	15 years	54 ± 1
Chlorella sp. [57]	Cryoprotectant: 10% (*v*/*v*) DMSO Cooling-freezing-cryopreservation: gradual decrease of temperature from +25 °C to −30 °C. with rate of 1 °C min^{-1}.	12 months	77.5
	Cryoprotectant: 5% (*v*/*v*) methanol. Direct freezing in liquid nitrogen (−196 °C).	6 months	80
H. ostrearia NCC-Jμ1 [53]	Immobilization of cells in Ca-alginate gel and dehydration in the presence of 0.7 M sucrose solution with further freezing in liquid nitrogen (−196 °C).	48 h	57.4 ± 3.9
	Immobilization of cells in Ca-alginate gel and dehydration in the presence of 0.7 M sucrose solution. Two-step freezing: freezing for 1 h at −80 °C and then quick transfer into liquid nitrogen (−196 °C).		76.9 ± 3.3
Oocystis sp. [58]	Encapsulation in Ca-alginate microbeads with addition of cryoprotectants supplemented with 100 (μM) glutathione at room temperature. The algal-encapsulated microbeads were exposed to equilibration solution (15% (*w*/*v*) glycerol, 7.5% (*w*/*v*) ethylene glycol, 7.5% (*w*/*v*) dimethyl sulfoxide) followed by vitrification solution (30% (*w*/*v*) glycerol, 15% (*w*/*v*) ethylene glycol, 15% (*w*/*v*) DMSO) at 27 °C for 0.5 h and 0.25 h respectively. Freezing in liquid nitrogen (−196 °C).	14 days	79 ± 1.6

Table 4. *Cont.*

Strain [Reference]	Cryopreservation of Biomass *	Storage Time	Cell Viability, %
Rivularia aquatica and *Gloeotrichia echinulata* [54]	Cells were entrapped in Ca-alginate gel and stored at −20 °C.	1 year	47 and 49, respectively
C. vulgaris [Beier.] rsemsu Chv-20/11 [This work]	Mixing with PVA solution and freezing at –70 °C.	18 month	90 ± 5

* EG—ethylene glycol, DMSO—dimethyl sulfoxide.

An essential feature of the approach to cryoimmobilization and cryoconservation of microalgal cells that we tested is that freezing is performed in the same stage as the entrapment of the phototrophic microorganism cells in the polymeric cryogel, the latter playing the role of cryoprotectant at the same time [27].

It was established via a comparison of various techniques for cryoconservation of microalgal cells that the viability of cells cryoimmobilized in PVA cryogel was essentially higher than that used in many other techniques summarized in Table 4.

The results observed in this study, as well as the known data on the microalgal biomass accumulation upon cultivation in industrial wastewater of various types with the use of both suspended and immobilized inoculum cells are summarized in (Table 5). The productivity of immobilized cells in terms of biomass accumulated in wastewater was 30–70 mg CDW/L/d, which is 4–11 times lower than in the case of cells cryoimmobilized in PVA cryogel (311.3–407.3 mg CDW/L/d), which was used in the study (Table 5). Thus the inoculum based on *C. vulgaris* cells cryoimmobilized in the PVA cryogel was found to be effective for biomass accumulation in various media. Note that the cell growth rate was almost unchanged after at least 5 reuses of the inoculum granules.

Table 5. Biomass productivity of various species of *Chlorella* during cultivation of free and immobilized cells in wastewater from various sources.

Strain	Cell Form	Wastewater Source [Reference]	Biomass Productivity, mg CDW/L/d
C. saccharophila	Suspended	Water from carpet industry with municipal sewage [59]	23.0
C. sorokiniana UTEX1230		Municipal and domestic with water supplemented CO_2 [60]	82.5
C. vulgaris		Aquaculturing [61]	42.6
C. vulgaris SAG 211-11b		Poultry litter [62]	127.0
Chlorella sp.		Manure [63]	90.0
		Artificial [64]	130.0
	* Immobilized in polymeric carrier NaCS–DMDAAC		55.0
C. vulgaris	Immobilized on poly(vinylidene fluoride) hollow-fiber membrane	Municipal secondary effluent [65]	72.0
		Municipal water [66]	39.9
		Domestic secondary effluent [67]	50.7
		Aquaculturing [61]	42.6
C. vulgaris AG30007 + activated sludge	Immobilized in Ca-alginate gel	Municipal wastes [68]	50.0

Table 5. *Cont.*

Strain	Cell Form	Wastewater Source [Reference]	Biomass Productivity, mg CDW/L/d
C. vulgaris ATCC 13482	Immobilized on commercial thin film composite membrane	Synthetic wastewater [69]	31.0
Chlorella sp. ADE4	Immobilized on high-density polyethylene hollow fiber microfiltration membrane	Treated sewage effluent [70]	55.0
C. vulgaris [Beier.] rsemsu Chv-20/11	Immobilized in PVA cryogel	Horticultural wastes	335.3 ± 16.7
		Municipal wastes	311.3 ± 15.5
		Milk plant wastes [this work]	407.3 ± 20.3

* NaCS–DMDAAC—sodium cellulose sulphate/poly(dimethyl-diallyl-ammonium chloride).

Since it is known that microalgal cells can be used for continuous bioindicative monitoring of the presence of pollutants in waters to be discharged into environmental water systems, for example, after aquaculturing [6], then the use of microalgal cells immobilized in mechanically strong granules of PVA cryogel for these purposes can allow such control in the flow-through systems by analogy with previously developed bioanalytical sensors based on cells of luminous bacteria cells included in the PVA cryogel that have a bioluminescent reaction to ecotoxicants [71,72].

Phototrophic microorganisms are often the objects of research in connection with the study of their behavior during the blooming of natural reservoirs [73,74]. For such work, among other things, it is necessary to assess the concentrations of cells preceding a sharp jump in biomass accumulation, as well as the functional and metabolic activities of microalgae and cyanobacteria cells in high-density populations associated with the synthesis and secretion of toxic substances by these cells [75,76].

This phenomenon provokes not only a reduction in water quality, but also causes direct damage to the environment or the creation of a serious environmental threat to various living objects.

The use of artificially immobilized inoculates in such studies can make it possible to simulate and help to study natural processes under laboratory conditions in order to predict and manage them, including preventing their possible ecological development with some negative impacts on the sustainable state of ecosystems. Thus, stable cryoimmobilized inocula providing accumulation of microalgal cells with reproducible characteristics can be useful in this case.

5. Conclusions

Thus, it was shown that due to the cryoimmobilization of cells of phototrophic microorganisms by including them in the PVA cryogel, it is possible to provide:

- Their cryopreservation and long-term storage, without the loss of functional, metabolic activity and basic biochemical characteristics of the cells, including in hereditary phototrophic cells.
- The use of immobilized cells as an inoculum for the accumulation of phototrophic cell biomass, including after long-term storage, under mixotrophic conditions, that allows combining wastewater treatment with the growth of microalgae, and accumulation of the phototrophic cell biomass for its conversion to various products in the frame of green chemistry and nature-like processes.

The created samples of cryoimmobilized phototrophic cells confirmed that in the environment of polymers (in this work synthetic, such as PVA, and in nature, biopolymers), the structuring of which occurs under the influence of multiple weak chemical interactions, microalgal cells can tolerate freezing/thawing processes under environmental conditions and can sustainably maintain their viability and ability to grow and reproduce.

Author Contributions: Conceptualization, E.E.; methodology, O.S., O.M. and N.S.; validation, O.S. and N.S.; formal analysis, E.E. and O.M.; investigation, O.S. and N.S.; data curation, E.E.; writing—original draft preparation, O.S., N.S., O.M. and E.E.; writing—review and editing, O.S., N.S. and E.E.; visualization, N.S. and O.M.; supervision, E.E.; project administration, E.E. All authors have read and agreed to the published version of the manuscript.

Funding: This research was funded by State Task No.1201253312 of the Institute of Biochemical Physics Russian Academy of Science (Part 47.11 Chemical aspects of energetic) in a part of obtaining cryoimmobilized microalgal cells. The research was funded by State Task of Lomonosov Moscow State University (AAAA-A21-121041500039-8) in a part of using wastewater various type of purification by the use of immobilized microalgal cells.

Institutional Review Board Statement: Not applicable.

Informed Consent Statement: Not applicable.

Data Availability Statement: The data presented in this study are available by request.

Acknowledgments: This research was performed according to the Development program of the Interdisciplinary Scientific and Educational School of Lomonosov Moscow State University "The future of the planet and global environmental change". This study was made possible through the support of the Applied Genetics Resource Facility of the Moscow Institute of Physics and Technology (MIPT).

Conflicts of Interest: The authors declare no conflict of interest.

References

1. Tan, J.S.; Lee, S.Y.; Chew, K.W.; Lam, M.K.; Lim, J.W.; Ho, S.-H.; Show, P.L. A review on microalgae cultivation and harvesting, and their biomass extraction processing using ionic liquids. *Bioengineered* **2020**, *11*, 116–129. [CrossRef] [PubMed]
2. Khan, M.I.; Shin, J.H.; Kim, J.D.; Khan, M.I.; Shin, J.H.; Kim, J.D. The promising future of microalgae: Current status, challenges, and optimization of a sustainable and renewable industry for biofuels, feed, and other products. *Microb. Cell Factories* **2018**, *17*, 36. [CrossRef]
3. Sathasivam, R.; Radhakrishnan, R.; Hashem, A.; Abd_Allah, E.F. Microalgae metabolites: A rich source for food and medicine. *Saudi J. Biol. Sci.* **2019**, *26*, 709–722. [CrossRef]
4. Efremenko, E.N.; Nikolskaya, A.B.; Lyagin, I.V.; Sen'Ko, O.V.; Makhlis, T.A.; Stepanov, N.A.; Maslova, O.V.; Mamedova, F.; Varfolomeev, S.D. Production of biofuels from pretreated microalgae biomass by anaerobic fermentation with immobilized Clostridium acetobutylicum cells. *Bioresour. Technol.* **2012**, *114*, 342–348. [CrossRef]
5. Kokkinos, K.; Karayannis, V.; Moustakas, K. Optimizing Microalgal Biomass Feedstock Selection for Nanocatalytic Conversion Into Biofuel Clean Energy, Using Fuzzy Multi-Criteria Decision Making Processes. *Front. Energy Res.* **2021**, *8*, 622210. [CrossRef]
6. O'Neill, E.A.; Neil, J.R. Microalgae as a natural ecological bioindicator for the simple real-time monitoring of aquaculture wastewater quality including provision for assessing impact of extremes in climate variance: A comparative case study from the Republic of Ireland. *Sci. Total Environ.* **2022**, *802*, 149800. [CrossRef]
7. van den Broek, L.A.M.; Wagemakers, M.J.M.; Verschoor, A.M.; Frissen, A.E.; van Haveren, J.; Blaauw, R.; Mooibroek, H. Microalgae as Renewable Raw Material for Bioproducts: Identification and Biochemical Composition of Microalgae from a Raceway Pond in The Netherlands. In *Biomass as Renewable Raw Material to Obtain Bioproducts of High-Tech Value*, 1st ed.; Popa, V., Volf, I., Eds.; Elsevier: Amsterdam, The Netherlands, 2018; pp. 39–68. [CrossRef]
8. Udaiyappan, A.F.M.; Abu Hasan, H.; Takkriff, M.S.; Abdullah, S.R.S. A review of the potentials, challenges and current status of microalgae biomass applications in industrial wastewater treatment. *J. Water Process Eng.* **2017**, *20*, 8–21. [CrossRef]
9. Lam, M.K.; Yusoff, M.I.; Uemura, Y.; Lim, J.-W.; Khoo, C.G.; Lee, K.T.; Ong, H.C. Cultivation of Chlorella vulgaris using nutrients source from domestic wastewater for biodiesel production: Growth condition and kinetic studies. *Renew. Energy* **2017**, *103*, 197–207. [CrossRef]
10. Ahmad, A.; Bhat, A.H.; Buang, A. Immobilized Chlorella vulgaris for efficient palm oil mill effluent treatment and heavy metals removal. *Desalinisation Water Treat.* **2017**, *81*, 105–117. [CrossRef]
11. Makhlis, T.A.; Senko, O.V.; Mamedova, F.T.; Efremenko, E.N. Immobilization of cells as approach to their long-term storage. In *Immobilized Cells: Biocatalysts and Processes*; Efremenko, E.N., Ed.; RIOR: Moscow, Russia, 2018; pp. 97–122.
12. Vasilieva, S.G.; Lobakova, E.S.; Lukyanov, A.A.; Solovchenko, A.E. Immobilized microalgae in biotechnology. *Mosc. Univ. Biol. Sci. Bull.* **2016**, *71*, 170–176. [CrossRef]
13. Xiao, R.; Zheng, Y. Overview of microalgal extracellular polymeric substances (EPS) and their applications. *Biotechnol. Adv.* **2016**, *34*, 1225–1244. [CrossRef] [PubMed]

14. Ray, A.; Banerjee, S.; Das, D. Microalgal bio-flocculation: Present scenario and prospects for commercialization. *Environ. Sci. Pollut. Res.* **2021**, *28*, 26294–26312. [CrossRef]
15. Osorio, J.H.M.; Pollio, A.; Frunzo, L.; Lens, P.N.L.; Esposito, G. A Review of Microalgal Biofilm Technologies: Definition, Applications, Settings and Analysis. *Front. Chem. Eng.* **2021**, *3*, 737710. [CrossRef]
16. Caldwell, G.S.; In-Na, P.; Hart, R.; Sharp, E.; Stefanova, A.; Pickersgill, M.; Walker, M.; Unthank, M.; Perry, J.; Lee, J. Immobilising Microalgae and Cyanobacteria as Biocomposites: New Opportunities to Intensify Algae Biotechnology and Bioprocessing. *Energies* **2021**, *14*, 2566. [CrossRef]
17. Ng, F.-L.; Phang, S.-M.; Vengadesh, P.; Periasamy, V.; Yunus, K.; Fisher, A.C. Enhancement of Power Output by using Alginate Immobilized Algae in Biophotovoltaic Devices. *Sci. Rep.* **2017**, *7*, 16237. [CrossRef]
18. Lozinsky, V.I. Cryostructuring of Polymeric Systems. 50. Cryogels and Cryotropic Gel-Formation: Terms and Definitions. *Gels* **2018**, *4*, 77. [CrossRef]
19. Podorozhko, E.A.; Buzin, M.I.; Golubev, E.K.; Shcherbina, M.A.; Lozinsky, V.I. A Study of Cryostructuring of Polymer Systems. 59. Effect of Cryogenic Treatment of Preliminarily Deformed Poly(vinyl alcohol) Cryogels on Their Physicochemical Properties. *Colloid J.* **2021**, *83*, 634–641. [CrossRef]
20. Senko, O.; Gladchenko, M.; Maslova, O.; Efremenko, E. Long-Term Storage and Use of Artificially Immobilized Anaerobic Sludge as a Powerful Biocatalyst for Conversion of Various Wastes Including Those Containing Xenobiotics to Biogas. *Catalysts* **2019**, *9*, 326. [CrossRef]
21. Stepanov, N.; Efremenko, E. "Deceived" Concentrated Immobilized Cells as Biocatalyst for Intensive Bacterial Cellulose Production from Various Sources. *Catalysts* **2018**, *8*, 33. [CrossRef]
22. Maslova, O.; Stepanov, N.; Senko, O.; Efremenko, E. Production of various organic acids from different renewable sources by immobilized cells in the regimes of separate hydrolysis and fermentation (SHF) and simultaneous saccharification and fermentation (SFF). *Bioresour. Technol.* **2019**, *272*, 1–9. [CrossRef]
23. Efremenko, E.N.; Tatarinova, N.Y. The effect of long-term preservation of bacterial cells immobilized in poly(vinyl alcohol) cryogel on their viability and biosynthesis of target metabolites. *Microbiology* **2007**, *76*, 336–341. [CrossRef]
24. Razumovsky, S.D.; Efremenko, E.N.; Makhlis, T.A.; Senko, O.V.; Bikhovsky, M.Y.; Podmaster'Ev, V.V.; Varfolomeev, S.D. Effect of immobilization on the main dynamic characteristics of the enzymatic oxidation of methane to methanol by bacteria Methylosinus sporium B-2121. *Russ. Chem. Bull.* **2008**, *57*, 1633–1636. [CrossRef]
25. Ali, P.; Fucich, D.; Shah, A.A.; Hasan, F.; Chen, F. Cryopreservation of Cyanobacteria and Eukaryotic Microalgae Using Exopolysaccharide Extracted from a Glacier Bacterium. *Microorganisms* **2021**, *9*, 395. [CrossRef]
26. Kapoore, R.V.; Huete-Ortega, M.; Day, J.G.; Okurowska, K.; Slocombe, S.P.; Stanley, M.S.; Vaidyanathan, S. Effects of cryopreservation on viability and functional stability of an industrially relevant alga. *Sci. Rep.* **2019**, *9*, 2093. [CrossRef]
27. Nowshari, M.A.; Brem, G. The protective action of polyvinyl alcohol during rapid-freezing of mouse embryos. *Theriogenology* **2000**, *53*, 1157–1166. [CrossRef]
28. Cunningham, C.J.; Ivshina, I.B.; Lozinsky, V.I.; Kuyukina, M.S.; Philp, J.C. Bioremediation of diesel-contaminated soil by microorganisms immobilised in polyvinyl alcohol. *Int. Biodeterior. Biodegrad.* **2004**, *54*, 167–174. [CrossRef]
29. Kumar, A. *Supermacroporous Cryogels: Biomedical and Biotechnological Application*; CRC Press: New York, NY, USA, 2016.
30. Stepanov, N.; Efremenko, E. Immobilised cells of Pachysolen tannophilus yeast for ethanol production from crude glycerol. *New Biotechnol.* **2017**, *34*, 54–58. [CrossRef]
31. Matsunaga, N.; Uehara, A.; Murase, N.; Kuriyama, A. Cryopreservation of Spirulina (Arthrospira) platensis NIES-46 by snap-freezing considering trichome morphology. *Cryobiol. Cryotechnol.* **2019**, *64*, 75–83. [CrossRef]
32. Shiraishi, H. Cryopreservation of the edible alkalophilic cyanobacterium Arthrospira platensis. *Biosci. Biotechnol. Biochem.* **2016**, *80*, 2051–2057. [CrossRef]
33. Prasad, R.N.; Sanghamitra, K.; Antonia, G.-M.; Juan, G.-V.; Benjamin, R.-G.; Luis, I.-M.J.; Guillermo, V.-V. Isolation, Identification and Germplasm Preservation of Different Native Spirulina Species from Western Mexico. *Am. J. Plant Sci.* **2013**, *04*, 65–71. [CrossRef]
34. Iwamoto, K.; Fukuyo, S.; Okuda, M.; Kobayashi, M.; Shiraiwa, Y. Cryopreservation of the Chlorophyll d-Containing Cyanobacterium Acaryochloris Marina. *Procedia Environ. Sci.* **2012**, *15*, 118–125. [CrossRef]
35. Mori, F.; Erata, M.; Watanabe, M.M. Cryopreservation of cyanobacteria and green algae in the NIES-collection. *Microbiol. Cult. Coll.* **2002**, *18*, 45–55.
36. Motham, M.; Peerapornpisal, Y.; Tongsriri, S.; Pumas, C.; Vacharapiyasophon, P. High subzero temperature preservation of Spirulina platensis (Spirulina fusiformis) and its ultrastructure. *Chiang Mai J. Sci.* **2012**, *39*, 554–561. Available online: https://www.thaiscience.info/journals/Article/CMJS/10905239.pdf (accessed on 4 January 2022).
37. Efremenko, E.N.; Senko, O.V.; Makhls, T.A.; Mamedova, F.T.; Holstov, A.V.; Varfolomeev, S.D. Method of Cryopreservation Phototrophic Microorganisms Cells. RU Patent 2508397, 27 February 2014.
38. Araujo, G.S.; Matos, L.J.; Fernandes, J.O.; Cartaxo, S.J.; Gonçalves, L.R.; Fernandes, F.A.; Farias, W.R. Extraction of lipids from microalgae by ultrasound application: Prospection of the optimal extraction method. *Ultrason. Sonochem.* **2013**, *20*, 95–98. [CrossRef] [PubMed]
39. ISO 6060:1989 Water Quality—Determination of the Chemical Oxygen Demand. Available online: https://www.iso.org/obp/ui/#iso:std:iso:6060:ed-2:v1:en (accessed on 4 January 2022).

40. ISO 11905-1:1997 Water Quality—Determination of Nitrogen—Part 1: Method Using Oxidative Digestion with Peroxodisulfate. Available online: https://www.iso.org/obp/ui/#iso:std:iso:11905:-1:ed-1:v1:en (accessed on 4 January 2022).
41. ISO 11923:1997 Water Quality—Determination of Suspended Solids by Filtration through Glass-Fibre Filters, APHA, AWWA, WEF. Standard Methods for Examination of Water and Wastewater. Available online: https://www.iso.org/obp/ui/fr/#iso:std:iso:11923:ed-1:v1:en (accessed on 4 January 2022).
42. ISO 6878:2004 Water Quality—Determination of Phosphorus-Ammonium Molybdate Spectrometric Method. Available online: https://www.iso.org/obp/ui/#iso:std:iso:6878:ed-2:v1:en (accessed on 4 January 2022).
43. Fursova, P.V.; Bobyrev, P.A.; Risnik, D.V.; Voronova, E.N.; Pogosyan, S.I. Bioindicational Potential of Biophysical and Hydrobiological Indicators of Phytoplankton in Experiments with Laboratory Algocenoses. *Biol. Bull. Rev.* **2020**, *10*, 193–201. [CrossRef]
44. Andres, A.I.; Petron, M.J.; Lopez, A.M.; Timon, M.L. Optimization of Extraction Conditions to Improve Phenolic Content and In Vitro Antioxidant Activity in Craft Brewers' Spent Grain Using Response Surface Methodology (RSM). *Foods* **2020**, *9*, 1398. [CrossRef]
45. Breuer, G.; Evers, W.A.C.; De Vree, J.H.; Kleinegris, D.M.M.; Martens, D.E.; Wijffels, R.H.; Lamers, P.P. Analysis of Fatty Acid Content and Composition in Microalgae. *J. Vis. Exp.* **2013**, e50628. [CrossRef]
46. Bohutskyi, P.; Kligerman, D.C.; Byers, N.; Nasr, L.K.; Cua, C.; Chow, S.; Su, C.; Tang, Y.; Betenbaugh, M.J.; Bouwer, E.J. Effects of inoculum size, light intensity, and dose of anaerobic digestion centrate on growth and productivity of Chlorella and Scenedesmus microalgae and their poly-culture in primary and secondary wastewater. *Algal Res.* **2016**, *19*, 278–290. [CrossRef]
47. Supraja, K.V.; Behera, B.; Balasubramanian, P. Performance evaluation of hydroponic system for co-cultivation of microalgae and tomato plant. *J. Clean. Prod.* **2020**, *272*, 122823. [CrossRef]
48. Prakash, J.; Kalia, V.C. Application of Quorum Sensing Systems in Production of Green Fuels. In *Quorum Sensing and Its Biotechnological Applications*; Kalia, V.P., Ed.; Springer: Singapore, 2018; pp. 155–166. [CrossRef]
49. Rincon, S.M.; Romero, H.M.; Aframehr, W.M.; Beyenal, H. Biomass production in Chlorella vulgaris biofilm cultivated under mixotrophic growth conditions. *Algal Res.* **2017**, *26*, 153–160. [CrossRef]
50. Herrera, N.; Echeverri, F. Evidence of Quorum Sensing in Cyanobacteria by Homoserine Lactones: The Origin of Blooms. *Water* **2021**, *13*, 1831. [CrossRef]
51. Chi, W.; Zheng, L.; He, C.; Han, B.; Zheng, M.; Gao, W.; Sun, C.; Zhou, G.; Gao, X. Quorum sensing of microalgae associated marine *Ponticoccus* sp. PD-2 and its algicidal function regulation. *AMB Express* **2017**, *7*, 59. [CrossRef]
52. Dow, L. How Do Quorum-Sensing Signals Mediate Algae–Bacteria Interactions? *Microorganisms* **2021**, *9*, 1391. [CrossRef]
53. Tanniou, A.; Turpin, V.; Lebeau, T. Comparison of cryopreservation methods for the long term storage of the marine diatom Haslea ostrearia (simonsen). *Cryobiology* **2012**, *65*, 45–50. [CrossRef]
54. Choudhary, K.K. Post-storage viability and metabolic stability of immobilized cyanobacteria. *Nova Hedwig.* **2010**, *90*, 215–226. [CrossRef]
55. Morschett, H.; Reich, S.; Wiechert, W.; Oldiges, M. Simplified cryopreservation of the microalga Chlorella vulgaris integrating a novel concept for cell viability estimation. *Eng. Life Sci.* **2016**, *16*, 36–44. [CrossRef]
56. Nakanishi, K.; Deuchi, K.; Kuwano, K. Cryopreservation of four valuable strains of microalgae, including viability and characteristics during 15 years of cryostorage. *J. Appl. Phycol.* **2012**, *24*, 1381–1385. [CrossRef]
57. Saadaoui, I.; Al Emadi, M.; Bounnit, T.; Schipper, K.; Al Jabri, H. Cryopreservation of microalgae from desert environments of Qatar. *J. Appl. Phycol.* **2015**, *28*, 2233–2240. [CrossRef]
58. Kumari, N.; Gupta, M.K.; Singh, R.K. Open encapsulation-vitrification for cryopreservation of algae. *Cryobiology* **2016**, *73*, 232–239. [CrossRef]
59. Chinnasamy, S.; Bhatnagar, A.; Hunt, R.W.; Das, K.C. Microalgae cultivation in a wastewater dominated by carpet mill effluents for biofuel applications. *Bioresour. Technol.* **2010**, *101*, 3097–3105. [CrossRef]
60. Lizzul, A.M.; Hellier, P.; Purton, S.; Baganz, F.; Ladommatos, N.; Campos, L. Combined remediation and lipid production using Chlorella sorokiniana grown on wastewater and exhaust gases. *Bioresour. Technol.* **2014**, *151*, 12–18. [CrossRef]
61. Gao, F.; Li, C.; Yang, Z.-H.; Zeng, G.-M.; Feng, L.-J.; Liu, J.-Z.; Liu, M.; Cai, H.-W. Continuous microalgae cultivation in aquaculture wastewater by a membrane photobioreactor for biomass production and nutrients removal. *Ecol. Eng.* **2016**, *92*, 55–61. [CrossRef]
62. Markou, G. Fed-batch cultivation of Arthrospira and Chlorella in ammonia-rich wastewater: Optimization of nutrient removal and biomass production. *Bioresour. Technol.* **2015**, *193*, 35–41. [CrossRef]
63. Wang, L.; Li, Y.; Chen, P.; Min, M.; Chen, Y.; Zhu, J.; Ruan, R.R. Anaerobic digested dairy manure as a nutrient supplement for cultivation of oil-rich green microalgae *Chlorella* sp. *Bioresour. Technol.* **2010**, *101*, 2623–2628. [CrossRef]
64. Zeng, X.; Danquah, M.K.; Zheng, C.; Potumarthi, R.; Chen, X.D.; Lu, Y. NaCS–PDMDAAC immobilized autotrophic cultivation of *Chlorella* sp. for wastewater nitrogen and phosphate removal. *Chem. Eng. J.* **2012**, *187*, 185–192. [CrossRef]
65. Gao, F.; Yang, Z.-H.; Li, C.; Zeng, G.-M.; Ma, D.-H.; Zhou, L. A novel algal biofilm membrane photobioreactor for attached microalgae growth and nutrients removal from secondary effluent. *Bioresour. Technol.* **2015**, *179*, 8–12. [CrossRef]
66. Gao, F.; Yang, Z.-H.; Li, C.; Wang, Y.-J.; Jin, W.-H.; Deng, Y.-B. Concentrated microalgae cultivation in treated sewage by membrane photobioreactor operated in batch flow mode. *Bioresour. Technol.* **2014**, *167*, 441–446. [CrossRef]

67. Gao, F.; Li, C.; Yang, Z.-H.; Zeng, G.-M.; Mu, J.; Liu, M.; Cui, W. Removal of nutrients, organic matter, and metal from domestic secondary effluent through microalgae cultivation in a membrane photobioreactor. *J. Chem. Technol. Biotechnol.* **2016**, *91*, 2713–2719. [CrossRef]
68. Mujtaba, G.; Lee, K. Treatment of real wastewater using co-culture of immobilized Chlorella vulgaris and suspended activated sludge. *Water Res.* **2017**, *120*, 174–184. [CrossRef]
69. Praveen, P.; Loh, K.-C. Nitrogen and phosphorus removal from tertiary wastewater in an osmotic membrane photobioreactor. *Bioresour. Technol.* **2016**, *206*, 180–187. [CrossRef]
70. Boonchai, R.; Seo, G. Microalgae membrane photobioreactor for further removal of nitrogen and phosphorus from secondary sewage effluent. *Korean J. Chem. Eng.* **2015**, *32*, 2047–2052. [CrossRef]
71. Senko, O.; Stepanov, N.; Maslova, O.; Akhundov, R.; Ismailov, A.; Efremenko, E. Immobilized Luminescent Bacteria for the Detection of Mycotoxins under Discrete and Flow-Through Conditions. *Biosensors* **2019**, *9*, 63. [CrossRef]
72. Efremenko, E.N.; Maslova, O.V.; Kholstov, A.V.; Senko, O.V.; Ismailov, A.D. Biosensitive element in the form of immobilized luminescent photobacteria for detecting ecotoxicants in aqueous flow-through systems. *Luminescence* **2016**, *31*, 1283–1289. [CrossRef] [PubMed]
73. Khesina, Z.B.; Karnaeva, A.E.; Pytskii, I.S.; Buryak, A.K. The mysterious mass death of marine organisms on the Kamchatka Peninsula: A consequence of a technogenic impact on the environment or a natural phenomenon? *Mar. Pollut. Bull.* **2021**, *166*, 112175. [CrossRef] [PubMed]
74. Zohdi, E.; Abbaspour, M. Harmful algal blooms (red tide): A review of causes, impacts and approaches to monitoring and prediction. *Int. J. Environ. Sci. Technol.* **2019**, *16*, 1789–1806. [CrossRef]
75. Zingone, A.; Escalera, L.; Aligizaki, K.; Fernández-Tejedor, M.; Ismael, A.; Montresor, M.; Mozetič, P.; Taş, S.; Totti, C. Toxic marine microalgae and noxious blooms in the Mediterranean Sea: A contribution to the Global HAB Status Report. *Harmful Algae* **2021**, *102*, 101843. [CrossRef] [PubMed]
76. Tester, P.A.; Litaker, R.W.; Berdalet, E. Climate change and harmful benthic microalgae. *Harmful Algae* **2020**, *91*, 101655. [CrossRef] [PubMed]

sustainability

MDPI

Article

Effect of the Co-Application of *Eucalyptus Wood* Biochar and Chemical Fertilizer for the Remediation of Multimetal (Cr, Zn, Ni, and Co) Contaminated Soil

Subhash Chandra [1,2], Isha Medha [2,3], Jayanta Bhattacharya [1,3,*], Kumar Raja Vanapalli [1] and Biswajit Samal [1]

1 Indian Institute of Technology Kharagpur, School of Environmental Science and Engineering, West Bengal 721302, India; subhash.civil.iitkgp@gmail.com (S.C.); kumarraja.vanapalli@gmail.com (K.R.V.); biswa062@gmail.com (B.S.)
2 Department of Civil Engineering, Vignan's Institute of Information Technology, Visakhapatnam 530049, India; medhaisha@gmail.com
3 Department of Mining Engineering, Indian Institute of Technology Kharagpur, West Bengal 721302, India
* Correspondence: jayantab@mining.iitkgp.ac.in or jayantaism@gmail.com; Tel.: +91-3222-283702; Fax: +91-3222-282282

Abstract: Contamination of soil with heavy metals is a worldwide problem, which causes heavy metals to release into the environment. Remediation of such contaminated soil is essential to protect the environment. The aims of this study are: first, to compare the effect of biochar and the joint application of biochar with fertilizer for the phytoremediation of heavy metals-contaminated soil using *Acacia auriculiformis*; second, to study the effect of the application rate of biochar in improving the physicochemical properties of the soil. The soil samples were collected from an active coal mine dump and assessed for their physicochemical properties and heavy metals toxicity. Initial results indicated that the soil has poor physicochemical properties and was contaminated with the presence of heavy metals such as Zn, Ni, Cu, Cr, and Co. Later, the heavy metals-contaminated soil was mixed with the 400 and 600 °C biochar, as well as the respective biochar–fertilizer combination in varying mixing ratios from 0.5 to 5% (*w*/*w*) and subjected to a pot-culture study. The results showed that the application of both varieties of biochar in combination with fertilizer substantially improved the physicochemical properties and reduced the heavy metals toxicity in the soil. The biochar and fertilizer joint application also substantially improved the soil physiochemical properties by increasing the application rate of both varieties of biochar from 0.5 to 5%. The soil fertility index (SFI) of the biochar and biochar–fertilizer amended soil increased by 49.46 and 52.22%, respectively. The plant's physiological analysis results indicated a substantial increase in the plant's shoot and root biomass through the application of biochar and biochar–fertilizer compared to the control. On the other hand, it significantly reduced the heavy metals accumulation and, hence, the secretion of proline and glutathione hormones in the plant cells. Therefore, it can be concluded that the joint application of biochar with the application rate varying between 2.5 to 5% (*w*/*w*) with the fertilizer significantly improved the physicochemical properties of the soil and reduced the heavy metals toxicity compared to the controlled study.

Keywords: biochar; remediation; fertilizer; contaminated soil; heavy metals; and soil fertility index

Citation: Chandra, S.; Medha, I.; Bhattacharya, J.; Vanapalli, K.R.; Samal, B. Effect of the Co-Application of *Eucalyptus Wood* Biochar and Chemical Fertilizer for the Remediation of Multimetal (Cr, Zn, Ni, and Co) Contaminated Soil. *Sustainability* **2022**, *14*, 7266. https://doi.org/10.3390/su14127266

Academic Editors: Sunil Kumar, Pooja Sharma and Deblina Dutta

Received: 26 April 2022
Accepted: 8 June 2022
Published: 14 June 2022

1. Introduction

Heavy metals toxicity is the presence of toxic inorganic elements in the soil in a quantity in excess of the permissible limit [1]. In the last few decades, heavy metal toxicity in the soil reportedly increased worldwide due to the increase in anthropogenic activities, such as mining activities, refinery operations, and other industrial activities [2]. These anthropogenic activities reportedly release various heavy metals, namely, Cd, Cr, Pb, Cu, and Zn, into the environment [3–6]. Excessive release of heavy metals into the environment and ecosystem poses a threat to land productivity, ecological balance, and soil microbes. Moreover, the leaching of heavy metals reportedly increases metal toxicity in the plants, which ultimately, through various food chain pathways, can pose a threat to human health [7–13]. The contaminated soils, which are generated through the mining activities, are characterized by variable pH, low organic matter, poor nutrients, low cation exchange capacity (CEC), and heavy metals toxicity [14,15]. The release of heavy metals from such soils is reportedly reduced or prevented by in situ soil remediation techniques [16,17]. Among these, the revegetation technique using a suitable plant species plays an important role in enhancing the physical, chemical, and biological characteristics of the contaminated soil [18,19]. The metal-tolerant species bioaccumulate the heavy metals in the root and shoot parts, thereby reducing their availability and leaching through the soil [20]. However, there is still a chance that the consumption of the metal-tolerant species by the grazing cattle in the area can lead to the biomagnification of the heavy metals within the food chain. Various studies have suggested the role of phytoremediation as a cost-effective technique for the remediation of contaminated soils using a suitable plant species [21–24].

However, to sustain the vegetation growth during the in situ remediation process, the use of soil ameliorants for the amendment of the contaminated soils plays a vital role [25–27]. The various soil-amending materials reportedly used in the in situ remediation of the contaminated soils comprised manure, bonemeal [Ca_{10} $(PO_4)_6$ OH_2], fly ash, and lime [25,27–30]. However, the above-reported ameliorants have certain limitations; for example, lime can only be used in acidic soils to increase the pH, fly ash application could increase the heavy metals content in the soil [31], and the use of manure or bio-solids may increase the mobility of the heavy metal due to their high dissolved organic matter content [31–34].

In recent years, biochar has gained much attention due to its beneficial soil properties, porous structure, presence of functional groups, and ability to immobilize the heavy metals within the soil matrix, thereby possible limiting the bioaccumulation of the heavy metals within the plant's tissue [25,35]. Biochar is a carbon material derived through the pyrolysis of biomass waste [36,37]. Biochar's physicochemical properties reportedly vary with the change in pyrolysis parameters, such as pyrolysis temperature and feedstock [38–40]. Several studies reported the beneficial effect of biochar on the soil's physicochemical and biological properties, as well as on the growth of the plants [41–43]. Abdelhafez [41] reported the advantageous effect of biochar on the soil physicochemical properties and remediation of Pb contaminated soil. Medyńska-Juraszek [44] reported the effect of biochar in reducing the mobility of the Cu-, Pb-, Zn-, and Cd-contaminated soil. In another study, Lu [45] reported that the application of rice straw biochar at a rate of 5% (*w*/*w*) reduced the pool of Cd, Cu, Pb, and Zn by 11, 17, 34, and 6%, respectively, in the soil. However, biochar has certain limitations to act as a fertilizer in the soil due to having limited availability of nitrogen and phosphate sources [46]. In this regard, the application of biochar in combination with nitrogen and phosphorous-enriched material could enhance its heavy metals sorption capacity and agronomic benefits in the soil. For example, Liang [47] reported that the co-application of biochar and compost in the contaminated wetland soil reportedly reduced the exchangeable fractions of Cd and Zn by 67 and 37.5%, respectively. Karami [48] reported that the application of biochar in combination with the green compost significantly reduced the Pb levels in plant shoot part by 63% compared to the single application of biochar. Zeng [49] reported that the application of biochar in combination with compost reduced the availability of Cd by 13.3% in the soil compared to the single application of the biochar. From the above-reported results, it was implied that the joint application of biochar in combination with the compost or other biofertilizer has effectively reduced the availability of heavy metals in soil.

Until now, various studies also reported the beneficial effect of the single application of biochar and the joint effect of the combined biochar and compost in the metal-contaminated soil [50–53]. Additionally, very few studies have reported the effect of different temperatures of biochar on the physicochemical characteristics of the heavy metals-contaminated soil [54–56]. However, none of the studies comprehensively reported the effect of joint application of different pyrolytic temperature *Eucalyptus* wood biochar with the NPK fertilizer for the remediation of multimetal-contaminated soil using plant species, namely *Acacia auriculiformis*.

Briefly, this study reports the effect of the application of low (400 °C) and high-temperature (600 °C) biochar and the joint application of respective biochar with fertilizer on the physicochemical properties of the soil based on a pot-culture study, variations in the soil enzymatic activities, and changes in the physiological and biochemical properties of *A. auriculiformis*. The reason behind the combined application of biochar with fertilizer lies in the fact that the biochar lacks a sufficient intrinsic pool of nitrogen and phosphorous. Hence, the joint application with fertilizer could enhance biochar's nutrient retention capacity in the soil [40], which would be otherwise lost in the form of leachate. Additionally, the presence of nitrogen, potassium, and phosphorous ions promotes the adsorption of heavy metals within the soil through the precipitation and ion exchange processes [57]. This study also investigated the effect of biochar and the combined effect of the biochar–fertilizer mixture in reducing the availability of the heavy metals in the plant tissues and changes in the plant's biochemical hormones (indicators of abiotic and toxicity stress). Finally, this study calculated the soil fertility index (SFI) as the pointer toward the increase in the soil fertility level after the co-application of biochar and fertilizer.

2. Materials and Methods

2.1. Study Area and Soil Sample Collection

The study area is located at the geographical coordinates of longitude 86°25′49.41″ E and latitude 23°45′41.82″ N in Bastacolla area, Dhanbad district, Jharkhand, India is known as Bera opencast project as shown in Figure S1. The rest of the details of the study area is given in the Supplementary Material. The soil samples were collected using a randomized design covering the entire area of the dump to represent the aggregate sampling. The samples were collected from a depth of 0–30 cm by forming a regular square grid of 5 × 5 feet, followed by the coning and quartering method to reduce sample volume. Subsequently, the soil samples were stored in airtight polystyrene bags for further use in the laboratory.

2.2. Biochar Production and Characterization

Biochar was produced through the pyrolysis of *Eucalyptus wood* waste at 400 °C and 600 °C for 150 min using a slow pyrolysis unit reported in the author's previous study [58] and denoted as 400 and 600 EB. *Eucalyptus wood* biochar (EB) was crushed and sieved through 1 mm and stored for further characterization. The details of the biochar characterization techniques are given in the Supplementary Material.

2.3. Pot Experiment Using Biochar and Fertilizer as Amending Materials

A pot experiment (Figure S2) was done in a poly-house to reckon the efficacy of biochar and the co-application of biochar and fertilizer for the remediation of the multimetal-contaminated soil planted with the *A. auriculiformis* seeds [59,60]. The details of the pot study are given in Supplementary Materials.

The design of the experiment for the pot-culture study was divided into four sets: soil with biochar, soil with biochar and fertilizer, soil with fertilizer (control 1), and only soil (control 2). A third control (control 3) representing the field sample before the pot-culture study was also included in the study. Moreover, the application of biochar with the soil was further divided into two sets, i.e., soil with 400 °C EB and 600 °C EB, to determine the effect of the pyrolysis temperature on the remediation potential of the biochar. Moreover, the 400 °C EB-fertilizer and 600 °C EB-fertilizer mixture were incorporated into the soil to estimate the co-effect of both biochar and fertilizer for the remediation of the soil. The 400 and 600 °C EB were mixed with 5 kg of the soil in the pots in a weight percentage of 0.5, 1, 2, and 5% to represent the 6, 12, 24, and 60 tonnes per hectare application of biochar in the top layer (up to 20 cm) of the soil layer. The fertilizer (containing nitrogen:phosphorous:potassium in a weight ratio of 10:8:10) was mixed with the 5 kg of the soil in the pot along with the biochar in a dose an equivalent to 75 kg of N ha^{-1}, 30 kg of P ha^{-1}, and 90 kg of K ha^{-1}, equivalent to the dose of fertilizer recommended for the reclamation of sandy soil [61,62].

2.4. Characterization of Soil

Pre- and Post-Soil Physicochemical Characterization

The soil was characterized for the physicochemical properties before and after the pot culture study to evaluate the effect of biochar and joint effect of biochar–fertilizer amendments on the properties of the soil. The pH of the soil was determined using the method described in earlier studies [63,64]. The soil organic carbon was determined using the Walkley–Black method, as reported in detail in the earlier studies [65,66]. The soil organic matter was determined using the loss-on-ignition method [67,68]. The plant-available nutrients in the soil (Na, K, Ca, and Mg) were determined by extracting the soil samples with 1 M ammonium acetate with a weight to volume ratio of 1:5, followed by N analysis on ion chromatography (IC) (Model: Dionex ICS 2100, Thermo Scientific, Sunnyvale, CA, USA). The plant-available phosphorous (H_2PO_4) in the soil was determined using Bray's method as reported in earlier studies [63,69]. The plant-available nitrogen (NH_4-N) in the soil was determined using the alkaline permanganate method on the Kjeldahl apparatus [63,70]. Soil cation exchange capacity was determined using the sodium acetate (buffered at pH 8.2) and ammonium acetate (buffered at 7.0) method [63,71]. Acid extractable heavy metals in the mine soil were determined using a method reported in an earlier study [72]. Briefly, 0.5 g of soil sample was digested using the acids HNO_3 (concentrated 69%) and HCl (strength 37%) in a 3:1 (v/v) ratio in a microwave digester (Model No. SK 10/HPR-GE-12, Milestone Technologies, Fremont, CA, USA) at 200 °C. The diethylenetriaminepentaacetic acid (DTPA) extractable heavy metals in the mine soil were determined by the method described in an earlier study [73]. Soil catalase activity was determined using the method reported in an earlier study [74,75]. The β-glucosidase in the soil was determined using the colorimetric method for the estimation of p-nitrophenol (pNP) formed by the hydrolysis of p-nitrophenyl-β-D-glucopyranoside [75,76]. Soil urease activity in the soil was determined using the method described in an earlier study [77]. The soil fertility index for pre- and the post-pot culture study was calculated using Equation (1) [78,79], to determine the effect of biochar and co-application of biochar–fertilizer on the fertility level of the soil.

$$\begin{aligned}\text{Soil fertility index (SFI)} &= \text{pH} + \text{Organic matter (\%)} + \text{available P (mg kg}^{-1}\text{)} + \\ &\text{exchangeable K (cmol kg}^{-1}\text{)} + \text{exchangeable Ca (cmol kg}^{-1}\text{)} + \text{exchangeable} \\ &\text{Mg (cmol kg}^{-1}\text{)} - \text{exchangeable Al (cmol kg}^{-1}\text{)}\end{aligned} \quad (1)$$

2.5. Plant Analysis

The plant biomass from the shoot and root parts was collected from the top, bottom, and middle parts to represent the whole plant and cleaned with distilled water to remove dust. Subsequently, the cleaned shoot and root biomass were subjected to the ultrasonic bath using an ultrasonicator (Phoenix instrument) to remove metals, if any, present in the leaf and root tissues that might have loosely bound within the pores through the dust particles deposited on them. The cleaning of the shoot and root parts through ultrasonication is essential, as otherwise, it may lead to variations in the concentration of heavy metals during analysis. Furthermore, the cleaned shoot and root biomass were oven-dried at 70 °C, until constant weight was obtained, and the shoot and root biomass was recorded. The lengths of the plant's shoot and root in all the pots were measured using a digital measuring tape to ascertain the plant growth above and below the soil layer.

Total glutathione content in freshly collected plant leaves was determined using the method described in an earlier study [80]. Proline content in the plant leaves was determined using the acid-ninhydrin method [81]. The chlorophyll a, chlorophyll b, and total carotene in the fresh plant leaves were determined using the method described in an earlier study [82]. The content of chlorophyll a, chlorophyll b, and total carotene was calculated using the Equations (2)–(4), respectively [83]:

$$C_a = 10.05A_{662} - 0.766A_{644} \quad (2)$$

$$C_b = 16.37A_{644} - 3.140A_{662} \quad (3)$$

$$C_T = 1000A_{470} - 1.280C_a - 56.7C_b/230 \quad (4)$$

where C_a is chlorophyll a, C_b is chlorophyll b, C_T is the total carotene, A_{662} is the absorbance at 662 nm, A_{644} is the absorbance at 644 nm, and A_{470} is the absorbance at 470 nm.

The heavy metals content in the plant's shoot and root parts was determined using the method reported in an earlier study [84]. The digested shoot and root samples were diluted and analyzed for heavy metals (Zn, Ni, Co, Cr, Fe, Pb, and Mn) using an atomic absorption spectrophotometer (AAS ICE 3000 series, Thermo Scientific, Waltham, MA, USA). The bioaccumulation factor (BAF) in shoot and root parts was calculated using Equations (5) and (6), respectively [85]. The translocation factor in plants was calculated using Equation (7) [85,86]:

$$BAF_{shoot} = \frac{(\text{heavy metals concentration in the shoot})}{(\text{heavy metals concentration in soil})} \quad (5)$$

$$BAF_{root} = \frac{(\text{heavy metals concentration in root})}{(\text{heavy metals concentration in soil})} \quad (6)$$

$$TF = \frac{(\text{heavy metals concentration in the shoot})}{(\text{heavy metals concentration in root})} \quad (7)$$

where BF_{shoot} is the bioaccumulation factor for the shoot, BF_{root} is the bioaccumulation factor for root, and TF is the translocation factor.

2.6. Statistical Analysis

The pot culture study in the polyhouse using biochar and biochar–fertilizer mixture as the soil-amending material was conducted in a completely randomized design. All the data generated from the soil and plant analysis were tested for normal distribution using SPSS 20 for the calculation of Skewness Z-score and Shapiro-Wilks *p*-value as shown in Table S7 before being analyzed using one-way ANOVA statistical. It can be observed from Table S7 that all the soil data have shown normal distribution due to having a Skewness z-score between ± 1.96 and the Shapiro-Wilks test *p*-value was greater than 0.05, which showed the acceptance of the null hypothesis that the data are normally distributed. Hence, all the data were further subjected to the one-way ANOVA analysis at a 95% confidence interval using Origin pro-2020 software (Copyright, Origin Lab, Northampton, MA, USA) followed by a post-hoc Tukey test to find the significant mean difference among the cases (biochar and biochar–fertilizer mixture amendment) and controls (control 1, control 2, and control 3). The correlation plots among the various groups of data were plotted using the corrplot function in the R stats (version 3.6.3) statistical package.

3. Results and Discussion

3.1. Soil and Biochar Physicochemical Characterization

3.1.1. Biochar Characterization Results

The details of the biochar characterization results are given in the Supplementary Material in the results and discussion Section 3.1.1.

3.1.2. Pre-Soil Characterization

Initial evaluation of the physicochemical properties of the soil samples collected from the two-year-old overburden dump was done to ascertain the current status of the soil physicochemical properties before being implemented in the pot-culture study. The initial characterization results showed that the soil has a sandy loam texture (Table 1) having acidic pH (5.86 ± 0.178). The soil was low in both organic carbon (<0.52%) and organic matter content (<1.03%) along with having meagre availability of the nutrients, which makes it imperative to categorize the entire area of the dump as a degraded land [87]. The degraded lands reportedly require a very long time (>30 years) to reach the complete reclamation stage [88]. The CEC of the soil were also very low (Table 1), hence making it difficult to hold the applied nutrients within the soil matrix. Apart from having degraded characteristics, the soil was also contaminated with the presence of multiple heavy metals (Ni, Zn, Co, and Cr) (Table 1). It can be observed that the total concentration of Ni (60.03 mg/kg), Co (63.60 mg/kg), Zn (52.62 mg/kg), and Cr (139.66 mg/kg) were higher than their respective permissible limit of (as per ecological and health risk guidelines) in the soil [1,89]. Moreover, the plant-available concentration of heavy metals, namely, Cr (17.89 mg/kg), Cu (9.31 mg/kg), and Ni (11.25 mg/kg), were also higher than their respective permissible limits in the soil [1]. The microbial activities comprising soil catalase, β–glucosidase activity, and urease activities were very low in the soil, which might be limited due to the absence of organic carbon, organic matter, nutrients, acidic pH, and toxicity of heavy metals in the soil. Such conditions of the soil can be improved using the remediation process, which utilizes the soil amendment and phytoremediation technique [90].

Table 1. Pre-pot culture soil physicochemical properties (mean ± S.D, n = 3).

Soil Parameter		**Values**	
Particle size	Sand	49.50%	Sandy loam texture
	Silt	46.79%	
	Clay	3.80%	
Water holding capacity, WHC (%)		13.18 ± 1.53	
pH		5.86 ± 0.17	
Total organic carbon (%)		0.516 ± 0.12	
Organic Matter (%)		1.03 ± 0.18	
Exchangeable Na (mg/kg)		131.45 ± 10.58	
Exchangeable K (mg/kg)		95.88 ± 7.11	
Exchangeable Ca (mg/kg)		146.68 ± 6.08	
Exchangeable Mg (mg/kg)		16.21 ± 3.27	
Available P (mg/kg)		0.70 ± 0.11	
Available N (mg/kg)		194.05 ± 8.37	
Cation exchange capacity, CEC (cmol/kg)		6.16 ± 0.74	
Nickel, Ni (mg/kg)		40.03 ± 9.25	
Copper, Cu (mg/kg)		34.60 ± 5.56	
Zinc, Zn (mg/kg)		52.62 ± 11.16	
Cobalt, Co (mg/kg)		63.60 ± 11.03	
Chromium, Cr (mg/kg)		139.66 ± 13.04	
DTPA-extractable Ni (mg/kg)		11.25 ± 1.38	
DTPA-extractable Cu (mg/kg)		9.31 ± 1.50	
DTPA-extractable Zn (mg/kg)		8.55 ± 2.78	
DTPA-extractable Co (mg/kg)		12.33 ± 1.45	
DTPA-extractable Cr (mg/kg)		17.89 ± 2.81	
Soil catalase (0.1 mol KMnO4 g-1 of soil)		0.59 ± 0.05	
β-glucosidase (mol PNF g-1 h-1)		0.76 ± 0.08	
Urease (μg N-NH4 kg-1 h-1)		0.30 ± 0.04	

3.1.3. Post-Soil Characterization

Effect of Biochar and Fertilizer Amendment on the Physicochemical Properties of the Soil

Here, the post-soil is represented by the soil collected after the completion of the pot-culture study. Representative post-soil characterization was done to evaluate the effectiveness of the biochar and the joint biochar–fertilizer amendments on its physicochemical properties. It can be observed from the initial characterization of the soil (Table 1) that the soil has acidic pH and insufficiency of nutrients and organic matter to support revegetation. The results of the post-soil analysis showed that the pH of the soil that was amended with 400 and 600 EB and the biochar–fertilizer mixture was significantly ($p < 0.05$) improved (Figure 1a) compared to the control 1, control 2, and control 3. Additionally, it can be concluded from Figure 1a that increasing its application rate from 0.5 to 5% significantly increased the pH by 0.8 units.

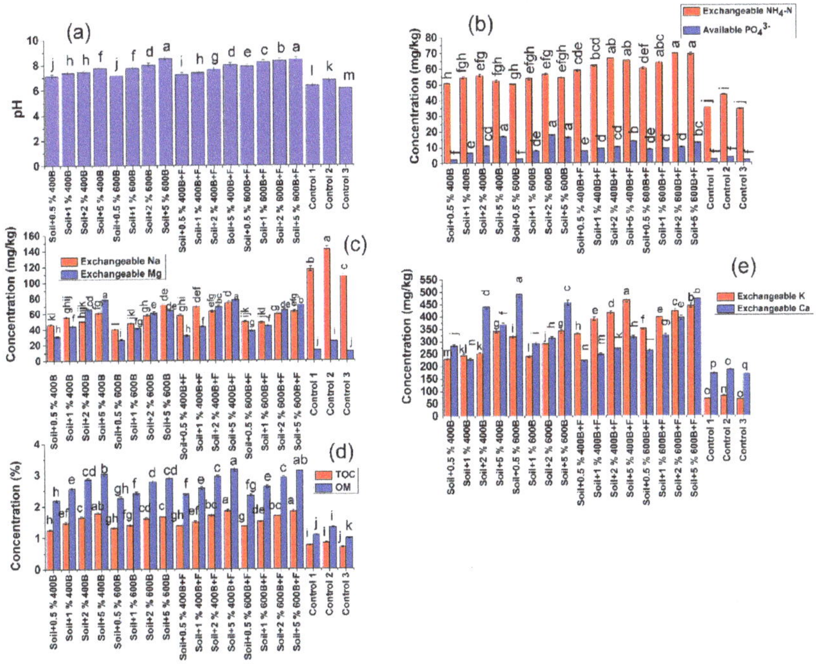

Figure 1. Effect of biochar and fertilizer amendment on the soil properties: (**a**) change in the pH values of the soil; (**b**) change in the concentration of available NH_4-N and P; (**c**) change in the exchangeable Na and Mg; (**d**) change in the total organic carbon and organic matter; (**e**) change in the concentration of exchangeable K and Ca. Lower letters above the bar graph indicate the statistically significant difference among the cases.

The joint application of 400 EB with fertilizer did not show much of an effect on the pH of the soil as compared to the sole biochar, except at the application rates of 2 and 5%. However, the joint application of 600 EB with fertilizer substantially ($p < 0.05$) improved the pH of the soil at the application rate of 1 and 2% by 0.44 and 0.33 units, respectively. A similar observation of the increase in the soil pH after the application of biochar and biochar with compost was reported in earlier studies [91,92]. An increase in the pH of the soil after the completion of the pot-culture study might be related to the liming effect of the biochar (Table S2) and the release of carbonates into the soil. Additionally, the joint application of biochar with fertilizer (pH∼7.80) may have increased the liming effect, and hence, showed a considerable increase in the soil pH [92].

The total organic carbon (TOC), organic matter (OM), available nutrients (NH_4-N, K, Ca, Mg, P), and cation exchange capacity (CEC) of the soil were significantly increased ($p < 0.05$, Figures 1 and 2) compared to the controls (C1, C2, and C3) at the end of the pot-culture study. The TOC and OM contents in the soil amended with 400 EB were significantly increased ($p < 0.05$, Figure 1d) by 29.77 and 27.96%, respectively, as the biochar's application rate was increased up to 5% (*w/w*). Likewise, the TOC and OM content in the soil amended with 600 EB was increased by 20.95 and 21.18%, respectively. It can be observed that the increase in the TOC and OM in the soil amended with 400 EB was relatively higher compared to 600 EB. This can be attributed to the increase in the presence of non-liable carbon in the biochar with the rise in the temperature [93,94]. The joint application of biochar with the fertilizer at a low application rate (up to 2%) has shown a significant change in TOC and OM in the soil. A similar increase in the soil organic carbon and organic matter after the application of biochar and biochar with compost or cow dung were reported in the earlier studies [95–99]. The increase in the soil's organic carbon and organic matter can be correlated with the change in the soil's physicochemical properties, (Figures 1 and 2a). The improvement in the soil physicochemical properties is positively correlated with the soil microbial activities ($r^2 > 0.90$, Figure S5), which might have increased the humification process in the soil. A similar mechanism was reported in a study by Jien and Wang [97], which stated that the soil environment could accelerate the self-humification of biochar and release of complex organic carbons into the soil. Another study by Liang et al. [98] reported that biochar is more prone to oxidation of carbon rings on its surface than on the inner core surface, which might have made it possible to break the complex organic carbon into simpler forms to be utilized by the microbes, and hence, had increased the soil organic carbon.

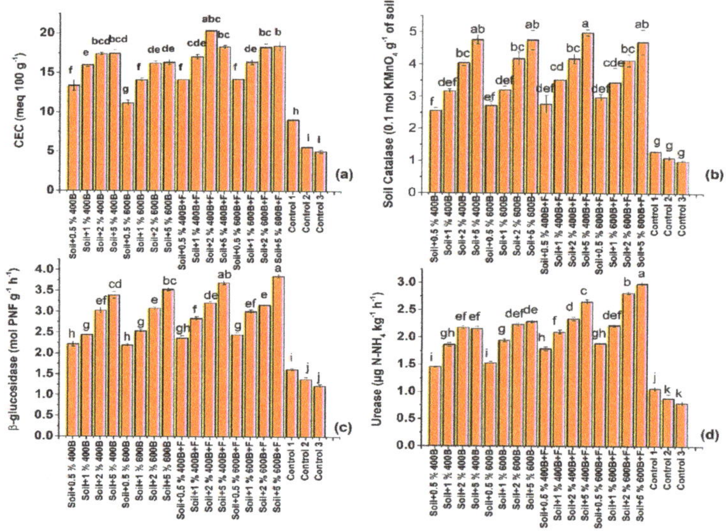

Figure 2. Effect of biochar and fertilizer amendment on the mine soil properties: (**a**) change in the CEC; (**b**) change in the soil catalase activity; (**c**) change in the β-glucosidase activity; (**d**) change in the soil urease activity. Lower letters above the bar graph indicate the statistically significant difference among the cases.

The exchangeable nutrients (nutrients that are available for plant's uses in the soil through the cation or anion exchange process) in the soil amended with 400 and 600 EB were significantly higher than those of controls (Figure 1). Additionally, the availability of the nutrients was increased ($p < 0.05$, Figure 1b,c,e) with the increase in the application rate of the EB from 0.5 to 5% in the soil. The joint application of 400 and 600 EB with fertilizer in the soil marked a noticeable increase ($p < 0.05$, Figure 1c–e) in the availability of nutrients compared to the merely applied biochar. Such an increase in the plant-available nutrient can be linked to the reduction in nutrient leaching and increased availability [40,99,100].

Moreover, at a high application rate (2 and 5%) of biochar, the soil amended with the 600 EB showed higher nutrient content compared to the 400 EB. Such variation can be linked to the higher mineral matter content and CEC in 600 EB compared to the 400 EB [37,93]. The cationic nutrients are usually present in the biochar either in the electrostatically bonded form to the negatively charged surface of the biochar or on its surface through the weak van der Waals force. A similar observation of an increase in the availability of the applied nutrients (NH_4^+, NO_3^-, and PO_4^{3-}) in the sandy loam soil amended with the rice straw biochar was reported in our earlier study [40]. The availability of nutrients in the soil can also be correlated ($r^2 > 0.80$, Figure S5) with the CEC, which was also significantly increased ($p < 0.05$, Figure 2a) in the soil amended with 400 and 600 EB. The CEC of the soil amended with the biochar–fertilizer mixture (Figure 2a) was significantly higher at the application rate greater than 0.5% (*w*/*w*) compared to the soil merely mixed with the 400 and 600 EB. A similar observation of the increase in the CEC of the soil amended with biochar and compost was reported in the earlier studies [91,92]. The increase in the CEC of the soil can be associated with the increase in the organic matter, charge density, and pH of the soil [92]. Moreover, the high specific surface area of the biochar and increase in the surface negative chare due to the surface oxidation of aromatic carbon to form carboxylate groups can also be linked to the increase in the CEC of the soil [92,98,101].

Soil microbial activities, i.e., catalase, β-glucosidase, and urease, are the indicators of an increase in the aerobic microbial activity, microbial activity related to the carbon cycle, and conversion of nitrogen to ammonium form, respectively, in the soil [92]. In the present study, the soil microbial activities were evinced increasing with the application of 400 and 600 EB, as well as biochar–fertilizer mix compared to the controls (C1, C2, and C3). Nevertheless, the rate of application of biochar (from 0.5 to 5%) in the soil also significantly increased ($p < 0.05$, Figure 2b–d) the microbial activities. The increase in the soil microbial activities can be correlated ($r^2 > 0.75$, Figure S5) with the increase in the soil pH, CEC, available nutrients, and organic matter in the soil, which synergistically improved the local soil environment for their growth. The above finding of the association of microbial activities with the soil physicochemical properties is in line with the facts reported in earlier studies, which stated that an increase in the soil pH, CEC, and available nutrients are the key factors for the abundance of microbes in the soil [102,103]. Apart from this, another reason for the significant increase ($p < 0.05$) in the soil microbial activities could be the reduction in the availability of the toxic heavy metals due to their sorption on the biochar surface [101].

The soil fertility index (SFI), which is an indicator of overall soil fertility level, was substantially increased (Figure 3) for the soil amended with the 400 and 600 EB, as well as biochar and fertilizer compared to the controls (C1, C2, and C3). The SFI of the soil mixed with the biochar, as well as the biochar–fertilizer mixture at a higher mixing ratio, i.e., at 5% (*w*/*w*), was comparable to the SFI (29.80, SFI was calculated using the data reported in the paper) of the 7-year-old soil, as reported in an earlier study [104]. The increase in the SFI might be correlated (Figure S5) with the increase in the pH ($r^2 > 0.80$), available nutrients ($r^2 > 0.60$), OM ($r^2 > 0.70$), and CEC ($r^2 > 0.60$). Moreover, the SFI also significantly increased ($p < 0.05$, Figure 3) as the biochar application rate in the soil increased from 0.5 to 5 % (*w*/*w*) ratio. The SFI of the soil mixed with 600 EB biochar was significantly higher ($p < 0.05$) beyond the 2% application rate compared to the 400 EB, which might be due to the significantly higher ($p < 0.05$, Figure 2 and Table S2) availability of plant-available nutrients, liable organic carbon, organic matter, pH, and large surface area that might have helped in the growth of microbes in the soil. The SFI of the soil amended with the co-application of biochar–fertilizer was significantly higher ($p < 0.05$) than the sole application of 400 and 600 EB at the application rate of 2% (*w*/*w*) and above.

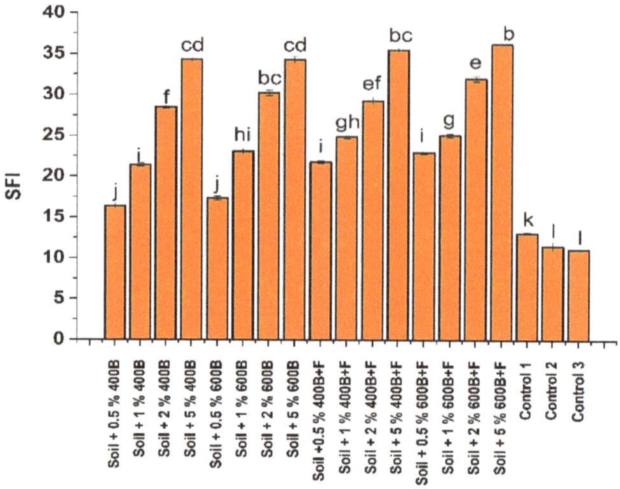

Figure 3. Changes in the soil fertility index (SFI) with the single application of 400 and 600 EB and joint application of biochar with fertilizer. Lower letters above the bar graph indicate the statistically significant difference among the cases.

Effect of Biochar and Fertilizer Amendment on the Availability of the Heavy Metals in the Soil

It has been discussed in the previous Section 3.1.1. that the soil collected from the overburden dump was found contaminated with multiple heavy metals (Ni, Co, Zn, Cu, and Cr). The joint application of the biochar and fertilizer was made to promote the remediation of the soil using the revegetation process, to reduce the metal toxicity to the plants and prevent the leaching of the metals from the soil. It can be observed from Table S6 that the concentrations of the acid-extractable heavy metals (Ni, Co, Cu, Zn, and Cr) in the soil amended with both 400 and 600 EB, as well as the respective biochar and fertilizer, were significantly reduced ($p < 0.05$) compared to the controls (C1 and C2). The application rate of the biochar also had a considerable effect on the sorption of the heavy metals in the soil. The content of total heavy metals (acid extractable) in the soil was significantly reduced ($p < 0.05$, Table S4) by increasing the mixing ratio of the biochar (EB 400 and 600) from 0.5 to 5% (w/w). Such a reduction in the heavy metals in the soil can be linked to the higher adsorption capacity of the EB that might have adsorbed the heavy metals through the formation of metal ion complexes, electrostatic attraction, and precipitation in the soil-biochar matrix. A similar observation was reported in an earlier study, where the acid extractable contents of the Cd, Cu, Zn, and Pb were significantly reduced with the increase in the application rate of bamboo and rice straw-derived biochar from 1 to 5% (w/w) [45]. Importantly, the soil amended with the same varying amount of biochar with the fixed-dose of fertilizer had a significantly ($p < 0.05$) lower content of acid-extractable heavy metals compared to the soil amended with 400 and 600 EB. Such a reduction in the concentration of acid extractable heavy metals can be linked to the fact that the jointly applied biochar–fertilizer in the soil might have promoted the higher adsorption of the metals within the soil [105]. The evidence of the adsorption of heavy metals onto the EB surface can be evinced from the post-pot culture biochar characterization study through SEM, EDAX mapping (Figure S8a), and XRD analysis (Figure S8b). It can be observed from the SEM image analysis and EDAX mapping (Figure S8a) that the heavy metals successfully adsorbed onto the biochar surface. This fact can further be verified through the post-XRD analysis results of biochar, which indicated an increase in the number of peaks of both 400 and 600 EB after the completion of the pot-culture study compared to the pre-biochar analysis. The change in the peaks can be marked at the 2 θ angles of 13°, 21°, 26.5°, 36°,

42°, 45°, 47°, 50°, 55°, 60°, 64°, 68°, and 79°, respectively (Figure S8b). Heavy metals are reportedly immobilized within the soil matrix by forming complex ionic compounds with the phosphate and carbonates to form precipitates, electrostatically bonding to the surface of the biochar, forming complexes with the deprotonated functional groups due to increasing pH, and reduction by accepting the π electrons from the biochar [105,106]. The detailed mechanism of heavy metals sorption onto the biochar within the soil matrix is diagrammatically shown in Figure 4.

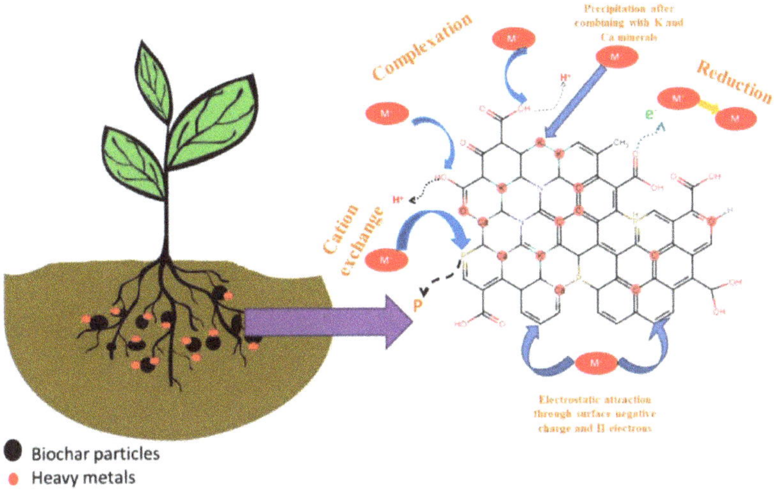

Figure 4. Mechanism of heavy metals sorption-desorption on the biochar present within the root–soil interaction zone.

The soil containing biochar–fertilizer had shown higher sorption capacity compared to the soil merely containing only biochar, and hence, has high adsorption efficiency for heavy metals. Furthermore, the soil containing 600 EB at the application rate greater than 1% had significantly lower ($p < 0.05$, Table S6) metal content compared to the 400 EB. A similar observation was reported in a study by Xiao et al. [107], who reported a high sorption capacity of Pb adsorbed using biochar produced at 600 °C compared to the low-temperature biochar (300 °C).

The DTPA-extractable heavy metals represent the available form of the metals, which are usually present in the soil organically bounded, and can be phytoextracted by the plants in their shoot and root parts [108]. It can be observed from Table S5 that the plant-available forms of heavy metals in the soil amended with both 400 and 600 EB, as well as the biochar–fertilizer, were significantly reduced ($p < 0.05$) compared to the controls (C1 and C2). Moreover, by increasing the application rate of 400 EB from 0.5 to 5% (w/w) in the soil, the availability of the heavy metals was substantially reduced by 64.48, 39.75, 33.73, 20.47, and 34.15%, for Ni, Cu, Zn, Co, and Cr, respectively (Table S3). Similarly, for the 600 EB the availability of the heavy metals was reduced by 59.89, 44.85, 36.2, 30, and 37.57%, respectively. The 600 EB showed a better reduction in the availability of heavy metals compared to the 400 EB. This might be due to the effective adsorption of heavy metals by 600 EB, owing to having high CEC, mineral matter content, and surface area [38]. The joint application of 400 and 600 EB with the fertilizer reduced the availability of the heavy metals at an application rate greater than 1% (w/w) (Table S5). It can be concluded from the above results that the co-application of both 400 and 600 EB up to 5% (application rate (w/w) along with the fertilizer further reduced the availability of Zn, Co, and Cr compared to the single application of biochar in the soil.

The decrease in the availability of the DTPA-extractable metals in the soil amended with both biochar and biochar–fertilizer can be linked to the increase in the soil pH, CEC, available nutrients, and high biochar surface area (Figures 1 and 2). With the increase in the pH of the soil and continuous oxidation with time, the abundance of the polar functional groups (such as –OH, –COOH, –CO, and phenolic groups) reportedly increased [109]. Due to the increased polar functional groups on the biochar surface, the adsorption of cationic metals might have occurred through the surface complexation process [106]. Additionally, the high surface area and the presence of carbonates and minerals on the biochar surface might have also facilitated the adsorption of heavy metals through the formation of metal-carbonate or metal ions precipitate on the biochar surface [107]. Perhaps the pH_{pzc} of the biochar also has a key role in the sorption of heavy metals. It has been reported that the surface of the biochar remains positively charged if the pH of the aqueous solution lies below the pH_{pzc} of the biochar; otherwise, it is negatively charged [40,110]. In the present study, the initial pH of the soil was acidic, with pH = 5.86, which is less than the pH_{pzc} of the biochar (Table S1); however, with time, due to the liming effect, the pH of the soil was increased up to 8.2 (Figure 1). Hence, the surface of the biochar, which was initially positively charged, might have progressively increased the surface negative charge to promote the sorption of the metals through electrostatic attraction. The application of biochar, as well as the biochar–fertilizer mixture, effectively reduced the availability of the metals below the permissible limit in the soil and reduced their toxicity to the plants.

3.2. Effect of Biochar and Fertilizer Amendment on the Plant's Physiological and Biochemical Properties

3.2.1. Effect of Biochar and Fertilizer Amendment on Plant's Growth and Biomass

The plant biomass is an important indicator to mark changes in the physiological characteristics of any vegetation. The plant biomass (*A. auriculiformis)* of both shoot and root parts was significantly increased ($p < 0.05$) with the incorporation of biochar and biochar–fertilizer in the soil compared to the controls (C1 and C2) (Figure 5). Moreover, the shoot biomass was significantly increased by 37.02 and 38.11% ($p < 0.05$, Figure 5) by increasing the application rate of both 600 and 400 EB up to 5% (w/w), respectively. The co-application of biochar and fertilizer significantly increased the shoot and root biomass beyond a 0.5% (w/w) application rate compared to the single application of the biochar (Figure 5). The increase in the plant biomass (shoot and root) in the pots accompanied by the co-application of biochar–fertilizer might be related to the increase in the availability of the nutrients, neutralization of acidic pH, high CEC and WHC, and reduction in the availability of the toxic metals in the soil. A similar observation in the increase in the plant biomass with the application of biochar was reported in the earlier studies [92,111]. Likewise, the shoot and root length in the *A. auriculiformis* also significantly increased ($p < 0.05$, Figure 5) with the increase in the application rate of biochar compared to the controls (C1 and C2). Additionally, the shoot and root lengths were significantly increased by increasing the application rate of 400 and 600 EB from 0.5 to 5% (w/w). The increase in the shoot and root length with the increase in the biochar application rate might be linked to the fact that the presence of biochar in large volumes in the soil may have increased the availability of major and minor nutrients to be utilized by the plants. This fact is in line with the previously reported studies, in which biochar has shown a high affinity towards the sorption when major and minor nutrients applied in the soil [40,112,113].

3.2.2. Effect of Biochar and Fertilizer Amendment on the Heavy Metals Toxicity in the Plant

Concerning the heavy metals in the shoot and root parts of the *A. auriculiformis*, the application of 400 and 600 EB, as well as the biochar and fertilizer significantly ($p < 0.05$, Table S6) reduced the heavy metals content compared to the controls (C1 and C2). Increasing the application rate of 400 and 600 EB, as well as the biochar–fertilizer mixture up to 5% (w/w) significantly ($p < 0.05$, Table S8) reduced the heavy metals in both the shoot and root parts of the plant. A similar observation of a reduction in the heavy metals toxicity in the plant's shoot and root biomass with the application of the biochar was reported in the earlier studies [113,114]. The reduction in the heavy metals toxicity in the plant's shoot and root biomass might be associated with the increase in the sorption of the metals within the biochar–soil interface following the complex adsorption mechanism, as shown in Figure 4 [103,112]. The decrease in the heavy metals in shoot and root biomass can also be linked to the decrease in the content of DTPA-extractable heavy metals in the soil (Table S5).

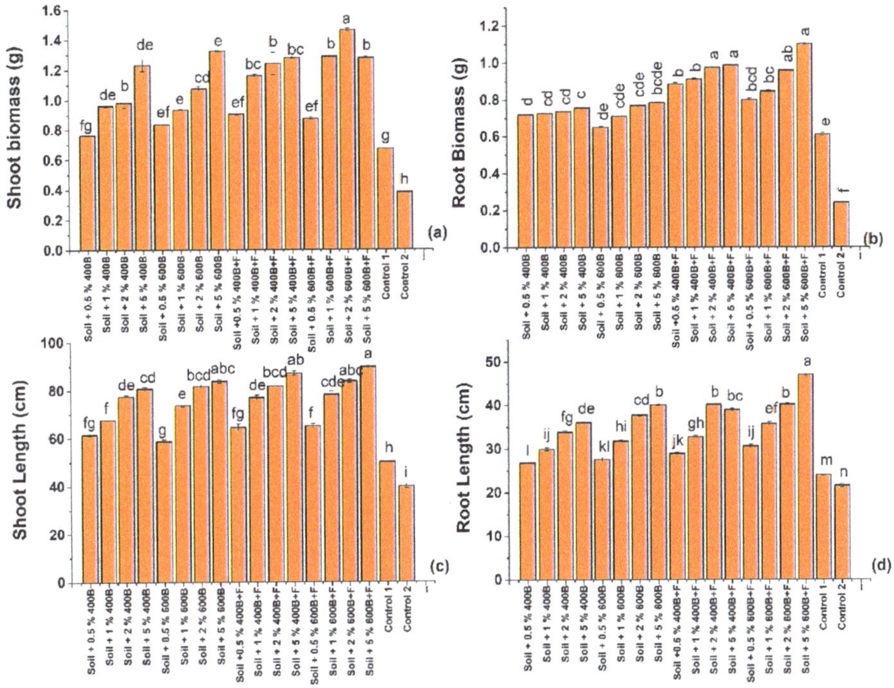

Figure 5. Effect of biochar and biochar–fertilizer mixture on *A. auriculiformis*: (**a**) shoot biomass; (**b**) root biomass, (**c**) shoot length, and (**d**) root length. Lower letters above the bar graph indicate the statistically significant difference among the cases.

The translocation factor (TF) is an important index for the phytoavailability from the root to the shoot part of the plant, whereas the bioaccumulation factor (BAF) is an index for the translocation of metals from the soil to the plant [114,115]. A TF factor value higher than one indicates translocation of the heavy metals from the root to the shoot part of the plant, whereas a TF value less than one indicates the root part has a higher metal concentration than the shoot part [85]. It can be observed from Figure 6a that the TF values for the heavy metals in the plants that grew in the soil amended with biochar and fertilizer were less than the TF values of controlled studies indicating effective sorption of heavy metals within the soil root zone. The TF values for Zn, Ni, Cr, and Cu, which were less than unity, were further decreased with the increase in the application rate of 400 and 600 EB up to 5% (w/w) in the soil. The TF values for Zn, Ni, Cr, and Cu were further reduced with the joint application of biochar and fertilizer in the soil. This indicated that the application of the biochar and biochar–fertilizer effectively immobilized them within the soil-root zone through the various sorption processes [103].

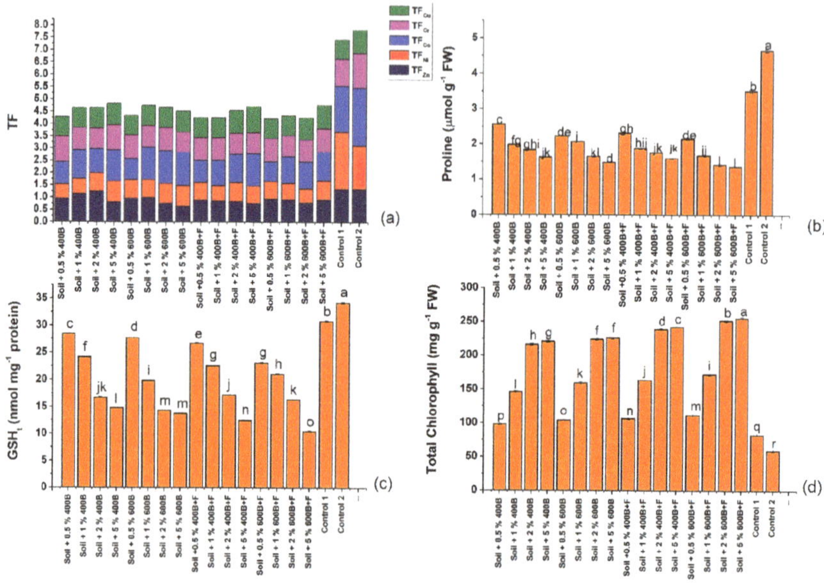

Figure 6. Effect of the application rate of biochar and the biochar–fertilizer mixture on the (**a**) Translocation factor, (**b**) proline production, (**c**) GSH$_t$ production, and (**d**) total chlorophyll content in the plants (*A. Auriculiformis*) Lower letters above the bar graph indicate the statistically significant difference among the cases.

Instead, the TF value of Co was increased with the increase in the application rate, indicating that the excess of Co^{2+} ions was translocated from the root to the shoot parts due to their poor sorption in the biochar-soil matrix. Moreover, the co-application of biochar and fertilizer further increased the TF values due to the poor sorption of Co^{2+} ions in the soil matrix. Such a reduction in the sorption of Co^{2+} ions in the soil matrix amended with both biochar and fertilizer can be related to the reduction in the adsorption sites in the competitive environment.

The bioaccumulation factor (BAF) for the heavy metals was lower (<1) in the plants that grew in the soil amended with the biochar and biochar–fertilizer mixture compared to the controls (C1 and C2, Figure S7). Moreover, the BAF_{shoot} for heavy metals was further reduced in the plants grown in the soil amended with the co-applied biochar and fertilizer except for Zn. The reduction in the BAF_{shoot} due to the co-application of biochar and fertilizer can be linked to the increase in the CEC and the presence of functional groups on the biochar surface that might have immobilized the metals within the soil [116]. However, the existence of cations might also have reduced the relative affinity for Zn^{2+} ions within the soil matrix due to increased competence on the adsorption sites onto the biochar's surface. A similar observation of lower Zn sorption affinity for the biochar in a multimetal system was reported in an earlier study [117].

The BAF_{shoot} for Ni, Co, and Cu were increased with the increase in the application rate of both 400 and 600 EB from 0.5 to 5% (*w*/*w*) (Figure S7), whereas, for Zn and Cr, it was decreased by 31.92 and 14.28% and 13.95 and 16.67%, respectively. This indicated the higher translocation of Ni, Co, and Cu ions from the soil to the shoot part due to the presence of an excess of these ions within the soil matrix with the increase in the biochar's application rate. Such an increase in BAF_{shoot} for these heavy metals can be associated with a decrease in the sorption capacity due to an increase in the biochar volume in the soil. As the biochar also intrinsically contains these metals in its structure to some extent (Table S3), the existence of the metals might have increased with the increase in the application rate, which, in turn, may have reduced their sorption in the soil matrix. Conversely, the Zn and Cr ions, in which Cr ions mostly occur in Cr (VI) form ($HCrO_4^-$, CrO_4^{2-}, and $Cr_2O_7^{2-}$) under oxidizing conditions and at the acidic pH [118–120], might have adsorbed within the soil matrix through electrostatic attraction and abundance of cations and high organic matter content within the soil matrix.

The application of the 400 and 600 EB substantially reduced the BAF_{root} values for heavy metals in the root part of the plant (Table S8). This can be linked to the decrease in the concentration of the available form of the metals by the formation of metal complexes with the functional groups present on the biochar [103]. Additionally, the joint application of the biochar with fertilizer further reduced the BAF_{root} values for heavy metals in the root zone due to the presence of additional ions (K^+, PO_4^{3-}, and NO_3^-), as well as a relative increase in the pH compared to the single application of the biochar, which might have facilitated the precipitation of the metal ions.

3.2.3. Effect of Biochar and Fertilizer Amendment on a Plant's Biochemical Hormones

As discussed earlier, the contaminated soil was deprived of nutrients and had elevated levels of heavy metals. Plants that grew in such soil were subjected to the drought condition along with having the stress of metal toxicity [121]. Under such poor environmental conditions, overproduction of reactive oxygen species (ROS) occurs, which indicates damage to the plant cells [122]. Under such circumstances, a defense mechanism comes into action to prevent the plant cells from being damaged, which leads to the production of antioxidants such as proline and total glutathione (GSH_t) [123,124]. The excessive production of proline and GSH_t are the indicators of the plant being subjected to abiotic stress conditions. It can be observed from Figure 6 that the application of 400 and 600 EB, as well as the joint application of biochar and fertilizer in the soil, significantly ($p < 0.05$) reduced the level of proline and GSH_t in the plants compared to the controls (C1 and C2) due to the inhibition of the production of ROS. Moreover, with the increasing application rate of the 400 and 600 EB from 0.5 to 5% (w/w), the proline level in the plant cells was significantly ($p < 0.05$) reduced by 36.57 and 33.48%, respectively. Likewise, the GSH_t level was reduced by 47.78 and 50.32%, respectively. The co-application of biochar–fertilizer further significantly ($p < 0.05$, Figure 6) reduced the proline and GSH_t levels in the plant tissues at the various application rates (0.5 to 5% (w/w). A similar observation of the reduction in the proline and GSH_t due to the application of the biochar in the mung bean reported in an earlier study [125]. The reduction in the proline and GSH_t levels in the plant tissues with the application of 400 and 600 EB, as well as the co-applied biochar–fertilizer might be related to the increase in the plant-available nutrients, improvement in the soil pH, and a reduction in the plant-available heavy metals. The relation among proline, GSH_t, and soil physicochemical properties is evident in the correlation matrix Figure S6, where the proline and GSH_t levels in the plant tissues have shown a strong positive correlation ($r^2 > 0.85$) with the TF and BAC for the heavy metals, and strong negative correlation ($r^2 > 0.90$) with the soil pH. Chlorophyll is the green pigment present in the plant leaves primarily responsible for the photosynthesis in the plants. It can be observed from Figure 6 that the incorporation of 400 and 600 EB, as well as the jointly applied biochar–fertilizer, significantly ($p < 0.05$) improved the chlorophyll content in the plants compared to the controls (C1 and C2). Moreover, with the increase in the application rate of biochar in the soil from 0.5 to 5% (w/w), the chlorophyll content in the plant leaves was also significantly ($p < 0.05$) improved. The increase in the chlorophyll content in the plants that grew in the biochar and biochar–fertilizer amended soil might be associated with the increase in the availability of the major and minor nutrients, and a decrease in the metal toxicity as indicated by TF and BAF values discussed earlier (Figure S6).

4. Conclusions

The following conclusions can be obtained from the results of the present study:

- Incorporation of the biochar in the soil improved its physicochemical properties, such as an increase in pH, available nutrients, organic matter, and soil enzymatic activities.
- An increase in the rate of application of biochar up to 5% (*w*/*w*) significantly improved the physicochemical properties of the soil. This can be evinced by the increase in the pH, exchangeable nutrients, organic matter, CEC, and the soil's enzymatic activities. Simultaneously, it has reduced the plant-available content of the heavy metals in the soil.
- The soil fertility index was significantly increased with the application of both biochar and the biochar–fertilizer mixture compared to the controls. Furthermore, the co-application of the biochar with NPK fertilizer increased its efficacy to be used as a soil-amending material.
- High-temperature biochar (600 EB) showed better sorption of heavy metals in the soil compared to 400 EB indicating better efficacy of the 600 EB in reducing the metal toxicity in the soil.
- Plant analysis results showed that the co-application of biochar with fertilizer substantially reduced metal toxicity and water stress effect in the plants as evidenced by the BAF, TF, proline, and glutathione (GSH_t) data of the plant analysis.

Supplementary Materials: The following supporting information can be downloaded at: https://www.mdpi.com/article/10.3390/su14127266/s1, Materials and Methods Section [38,63,93,126–132]. Figure S1: Bera opencast coalmines, Bastacolla area, Dhanbad, Jharkhand, India. Figure S2: Pot-culture study in a polyhouse using biochar and fertilizer as the soil amending material in the soil planted with Accacia Auriculiformis. Figure S3: Correlation matrix for the physicochemical properties of the *Eucalyptus* wood biochar. Figure S4: (a) SEM image of the EB; (b) FTIR curve of the EB; (c) XRD curve of the EB produced at 400 and 600 °C. Figure S5: Correlation matrix among the soil physicochemical properties. Figure S6: Correlation matrix among proline, GSH, chlorophyll content, TF, BAC, and soil pH. Figure S7: Bioaccumulation factor (BAC) of the heavy metals in the shoot and root part of the *A. Auriculiformis*. Figure S8: Post *Eucalyptus wood* Biochar (EB) characterization: (a) SEM Image and heavy metals mapping adsorbed onto the EB surface; (b) XRD analysis of the post 400 and 600 EB. Table S1: Elemental and proximate analysis results of Eucalyptus wood biochar produced at 400 and 600 °C. Table S2: Physicochemical characteristics of eucalyptus wood biochar (mean ± S.D., n = 3). Table S3: Heavy metals content in the eucalyptus wood biochar (mean ± S.D., n = 3). Table S4: BET surface area and pore volume of the eucalyptus wood biochar. Table S5: DTPA extractable heavy metals in the mine soil (n = 3, mean ± S.D.) after pot-culture study. Table S6: Acid extractable heavy metals in the mine-soil after pot-culture study. Table S7: Statistical analysis to test the normal distribution of the data. Table S8: Heavy metals in the plant shoot and root parts (n = 3, mean ± S.D.).

Author Contributions: S.C.: Conceptualization, experimental design, method development, and original draft writing; I.M.: data curation, review and editing; J.B.: Investigation, supervision, project administration, review, editing, and resources; K.R.V.: Validation, review, and editing, B.S.: Validation. All authors have read and agreed to the published version of the manuscript.

Funding: This research received no external funding.

Institutional Review Board Statement: This study did not require ethical approval.

Informed Consent Statement: Not applicable.

Data Availability Statement: This study did not report any data for publication.

Acknowledgments: The authors sincerely acknowledge the Indian Institute of Technology Kharagpur for providing the research facility and MHRD and the government of India for providing the fellowship for the research work.

Conflicts of Interest: The authors declare no conflict of interest.

References

1. WHO. *Permissible Limits of Heavy Metals in Soil and Plants*; WHO: Geneva, Switzerland, 1996.
2. Dewangan, P.; Mishra, R.; Jhariya, D. Opencast Coal Mining at Large Depth in India-challenges ahead. *World J. Eng. Res. Technol.* **2017**, *3*, 201–211.
3. Loska, K.; Wiechuła, D.; Barska, B.; Cebula, E.; Chojnecka, A. Assessment of Arsenic Enrichment of Cultivated Soils in Southern Poland. *Pol. J. Environ. Stud.* **2003**, *12*, 187–192.
4. Raj, D.; Kumar, A.; Maiti, S.K. Evaluation of toxic metal(loid)s concentration in soils around an open-cast coal mine (Eastern India). *Environ. Earth Sci.* **2019**, *78*, 645. [CrossRef]
5. Kronbauer, M.A.; Izquierdo, M.; Dai, S.; Waanders, F.B.; Wagner, N.J.; Mastalerz, M.; Hower, J.C.; Oliveira, M.L.S.; Taffarel, S.R.; Bizani, D.; et al. Environment Geochemistry of ultra-fine and nano-compounds in coal gasification ashes: A synoptic view. *Sci. Total* **2013**, *457*, 95–103. [CrossRef] [PubMed]
6. Yenilmez, F.; Kuter, N.; Kemal, M.; Aksoy, A. International Journal of Coal Geology Evaluation of pollution levels at an abandoned coal mine site in Turkey with the aid of GIS. *Int. J. Coal Geol.* **2011**, *86*, 12–19. [CrossRef]
7. Liu, X.; Bai, Z.; Zhou, W.; Cao, Y.; Zhang, G. Changes in soil properties in the soil profile after mining and reclamation in an opencast coal mine on the Loess Plateau, China. *Ecol. Eng.* **2017**, *98*, 228–239. [CrossRef]
8. Loupasakis, C.; Angelitsa, V.; Rozos, D.; Spanou, N. Mining geohazards—Land subsidence caused by the dewatering of opencast coal mines: The case study of the Amyntaio coal mine, Florina, Greece. *Nat. Hazards* **2014**, *70*, 675–691. [CrossRef]
9. Bai, Z.K.; Fu, M.C.; Zhao, Z.Q. On soil environmental problems in mining area. *Ecol. Environ.* **2006**, *15*, 1122–1125.
10. Liu, X.; Shi, H.; Bai, Z.; Zhou, W.; Liu, K.; Wang, M.; He, Y. Heavy metal concentrations of soils near the large opencast coal mine pits in China. *ECSN* **2019**, *244*, 125360. [CrossRef]
11. Park, B.; Lee, J.; Ro, H.; Ho, Y. Effects of heavy metal contamination from an abandoned mine on nematode community structure as an indicator of soil ecosystem health. *Appl. Soil Ecol.* **2011**, *51*, 17–24. [CrossRef]
12. Wang, J.; Liu, W.; Yang, R.; Zhang, L.; Ma, J. Assessment of the potential ecological risk of heavy metals in reclaimed soils at an opencast coal mine. *Disaster Adv.* **2013**, *6*, 366–377.
13. Zhai, X.; Li, Z.; Huang, B.; Luo, N.; Huang, M.; Zhang, Q.; Zeng, G. Environment Remediation of multiple heavy metal-contaminated soil through the combination of soil washing and in situ immobilization. *Sci. Total Environ.* **2018**, *635*, 92–99. [CrossRef] [PubMed]
14. Maiti, S.K. Bioreclamation of coalmine overburden dumps—With special emphasis on micronutrients and heavy metals. *Environ. Monit. Assess.* **2007**, *125*, 111–122. [CrossRef]
15. Vega, F.A.; Covelo, E.F.; Andrade, M.L. Competitive sorption and desorption of heavy metals in mine soils: Influence of mine soil characteristics. *J. Colloid Interface Sci.* **2006**, *298*, 582–592. [CrossRef] [PubMed]
16. Peng, J.; Song, Y.; Yuan, P.; Cui, X.; Qiu, G. The remediation of heavy metals contaminated sediment. *J. Hazard. Mater.* **2009**, *161*, 633–640. [CrossRef] [PubMed]
17. Montinaro, S.; Concas, A.; Pisu, M.; Cao, G. Remediation of heavy metals contaminated soils by ball milling. *Chemosphere* **2007**, *67*, 631–639. [CrossRef]
18. Sheoran, V.; Sheora, A.S.; Poonia, P. Soil Reclamation of Abandoned Mine Land by Revegetation: A Review. *Int. J. Soil Sediment Water* **2010**, *3*, 13.
19. Yang, S.; Liao, B.; Yang, Z.; Chai, L.; Li, J. Revegetation of extremely acid mine soils based on aided phytostabilization: A case study from southern China. *Sci. Total Environ.* **2016**, *562*, 427–434. [CrossRef]
20. Wong, M. Ecological restoration of mine degraded soils, with emphasis on metal contaminated soils. *Chemosphere* **2003**, *50*, 775–780. [CrossRef]
21. Novo, L.A.B.; Castro, P.M.L.; Alvarenga, P.; da Silva, E.F. *Plant Growth-Promoting Rhizobacteria-Assisted Phytoremediation of Mine Soils*; Elsevier Inc.: Amsterdam, The Netherlands, 2018; ISBN 9780128129876.
22. Palutoglu, M.; Akgul, B.; Suyarko, V.; Yakovenko, M.; Kryuchenko, N.; Sasmaz, A. Phytoremediation of Cadmium by Native Plants Grown on Mining Soil. *Bull. Environ. Contam. Toxicol.* **2018**, *100*, 293–297. [CrossRef]
23. Mahar, A.; Wang, P.; Ali, A.; Awasthi, M.K.; Lahori, A.H.; Wang, Q.; Li, R.; Zhang, Z. Challenges and opportunities in the phytoremediation of heavy metals contaminated soils: A review. *Ecotoxicol. Environ. Saf.* **2016**, *126*, 111–121. [CrossRef] [PubMed]
24. Acosta, J.A.; Abbaspour, A.; Zornoza, R.; Faz, A.; Abbaspour, A.; Faz, A. Phytoremediation of mine tailings with Atriplex halimus and organic/inorganic amendments: A five-year field case study. *Chemosphere* **2018**, *204*, 71–78. [CrossRef] [PubMed]
25. Palansooriya, K.N.; Shaheen, S.M.; Chen, S.S.; Tsang, D.C.W.; Hashimoto, Y.; Hou, D.; Bolan, N.S.; Rinklebe, J.; Ok, Y.S. Soil amendments for immobilization of potentially toxic elements in contaminated soils: A critical review. *Environ. Int.* **2020**, *134*, 105046. [CrossRef] [PubMed]
26. Wong, V. Remediation and restoration of contaminated soils for plant growth and establishment. In Proceedings of the EGU General Assembly Conference, Vienna, Austria, 27 April–2 May 2014; Volume 16, p. 16586.
27. Hodson, M.E.; Valsami-Jones, E.; Cotter-Howells, J.D. Bonemeal additions as a remediation treatment for metal contaminated soil. *Environ. Sci. Technol.* **2000**, *34*, 3501–3507. [CrossRef]
28. Abbott, D.E.; Essington, M.E.; Mullen, M.D.; Ammons, J.T. Fly Ash and Lime-Stabilized Biosolid Mixtures in Mine Spoil Reclamation: Simulated Weathering. *J. Environ. Qual.* **2001**, *30*, 608–616. [CrossRef]

29. Ciccu, R.; Ghiani, M.; Serci, A.; Fadda, S.; Peretti, R.; Zucca, A. Heavy metal immobilization in the mining-contaminated soils using various industrial wastes. *Miner. Eng.* **2003**, *16*, 187–192. [CrossRef]
30. Tandy, S.; Healey, J.R.; Nason, M.A.; Williamson, J.C.; Jones, D.L. Remediation of metal polluted mine soil with compost: Co-composting versus incorporation. *Environ. Pollut.* **2009**, *157*, 690–697. [CrossRef]
31. Singh, S.N.; Kulshreshtha, K.; Ahmad, K.J. Impact of fly ash soil amendment on seed germination, seedling growth and metal composition of *Vicia faba* L. *Ecol. Eng.* **1997**, *9*, 203–208. [CrossRef]
32. Kiikkilä, O.; Pennanen, T.; Perkiömäki, J.; Derome, J.; Fritze, H. Basic and Applied Ecology Organic material as a copper immobilising agent: A microcosm study on remediation. *Basic Appl. Ecol.* **2002**, *253*, 245–253. [CrossRef]
33. Wu, L.; Ma, L.Q.; Martinez, G.A. Comparison of Methods for Evaluating Stability and Maturity of Biosolids Compost. *J. Environ. Qual.* **2000**, *29*, 424–429. [CrossRef]
34. Beecher, N.; Harrison, E.; Goldstein, N.; McDaniel, M.; Field, P.; Susskind, L. Risk Perception, Risk Communication, and Stakeholder Involvement for Biosolids Management and Research. *J. Environ. Qual.* **2005**, *34*, 122–128. [CrossRef] [PubMed]
35. Novak, J.M.; Busscher, W.J.; Laird, D.L.; Ahmedna, M.; Watts, D.W.; Niandou, M.A.S. Impact of biochar amendment on fertility of a southeastern coastal plain soil. *Soil Sci.* **2009**, *174*, 105–112. [CrossRef]
36. Zhang, J.; Liu, J.; Liu, R. Effects of pyrolysis temperature and heating time on biochar obtained from the pyrolysis of straw and lignosulfonate. *Bioresour. Technol.* **2015**, *176*, 288–291. [CrossRef] [PubMed]
37. Khanmohammadi, Z.; Afyuni, M. Effect of pyrolysis temperature on chemical and physical properties of sewage sludge biochar. *Waste Manag. Res.* **2015**, *33*, 275–283. [CrossRef]
38. Gai, X.; Wang, H.; Liu, J.; Zhai, L.; Liu, S.; Ren, T.; Liu, H. Effects of feedstock and pyrolysis temperature on biochar adsorption of ammonium and nitrate. *PLoS ONE* **2014**, *9*, e113888. [CrossRef]
39. Tang, J.; Zhu, W.; Kookana, R.; Katayama, A. Characteristics of biochar and its application in remediation of contaminated soil. *J. Biosci. Bioeng.* **2013**, *116*, 653–659. [CrossRef]
40. Chandra, S.; Medha, I.; Bhattacharya, J. Potassium-iron rice straw biochar composite for sorption of nitrate, phosphate, and ammonium ions in soil for timely and controlled release. *Sci. Total Environ.* **2020**, *712*, 136337. [CrossRef]
41. Abdelhafez, A.A.; Li, J.; Abbas, M.H.H. Feasibility of biochar manufactured from organic wastes on the stabilization of heavy metals in a metal smelter contaminated soil. *Chemosphere* **2014**, *117*, 66–71. [CrossRef]
42. Tripti; Kumar, A.; Usmani, Z.; Kumar, V.; Anshumali. Biochar and flyash inoculated with plant growth promoting rhizobacteria act as potential biofertilizer for luxuriant growth and yield of tomato plant. *J. Environ. Manag.* **2017**, *190*, 20–27. [CrossRef]
43. Alaboudi, K.A.; Ahmed, B.; Brodie, G. Effect of biochar on Pb, Cd and Cr availability and maize growth in artificial contaminated soil. *Ann. Agric. Sci.* **2019**, *64*, 95–102. [CrossRef]
44. Medyńska-Juraszek, A.; Ćwieląg-Piasecka, I. Effect of Biochar Application on Heavy Metal Mobility in Soils Impacted by Copper Smelting Processes. *Pol. J. Environ. Stud.* **2020**, *29*, 1749–1757. [CrossRef]
45. Lu, K.; Yang, X.; Gielen, G.; Bolan, N.; Ok, Y.S.; Niazi, N.K.; Xu, S.; Yuan, G.; Chen, X.; Zhang, X.; et al. Effect of bamboo and rice straw biochars on the mobility and redistribution of heavy metals (Cd, Cu, Pb and Zn) in contaminated soil. *J. Environ. Manag.* **2017**, *186*, 285–292. [CrossRef] [PubMed]
46. Kavitha, B.; Reddy, P.V.L.; Kim, B.; Lee, S.S.; Pandey, S.K.; Kim, K.H. Benefits and limitations of biochar amendment in agricultural soils: A review. *J. Environ. Manag.* **2018**, *227*, 146–154. [CrossRef] [PubMed]
47. Liang, J.; Yang, Z.; Tang, L.; Zeng, G.; Yu, M.; Li, X.; Wu, H.; Qian, Y.; Li, X.; Luo, Y. Changes in heavy metal mobility and availability from contaminated wetland soil remediated with combined biochar-compost. *Chemosphere* **2017**, *181*, 281–288. [CrossRef] [PubMed]
48. Karami, N.; Clemente, R.; Moreno-Jiménez, E.; Lepp, N.W.; Beesley, L. Efficiency of green waste compost and biochar soil amendments for reducing lead and copper mobility and uptake to ryegrass. *J. Hazard. Mater.* **2011**, *191*, 41–48. [CrossRef]
49. Zeng, G.; Wu, H.; Liang, J.; Guo, S.; Huang, L.; Xu, P.; Liu, Y.; Yuan, Y.; He, X.; He, Y. Efficiency of biochar and compost (or composting) combined amendments for reducing Cd, Cu, Zn and Pb bioavailability, mobility and ecological risk in wetland soil. *RSC Adv.* **2015**, *5*, 34541–34548. [CrossRef]
50. Karer, J.; Wawra, A.; Zehetner, F.; Dunst, G.; Wagner, M.; Pavel, P.B.; Puschenreiter, M.; Friesl-Hanl, W.; Soja, G. Effects of biochars and compost mixtures and inorganic additives on immobilisation of heavy metals in contaminated soils. *Water. Air. Soil Pollut.* **2015**, *226*, 342. [CrossRef]
51. Mohamed, B.A.; Ellis, N.; Kim, C.S.; Bi, X. The role of tailored biochar in increasing plant growth, and reducing bioavailability, phytotoxicity, and uptake of heavy metals in contaminated soil. *Environ. Pollut.* **2017**, *230*, 329–338. [CrossRef]
52. Gregory, S.J.; Anderson, C.W.N.; Camps Arbestain, M.; McManus, M.T. Response of plant and soil microbes to biochar amendment of an arsenic-contaminated soil. *Agric. Ecosyst. Environ.* **2014**, *191*, 133–141. [CrossRef]
53. Ahmad, M.; Lee, S.S.; Lee, S.E.; Al-Wabel, M.I.; Tsang, D.C.W.; Ok, Y.S. Biochar-induced changes in soil properties affected immobilization/mobilization of metals/metalloids in contaminated soils. *J. Soils Sediments* **2017**, *17*, 717–730. [CrossRef]
54. Novak, J.M.; Ippolito, J.A.; Lentz, R.D.; Spokas, K.A.; Bolster, C.H.; Sistani, K.; Trippe, K.M.; Phillips, C.L.; Johnson, M.G. Soil Health, Crop Productivity, Microbial Transport, and Mine Spoil Response to Biochars. *Bioenergy Res.* **2016**, *9*, 454–464. [CrossRef]
55. Qi-Kai, W.; Wen-Juan, G.; Guo-Hong, S.; Da-Song, L.; Ying-Ming, X.; Jing-Ru, L.; Shi-Lei, Y. Combined effects of biochar and fertilizer on cadmium contaminated soil remediation. *J. Agric. Resour. Environ.* **2015**, *32*, 583.

56. Netherway, P.; Reichman, S.M.; Laidlaw, M.; Scheckel, K.; Pingitore, N.; Gascó, G.; Méndez, A.; Surapaneni, A.; Paz-Ferreiro, J. Phosphorus-Rich Biochars Can Transform Lead in an Urban Contaminated Soil. *J. Environ. Qual.* **2019**, *48*, 1091–1099. [CrossRef] [PubMed]
57. Bradl, H.B. Adsorption of heavy metal ions on soils and soils constituents. *J. Colloid Interface Sci.* **2004**, *277*, 1–18. [CrossRef]
58. Chandra, S.; Bhattacharya, J. Influence of temperature and duration of pyrolysis on the property heterogeneity of rice straw biochar and optimization of pyrolysis conditions for its application in soils. *J. Clean. Prod.* **2019**, *215*, 1123–1139. [CrossRef]
59. Cheung, K.C.; Wong, J.P.K.; Zhang, Z.Q.; Wong, J.W.C.; Wong, M.H. Revegetation of lagoon ash using the legume species Acacia auriculiformis and Leucaena leucocephala. *Environ. Pollut.* **2000**, *109*, 75–82. [CrossRef]
60. Kasongo, R.K.; VanRanst, E.; Verdoodt, A.; Kanyankagote, P.; Baert, G. Impact of Acacia auriculiformis on the chemical fertility of sandy soils on the Batéké plateau, DR Congo. *Soil Use Manag.* **2009**, *25*, 21–27. [CrossRef]
61. Dudek, M.; Kloc, S.; Kręt, A. Wheat Straw Biochar and NPK Fertilization Efficiency in Sandy Soil Reclamation. *Agronomy* **2020**, *10*, 496.
62. Pietrzykowski, M.; Krzaklewski, W. Soil organic matter, C and N accumulation during natural succession and reclamation in an opencast sand quarry (southern Poland). *Arch. Agron. Soil Sci.* **2007**, *53*, 473–483. [CrossRef]
63. Maiti, S.K. *Ecorestoration of the Coalmine Degraded Lands*; Springer Science & Business Media: Berlin/Heidelberg, Germany, 2012; ISBN 9788578110796.
64. Mylavarapu, R.; Bergeron, J.; Wilkinson, N. Soil pH and Electrical Conductivity: A County Extension Soil Laboratory Manual 1 Solubility of Plant Nutrients. 2015. Available online: https://edis.ifas.ufl.edu/pdf/SS/SS11800.pdf (accessed on 25 April 2022).
65. Walkley, A.; Black, T.A. An examination of the Degtjareff method for determining soil organic matter, and a proposed modification of the chromic acid titration method. *Soil Sci.* **1934**, *37*, 29–38. [CrossRef]
66. Nelson, D.W.; Sommers, L.E. Total carbon, organic carbon, and organic matter. In *Methods of Soil Analysis. Part 2: Chemical and Microbiological Properties*; Page, A.L., Miller, R.H., Keeny, D.R., Eds.; American Society of Agronomy: Madison, WI, USA, 1982; pp. 539–579.
67. Schulte, E.E.; Hopkins, B.G. Estimation of Soil Organic Matter by Weight Loss-On-Ignition. *Soil Sci. Soc. Am.* **1996**, *049*, 21–31.
68. Wright, A.L.; Wang, Y.; Reddy, K.R. Loss-on-Ignition Method to Assess Soil Organic Carbon in Calcareous Everglades Wetlands. *Commun. Soil Sci. Plant Anal.* **2008**, *39*, 3074–3083. [CrossRef]
69. Bray, R.H.; Kurtz, L.T. Determination of total, organic, and available forms of phosphorus in soils. *Soil Sci.* **1945**, *59*, 39–46. [CrossRef]
70. Asija, G.L.; Subbiah, B.V. A rapid procedure for the estimation of available nitrogen in soils. *Curr. Sci.* **1956**, *25*, 259–260.
71. Sumner, M.E.; Miller, W.P. Cation exchange capacity and exchange coefficients. In *Methods of Soil Analysis. Part 3. Chemical Methods-SSSA Book Series No. 5*; Soil Science Society of America and American Society of Agronomy: Madison, WI, USA, 1994; pp. 1201–1229.
72. Element, C.A.S. Method 3051A microwave assisted acid digestion of sediments, sludges, soils, and oils. *Z. Für Anal. Chem.* **2007**, *111*, 362–366.
73. Lindsay, W.L.; Norvell, W.A. Equilibrium Relationships of Zn2+, Fe3+, Ca2+, and H+ with EDTA and DTPA in Soils. *Soil Sci. Soc. Am. J.* **1969**, *33*, 62–68. [CrossRef]
74. Roberge, M.R. Methodology of enzymes determination and extraction. In *'Soil Enzymes'*; Burns, R.G., Ed.; Academic Press: New York, NY, USA, 1978; pp. 341–352.
75. Ouyang, L.; Tang, Q.; Yu, L.; Zhang, R. Effects of amendment of different biochars on soil enzyme activities related to carbon mineralisation. *Soil Res.* **2014**, *52*, 706–716. [CrossRef]
76. Alef, K.; Nannipieri, P. Enzyme activities. In *Methods in Applied Soil Microbiology and Biochemistry*; Academic Press: London, UK, 1995; pp. 311–373.
77. Kandeler, E.; Gerber, H. Short-term assay of soil urease activity using colorimetric determination of ammonium. *Biol. Fertil. Soils* **1988**, *6*, 68–72. [CrossRef]
78. Panwar, P.; Pal, S.; Reza, S.K.; Sharma, B. Soil fertility index, soil evaluation factor, and microbial indices under different land uses in acidic soil of humid subtropical India. *Commun. Soil Sci. Plant Anal.* **2011**, *42*, 2724–2737. [CrossRef]
79. Lu, D.; Moran, E.; Mausel, P. Linking amazonian secondary succession forest growth to soil properties. *Land Degrad. Dev.* **2002**, *13*, 331–343. [CrossRef]
80. Anderson, M.E.; Orrenius, S.; Holmgren, A.; Mannervik, B.; Press, R. Determination of Glutathione and Glutathione Disulfide in Biological Samples. *Methods Enzymol.* **1985**, *113*, 548–555. [PubMed]
81. Bates, L.S.; Waldren, R.P.; Teare, I.D. Rapid determination of free proline for water-stress studies. *Plant Soil* **1973**, *39*, 205–207. [CrossRef]
82. Dere, Ş.; Güneş, T.; Sivaci, R. Spectrophotometric Determination of Chlorophyll-A, B and Total Carotenoid Contents of Some Algae Species Using Different Solvents. *Turk. J. Bot.* **1998**, *22*, 13–17.
83. Li, Y.; Sun, Y.; Jiang, J.; Liu, J. Spectroscopic determination of leaf chlorophyll content and color for genetic selection on Sassafras tzumu. *Plant Methods* **2019**, *15*, 73. [CrossRef]
84. Tuzen, M. Determination of heavy metals in soil, mushroom and plant samples by atomic absorption spectrometry. *Microchem. J.* **2003**, *74*, 289–297. [CrossRef]

85. Banerjee, R.; Goswami, P.; Pathak, K.; Mukherjee, A. Vetiver grass: An environment clean-up tool for heavy metal contaminated iron ore soil. *Ecol. Eng.* **2016**, *90*, 25–34. [CrossRef]
86. Wu, Q.; Wang, S. Phytostabilization Potential of *Jatropha curcas* L. in Polymetallic Acid Mine Tailings. *Int. J. Phytoremediation* **2011**, *13*, 788–804. [CrossRef]
87. Ghose, M.K. Land reclamation and protection of environment from the effect of coal mining operation. *Mine Technol.* **1989**, *10*, 35–39.
88. Daily, G.D. Restoring Value to the World's Degraded Lands. *Science* **1995**, *269*, 350–354. [CrossRef]
89. Tóth, G.; Hermann, T.; da Silva, M.R.; Montanarella, L. Heavy metals in agricultural soils of the European Union with implications for food safety. *Environ. Int.* **2016**, *88*, 299–309. [CrossRef]
90. Ali, H.; Khan, E.; Anwar, M. Phytoremediation of heavy metals—Concepts and applications. *Chemosphere* **2013**, *91*, 869–881. [CrossRef] [PubMed]
91. Mensah, A.K.; Frimpong, K.A. Biochar and/or Compost Applications Improve Soil Properties, Growth, and Yield of Maize Grown in Acidic Rainforest and Coastal Savannah Soils in Ghana. *Int. J. Agron.* **2018**, *2018*, 6837404. [CrossRef]
92. Khan, W.D.; Ramzani, P.M.A.; Anjum, S.; Abbas, F.; Iqbal, M.; Yasar, A.; Ihsan, M.Z.; Anwar, M.N.; Baqar, M.; Tauqeer, H.M.; et al. Potential of miscanthus biochar to improve sandy soil health, in situ nickel immobilization in soil and nutritional quality of spinach. *Chemosphere* **2017**, *185*, 1144–1156. [CrossRef] [PubMed]
93. Wu, W.X.; Yang, M.; Feng, Q.B.; McGrouther, K.; Wang, H.L.; Lu, H.H.; Chen, Y.X. Chemical characterization of rice straw-derived biochar for soil amendment. *Biomass Bioenergy* **2012**, *47*, 268–276. [CrossRef]
94. Harvey, O.R.; Kuo, L.J.; Zimmerman, A.R.; Louchouarn, P.; Amonette, J.E.; Herbert, B.E. An index-based approach to assessing recalcitrance and soil carbon sequestration potential of engineered black carbons (biochars). *Environ. Sci. Technol.* **2012**, *46*, 1415–1421. [CrossRef]
95. Bista, P.; Ghimire, R.; Machado, S.; Pritchett, L. Biochar Effects on Soil Properties and Wheat Biomass vary with Fertility Management. *Agronomy* **2019**, *9*, 623. [CrossRef]
96. Frimpong, K.A.; Amoakwah, E.; Osei, B.A.; Arthur, E. Changes in soil chemical properties and lettuce yield response following incorporation of biochar and cow dung to highly weathered acidic soils. *J. Org. Agric. Environ.* **2016**, *4*, 28–39.
97. Jien, S.H.; Wang, C.S. Effects of biochar on soil properties and erosion potential in a highly weathered soil. *Catena* **2013**, *110*, 225–233. [CrossRef]
98. Liang, B.; Lehmann, J.; Solomon, D.; Kinyangi, J.; Grossman, J.; O'Neill, B.; Skjemstad, J.O.; Thies, J.; Luizão, F.J.; Petersen, J.; et al. Black carbon increases cation exchange capacity in soils. *Soil Sci. Soc. Am. J.* **2006**, *70*, 1719–1730. [CrossRef]
99. Laird, D.; Fleming, P.; Wang, B.; Horton, R.; Karlen, D. Biochar impact on nutrient leaching from a Midwestern agricultural soil. *Geoderma* **2010**, *158*, 436–442. [CrossRef]
100. Troy, S.M.; Lawlor, P.G.; O'Flynn, C.J.; Healy, M.G. The impact of biochar addition on nutrient leaching and soil properties from tillage soil amended with pig manure. *Water. Air. Soil Pollut.* **2014**, *225*, 1900. [CrossRef]
101. Glaser, B.; Guggenberger, G.; Zech, W.; Ruivo, M.D.L. Soil organic matter stability in Amazonian Dark Earths. In *Amazonian Dark Earths*; Lehmann, J., Kern, D.C., Glaser, B., Wodos, W.I., Eds.; Springer: Dordrecht, The Netherlands, 2003; pp. 141–158.
102. Rillig, M.C.; Wagner, M.; Salem, M.; Antunes, P.M.; George, C.; Ramke, H.; Titirici, M.; Antonietti, M. Material derived from hydrothermal carbonization: Effects on plant growth and arbuscular mycorrhiza. *Appl. Soil Ecol.* **2010**, *45*, 238–242. [CrossRef]
103. Lehmann, J.; Rillig, M.C.; Thies, J.; Masiello, C.A.; Hockaday, W.C.; Crowley, D. Biochar effects on soil biota—A review. *Soil Biol. Biochem.* **2011**, *43*, 1812–1836. [CrossRef]
104. Mukhopadhyay, S.; Maiti, S.K.; Masto, R.E. Development of mine soil quality index (MSQI) for evaluation of reclamation success: A chronosequence study. *Ecol. Eng.* **2014**, *71*, 10–20. [CrossRef]
105. Li, H.; Dong, X.; da Silva, E.B.; de Oliveira, L.M.; Chen, Y.; Ma, L.Q. Mechanisms of metal sorption by biochars: Biochar characteristics and modifications. *Chemosphere* **2017**, *178*, 466–478. [CrossRef]
106. Yuan, J.H.; Xu, R.K.; Zhang, H. The forms of alkalis in the biochar produced from crop residues at different temperatures. *Bioresour. Technol.* **2011**, *102*, 3488–3497. [CrossRef]
107. Xiao, Y.; Xue, Y.; Gao, F.; Mosa, A. Sorption of heavy metal ions onto crayfish shell biochar: Effect of pyrolysis temperature, pH and ionic strength. *J. Taiwan Inst. Chem. Eng.* **2017**, *80*, 114–121. [CrossRef]
108. Miles, L.J.; Parker, G.R. DTPA Soil Extractable and Plant Heavy Metal Concentrations with Soil-added Cd Treatments. *Plant Soil* **1979**, *68*, 59–68. [CrossRef]
109. Ren, X.; Sun, H.; Wang, F.; Zhang, P.; Zhu, H. Effect of ageing in field soil on biochar's properties and its sorption capacity. *Environ. Pollut.* **2018**, *242*, 1880–1886. [CrossRef]
110. Jung, K.W.; Kim, K.; Jeong, T.U.; Ahn, K.H. Influence of pyrolysis temperature on characteristics and phosphate adsorption capability of biochar derived from waste-marine macroalgae (Undaria pinnatifida roots). *Bioresour. Technol.* **2016**, *200*, 1024–1028. [CrossRef]
111. Coumar, M.V.; Parihar, R.S.; Dwivedi, A.K.; Saha, J.K.; Rajendiran, S.; Dotaniya, M.L.; Kundu, S. Impact of pigeon pea biochar on cadmium mobility in soil and transfer rate to leafy vegetable spinach. *Environ. Monit. Assess.* **2016**, *188*, 31. [CrossRef] [PubMed]
112. Zhou, L.; Xu, D.; Li, Y.; Pan, Q.; Wang, J.; Xue, L.; Howard, A. Phosphorus and nitrogen adsorption capacities of biochars derived from feedstocks at different pyrolysis temperatures. *Water* **2019**, *11*, 1559. [CrossRef]

113. Yao, Y.; Gao, B.; Zhang, M.; Inyang, M.; Zimmerman, A.R. Effect of biochar amendment on sorption and leaching of nitrate, ammonium, and phosphate in a sandy soil. *Chemosphere* **2012**, *89*, 1467–1471. [CrossRef] [PubMed]
114. Wang, Y.; Zhong, B.; Shafi, M.; Ma, J.; Guo, J.; Wu, J.; Ye, Z.; Liu, D.; Jin, H. Effects of biochar on growth, and heavy metals accumulation of Moso bamboo (Phyllostachy pubescens), soil physical properties, and heavy metals solubility in soil. *Chemosphere* **2019**, *219*, 510–516. [CrossRef]
115. Maiti, S.K.; Nandhini, S. Bioavailability of Metals in Fly Ash and Their Bioaccumulation in Naturally Occurring Vegetation. *Environ. Monit. Assess.* **2006**, *116*, 263–273. [CrossRef]
116. Wang, Y.; Gu, K.; Wang, H.; Shi, B. Remediation of heavy-metal-contaminated soils by biochar: A review. In *Envionmental Geotechnics*; ICE Publishing: London, UK, 2019; pp. 1–14, ISBN 0000000264.
117. Koetlisi, A.; Muchaonyerwa, P. Sorption of Selected Heavy Metals with Different Relative Concentrations in Industrial Effluent on Biochar from Human Faecal Products and Pine-Bark. *Materials* **2019**, *12*, 1768. [CrossRef]
118. Shi, Z.; Peng, S.; Lin, X.; Liang, Y.; Lee, S.Z.; Allen, H.E. Predicting Cr(vi) adsorption on soils: The role of the competition of soil organic matter. *Environ. Sci. Process. Impacts* **2020**, *22*, 95–104. [CrossRef]
119. Fendorf, S.; Wielinga, B.W.; Hansel, C.M.; Fendorf, S.; Wielinga, B.W.; Hansel, C.M. Chromium Transformations in Natural Environments: The Role of Biological and Abiological Processes in Chromium (VI) Reduction Chromium Transformations in Natural Environments: The Role of Biological and Abiological Processes. *Int. Geol. Rev.* **2000**, *42*, 691–701. [CrossRef]
120. Richard, F.C.; Bourg, A.C.M. Aqueous Geochemistry of Chromium: A Review. *Water Res.* **1991**, *25*, 807–816. [CrossRef]
121. Chibuike, G.U.; Obiora, S.C. Heavy metal polluted soils: Effect on plants and bioremediation methods. *Appl. Environ. Soil Sci.* **2014**, *2014*, 752708. [CrossRef]
122. Quartacci, M.F.; Sgherri, C.; Cardklli, R.; Fantozzi, A. Biochar amendment reduces oxidative stress in lettuce grown under copper excess. *Agrochim. Pisa* **2015**, *59*, 188–202. [CrossRef]
123. Anderson, J.V.; Davis, D.G. Abiotic stress alters transcript profiles and activity of glutathione S -transferase, glutathione peroxidase, and glutathione reductase in Euphorbia esula. *Physiol. Plant.* **2004**, *120*, 421–433. [CrossRef] [PubMed]
124. Rennenberg, H. Processes involved in glutathione metabolism. In *Amino Acids and Their Derivatives in PlantsBiosynthesis and Metabolism*; Wallsgrove, R.M., Ed.; Cambridge University Press: Cambridge, UK, 1995; pp. 155–171.
125. Alam, M.Z.; McGee, R.; Hoque, M.A.; Ahammed, G.J.; Carpenter-Boggs, L. Effect of arbuscular mycorrhizal Fungi, Selenium and biochar on photosynthetic pigments and antioxidant enzyme activity under arsenic stress in Mung Bean (*Vigna radiata*). *Front. Physiol.* **2019**, *10*, 193. [CrossRef] [PubMed]
126. *ASTM D5142-09*; Standard Test Methods for Proximate Analysis of the Analysis Sample of Coal and Coke by Instrumental Procedures. ASTM International: West Conshohocken, PA, USA, 2009.
127. Zhang, H.; Chen, C.; Gray, E.M.; Boyd, S.E. Effect of feedstock and pyrolysis temperature on properties of biochar governing end use efficacy. *Biomass Bioenergy* **2017**, *105*, 136–146. [CrossRef]
128. Reck, I.M.; Paixão, R.M.; Bergamasco, R.; Vieira, M.F.; Vieira, A.M.S. Removal of tartrazine from aqueous solutions using adsorbents based on activated carbon and Moringa oleifera seeds. *J. Clean. Prod.* **2018**, *171*, 85–97. [CrossRef]
129. Igalavithana, A.D.; Park, J.; Ryu, C.; Lee, Y.H.; Hashimoto, Y.; Huang, L.; Kwon, E.E.; Ok, Y.S.; Lee, S.S. Slow pyrolyzed biochars from crop residues for soil metal(loid) immobilization and microbial community abundance in contaminated agricultural soils. *Chemosphere* **2017**, *177*, 157–166. [CrossRef]
130. Domingues, R.R.; Trugilho, P.F.; Silva, C.A.; Melo, I.C.N.A.A.; Melo, L.C.A.; Magriotis, Z.M.; Sánchez-Monedero, M.A. Properties of biochar derived from wood and high-nutrient biomasses with the aim of agronomic and environmental benefits. *PLoS ONE* **2017**, *12*, e0176884. [CrossRef]
131. EBC. Analysis of Biochar. Available online: https://www.european-biochar.org/en/analytical%20methods (accessed on 20 February 2018).
132. Yin, Q.; Wang, R.; Zhao, Z. Application of Mg–Al-modified biochar for simultaneous removal of ammonium, nitrate, and phosphate from eutrophic water. *J. Clean. Prod.* **2018**, *176*, 230–240. [CrossRef]

MDPI
St. Alban-Anlage 66
4052 Basel
Switzerland
Tel. +41 61 683 77 34
Fax +41 61 302 89 18
www.mdpi.com

Sustainability Editorial Office
E-mail: sustainability@mdpi.com
www.mdpi.com/journal/sustainability

www.ingramcontent.com/pod-product-compliance
Lightning Source LLC
Chambersburg PA
CBHW061505210926
51850CB00008B/1829
9783036582337